Cell Surface Glycolipids

Charles C. Sweeley, EDITOR
Michigan State University

Based on a symposium
sponsored by the
Division of Carbohydrate Chemistry
at the 178th Meeting of the
American Chemical Society,
Washington, D.C.
September 11–14, 1979.

ACS SYMPOSIUM SERIES 128

AMERICAN CHEMICAL SOCIETY
WASHINGTON, D. C.   1980

Library of Congress CIP Data

Cell surface glycolipids.
(ACS symposium series; 128 ISSN 0097-6156)
Bibliography: p.

Includes indexes.

1. Glycolipids—Congresses. 2. Membranes (Biology)
—Congresses.
I. Sweeley, Charles Crawford, 1930-   . II. American Chemical Society. Division of Carbohydrate Chemistry. III. Series: American Chemical Society. ACS symposium series; 128.
[DNLM: 1. Cell membrane—Congresses. 2. Glycolipids—Congresses. QU85 C393 1979]

QP572.G56C44   574.87'5   80-15283
ISBN 0-8412-0556-6 ACSMC8 128 1–504 1980

ACS Symposium Series

M. Joan Comstock, *Series Editor*

FOREWORD

The ACS Symposium Series was founded in 1974 to provide
a medium for publishing symposia quickly in book form. The
format of the Series parallels that of the continuing Advances
in Chemistry Series except that in order to save time the
papers are not typeset but are reproduced as they are submitted by the authors in camera-ready form. Papers are reviewed under the supervision of the Editors with the assistance
of the Series Advisory Board and are selected to maintain the
integrity of the symposia; however, verbatim reproductions of
previously published papers are not accepted. Both reviews
and reports of research are acceptable since symposia may
embrace both types of presentation.

CONTENTS

v

PREFACE

This collection of papers is a record of the proceedings of a symposium on the chemistry, metabolism, and biological functions of glycolipids. It is also a review of some topics presented at an earlier meeting on the same subject, held in Honolulu, Hawaii in October, 1977 under the auspices of the Japan–United States Science Exchange Program.

Both meetings were convened with a hope that aspects of glycolipid biochemistry at the cell surface would be highlighted, and to an extent these meetings have provided insight and inspiration to all of us who seek to understand the role of these interesting substances in nature. The reader will note substantial progress and new information about the chemistry and metabolism of the glycolipids, with especially comprehensive material on their separation and characterization. The antigenic behavior and the cell surface receptor role of glycolipids are discussed in some detail, providing a sound basis for future investigations of cell surface specificity to particular molecules and supramolecular systems.

I am indebted to the foreign speakers; Lars Svennerholm, Karl-Anders Karlsson, Guido Tettamanti, Robert Murray, and Yoshitaka Nagai generously consented to participate in this symposium with little financial support from the organizer or the society. Their contributions were an especially important part of the symposium. I am grateful as well to the American participants, who attended the meeting largely with their own funds and made the meeting a success. Although Professor Egge of the University of Bonn could not attend the meeting, he kindly provided an important chapter on high resolution proton NMR of glycosphingolipids, for which I am grateful.

Finally, I wish to thank Ms. Dorothy Byrne, Ms. Paula Allen, and Ms. Susan Leavitt for their dedicated assistance in the organization of the meeting, correspondence with speakers, and preparation of the final manuscripts for this book.

East Lansing, Michigan
December, 1979

CHARLES S. SWEELEY

Preparative and Analytical High Performance Liquid Chromatography of Glycolipids

R. H. MC CLUER

Eunice Kennedy Shriver Center, Waltham, MA 02154

M. D. ULLMAN

Center for Disease Control, Atlanta, GA 30333

High performance liquid chromatography (HPLC) implies the use of reusable columns, injection port sample application, the use of pumps for uniform solvent flow operated at high pressures if necessary and automatic on-line sample detection. The availability of a large variety of microparticulate column packing materials, efficient column packing techniques, high pressure-low volume pumping equipment and various types of highly sensitive detectors have led to the development of sensitive, rapid and quantitative methods, analogous to that available for volatile materials by gas-chromatography, for the isolation and analysis of a large variety of relatively high molecular weight substances of biological interest (1).

We have attempted to utilize these tools of modern liquid chromatography to develop rapid and highly sensitive methods for the analysis of glycolipids. Our early experience with HPLC techniques indicated that the analysis of glycolipids becomes interestingly sensitive and practical if derivatives are prepared that allow the use of ultraviolet detectors and that exhibit good chromatographic properties. We have primarily studied the preparation of the benzoyl derivatives of glycolipids for the development of analytical and preparative HPLC methods. The following is a concise review of studies with neutral glycosphingolipids with emphasis on recent work in which we have utilized p-dimethylaminopyridine (DMAP) as a catalyst to effect benzoylation with benzoic anhydride.

Preparation and analysis of benzoylated cerebrosides

We initially demonstrated that brain cerebrosides, galactosyl-ceramides containing hydroxy fatty acids (HFA) and nonhydroxy fatty acids (NFA), could be completely derivatized by reaction with 10% benzoyl chloride at 60°C for 1 hour (2). After removal of excess reagents by partition between hexane and alkaline aqueous methanol, the perbenzoyl derivatives were seen to separate into two completely resolved components (HFA and NFA

cerebrosides) by adsorption chromatography on a pellicular silica
column support (Zipax, E.I. DuPont) with methanol in pentane as
the eluting solvent and 280 nm detection. Hexane-ethyl acetate
was subsequently shown to be a superior eluting solvent because
regeneration of the column adsorbant activity was more reproducible (3). Attempts to recover the parent cerebrosides by
treatment of the benzoyl derivatives with mild alkali was
successful with the HFA-cerebrosides, but the NFA-cerebroside
derivative gave rise to benzoyl psychosine as well as the parent
NFA-cerebroside. This was demonstrated to result from the diacyl
amine structure of the perbenzoyl NFA-cerebroside. NMR studies
of the two cerebroside derivatives indicated the presence of six
benzoyl groups in each case and the presence of an amide proton
in the HFA-derivative which was absent in the NFA-cerebroside
derivative. As reported by Inch and Fletcher (4) for the
diacylamine derivatives of amino sugars, the N-acyl groups are
randomly removed during alkali hydrolysis thus cerebrosides and
other sphingolipids which contain non-hydroxy fatty acids or N-
acetyl amino sugars cannot be recovered in high yields because
alkaline hydrolysis of the perbenzoyl derivative results in the
formation of N-benzoyl compounds as well as the parent N-acyl
sphingolipid. Benzoylation of cerebrosides with 10% benzoic
anhydride in pyridine was shown to lead only to the formation of
O-acyl derivatives and the parent glycolipids could be recovered
after alkaline methanolysis; however, this reaction was sluggish
and required treatment at 110°C for 18 hours for completion.
Sulfatides were shown to be completely converted to benzoylated
cerebrosides during this anhydride reaction. We chose the benzoyl
chloride reaction for analytical purposes because reaction times
were shorter and sulfatides do not desulfate under conditions
required for cerebroside derivatization. Because sphingolipids
which contain only hydroxy fatty acids as N-acyl substituents
form the same derivative with either the benzoyl chloride or the
anhydride reaction, they can be easily distinguished from nonhydroxy fatty acid containing sphingolipids which form different
derivatives, distinquishable by HPLC, when benzoylated with the
chloride as compared to the anhydride reaction.

Analysis of ceramides

     A quantitative HPLC method for the analysis of sphingolipids
as their perbenzoyl derivatives was first developed for ceramides
(5). Ceramides can be conveniently derivatized with benzoic
anhydride in pyridine (3 hrs at 110°C) and the products formed
have been utilized for the quantitative analysis of NFA and HFA
ceramides in normal and Farber's disease tissue. Iwamori and
Moser also utilized this procedure for the analysis of ceramides
in Farber's disease urine (6). More recently Iwamori and Moser
(7) established that the ceramide derivatives formed by reaction
with benzoyl chloride or benzoic anhydride are analogous to those
formed with cerebrosides. They also characterized the behavior

of ceramides that contain phytosphingosine and described the use
of estrone as an internal standard. These published ceramide
methods utilized 280 nm detection and the sensitivity of the
procedures could easily be increased to the pmole level by
detection at 230 nm if a variable wave length detector is
utilized. Also, the speed of derivatization could undoubtedly be
greatly increased by the use of the DMAP as a catalyst as
described below for neutral glycosphingolipids. Samuelsson (8)
utilized elegent gas chromatograph-mass spectrometric (GC-MS)
methods for the analysis of ceramide molecular species, but HPLC
methods offer the advantages of non-destructive measurement so
that components can easily be collected for determination of
radioactivity or for further analysis.

Quantitative analysis of neutral glycosphingolipids

     An HPLC method for neutral glycosphingolipids was first
designed for the analysis of human plasma glycolipids (3), which
consist primarily of glucosylceramide, lactosylceramide, globotriaosylceramide and globotetraosylceramide (globoside).
Conditions for the simultaneous derivatization of this group of
compounds, which provided maximal yields for globoside, were
selected to be 37°C for 16 hours in 10% benzoyl chloride in
pyridine, slightly different from those previously utilized for
cerebrosides. Satisfactory chromatographic conditions, which
provided base line resolution of these four derivatives in a
minimum of time, were found with the Zipax column and a gradient
of ethyl acetate in hexane and 280 nm detection. With this
chromatographic system the standard glycolipid derivatives could
be separated and quantitated in less than 20 min and column
activity could be reproducibly regenerated in eight min. Less
than 20 nmole of each glycolipid could be easily quantitated with
this procedure.
     The utility of ethyl acetate in this chromatographic system
was excellent but prevented the use of detection below 260 nm.
Because the λmax of the benzoyl derivatives is at 230nm we sought
chromatographic solvents which could be utilized at this wavelength, still provide adequate chromatographic resolution and
allow rapidly re-equilibration of column adsorbant activity after
gradient elution. A dioxane-hexane solvent system proved adequate
except residual light absorption due to the dioxane produced an
undesirable rising base line during the gradient. The rising
base line was eliminated by directing the solvent flow through a
precolumn pre-injector high pressure reference cell. This path
generates a horizontal baseline with a negative and positive
deflection at the beginning and end of the gradient respectively.
With this system reliable quantitation of less than 50 pmoles of
each of the four major plasma glycosphingolipids can be obtained
(9). For the analysis of plasma glycolipids it is necessary to
first isolate a glycolipid fraction by solvent extraction,

chromatography on small Unisil columns and treatment with mild
alkali as originally described by Vance and Sweeley (10).
Consistent recoveries of the glycolipids is dependent upon
maintaining a fixed ratio between sample size and the quantity of
Unisil employed. Accuracy of the method is improved by the
utilization of an internal standard such as N-acetylpsychosine
which is added to the plasma samples prior to the initial lipid
extraction. One ml plasma samples are now routinely used for
glycolipid analysis although sensitivity of the HPLC procedure
theoretically should allow analysis of less than 0.1 ml. However,
the isolation of such small quantities of glycolipid prior to
derivatization present difficult recovery problems. The high
sensitivity of the detection system employed for the analysis of
pmole quantities also requires precautions so that UV absorbing
contaminates are not introduced during processing of the samples.
All glassware should be scrupulously clean and HPLC grade solvents
should be used for all steps in the isolation and chromatography.
This HPLC procedure has also been utilized for the analysis of
neutral glycosphingolipids from a variety of sources.

Human erythrocytes, peripheral leukocytes and liver have
been satisfactorily analyzed, but it should be recognized that
each different tissue source may require different extraction
conditions and modified solvent gradient elution in order to
obtain maximal recoveries and optimal chromatographic resolution
of the tissue characteristic glycosphingolipids. Fletcher,
Bremer and Schwarting (11) have optimized the procedure for the
analysis of erythrocyte glycolipids and demonstrated that
erythrocytes from blood group P_1 individuals contain more globotriaosylceramide and less lactosylceramide than erythrocytes from
blood group P_2 individuals. The dramatic sex difference in mouse
kidney glycolipids and the occurrence of large amounts of glycolipids in male mouse urine was readily demonstrated by these HPLC
methods. The light ear (le/le) mouse pigmentation mutant was
shown to have storage of glycolipids in their kidneys which is
apparently due to an abnormality in the secretion of multilamellar
lysosomal bodies that contain large amounts of glycosphingolipids
(12). Thus, the analytical HPLC method for glycolipids is proving
useful for a variety of studies related to glycosphingolipid
function and metabolism.

Other useful analytical HPLC procedures for the analysis of
derivatized glycolipids have been developed. Nanaka and Kishimoto
(13) have devised an HPLC procedure which allows the tissue levels
of NFA cerebroside, HFA cerebroside, NFA sulfatide, HFA sulfatide,
and monogalactosyl diglyceride to be determined simultaneously.
This procedure involves benzoylation of total lipid extracts,
desulfation with mild acid and subsequent chromatography with the
gradient of isopropanol in hexane. Susuki, Honda and Yamakawa
(14) prepared acetylated glycolipid which were subsequently
reacted with p-nitrobenzoyl chloride to form the O-acetyl-N-p-
nitrobenzoyl derivatives which have good chromatographic

properties and can be detected with high sensitivity with a single wavelength detector at 254 nm. While offering these advantages, this procedure cannot be utilized with glycolipids that contain only α-hydroxy fatty acids and no amino sugar. All of the benzoylated or O-acetyl-N-p-nitrobenzoyl derivatives can be usefully separated into molecular species by reverse phase chromatography (11,12).

We have recently shown that the use of the catalyst N-dimethylaminopyridine (DMAP) with benzoic anhydride greatly accelerates the derivatization with this reagent (13). Reaction with DMAP and the anhydride avoids amide acylation, forms single products with satisfactory chromatographic properties and parent glycosphingolipids can be regenerated by mild alkaline hydrolysis. For analytical purposes, this reaction has been utilized for the analysis of plasma neutral glycosphingolipids. The glycolipids were reacted with 20% benzoic acid anhydride, 5% DMAP in pyridine at 37°C for four hours. GlcCer, LacCer, GbOse$_3$Cer and GbOse$_4$ Cer each gave single reaction products with maximum yields with reaction times between 2 and 6 hours. Excess reagents were removed from the products by partition between hexane and aqueous alkaline methanol as described previously for the benzoyl chloride products. The products were than analyzed with the Zipax column and dioxane gradient also as previously described (3). The chromatographic analysis of the per-O-benzoylated glycosphingolipid standards and plasma glycosphingolipids are shown in Fig. 1 along with the elution pattern of plasma glycolipid derivatives obtained by reaction with benzoyl chloride.

The derivatives obtained by reaction with benzoic anhydride have longer retention times when compared to the benzoyl chloride products. We have previously shown that galactosylceramide which contains α-hydroxy fatty acids is not N-benzoylated with benzoyl chloride and reaction with benzoic anhydride or benzoyl chloride results in an identical product. Similar results have been obtained with anhydride in the presence of DMAP as illustrated in Fig. 2. The behavior of peak "b" which we have shown to be derived from α-hydroxy fatty acid containing glucosyl and galactosylceramides is illustrative. The UV response from each of the standard GSLs benzoylated by the anhydride and by the benzoyl chloride method were compared. The relative responses (chloride/anhydride) for the mono, di, tri and tetra-hexosyl ceramide were found to be 1.18, 1.15, 0.94 and 1.03 respectively. These values were not significantly different from calculated rations 1.20, 1.12, 1.09, and 1.15, based on the assumption that the anhydride method avoids amide benzoylation. The yields of the per-O-benzoylated products were similar to those obtained for the products of the benzoyl chloride method reported previously.

The parent GSLs can be regenerated from their per-O-benzoylated products by treatment with mild alkali. Globoside was benzoylated by both methods, the products subjected to HPLC, and the peaks collected and treated with 0.5N methanolic sodium

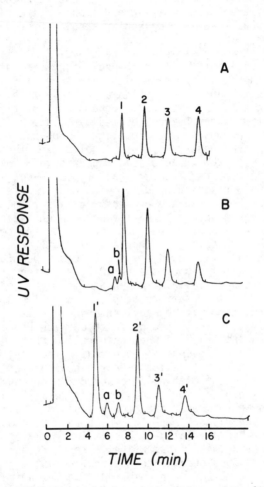

Figure 1.   HPLC of benzoylated standard and plasma glycosphingolipids

*The derivatized glycosphingolipids were injected onto a Zipex column (2.1 mm × 50 cm)
and eluted with a 13-min linear gradient of 2.5–25% dioxane in hexane with detection at
230 nm. A. Standard glycosphingolipids (GSL) per-O-benzoylated with benzoic anhydride
and 4-dimethylaminopyridine (DMAP). B. Plasma GSL per-O-benzoylated with benzoic
anhydride and DMAP. C. Plasma GSL perbenzoylated with benzoyl chloride. Glycosphingolipid peaks are identified as: (1) glycosylceramide, (2) lactosylceramide, (3)
galactosyl-lactosylceramide, (4) N-acetylgalactosaminylgalactosyllactosylceramide. Peak
A is unidentified, and peak B is hydroxy fatty acid containing galactosylceramide.*

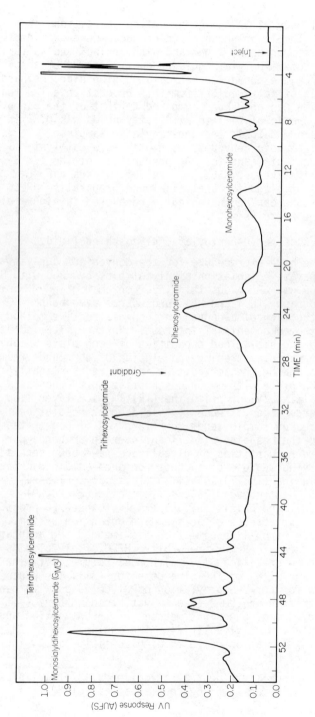

Figure 2.   Preparative isolation of liver glycolipids

A glycosphingolipid fraction (15 mg) was benzoylated with DMAP and benzoic anhydrydride, and the derivatives were chromatographed on a LiChrosorb SI 100 column with an ethyl acetate in hexane gradient as described in the text. Detection was at 280 nm. Components eluting at 7, 9, and 49 min are unidentified.

hydroxide for 1 hour at 37°C. After solvent partition, in
C/M/H$_2$0 (8/3/3) the lower phase lipid products were examined by
TLC and visualized under UV light and with orcinol and sulfuric
acid spray reagents. Only a single product, with no UV absorption
was obtained from the anhydride benzoylated globoside. Methanolysis
and GLC analysis of fatty acids from this product revealed that
their composition was unchanged compared to that of the parent
globoside. The ratio of 24:1 to 24:0 fatty acids was 0.51 in the
original sample and 0.50 in the debenzoylated sample.

    The use of benzoic anhydride with DMAP as a catalyst
provides a convenient means for the preparation of the per-O-
benzoylated derivatives of GSLs. These derivatives can
subsequently be utilized for analytical and preparative HPLC
because parent GSLs can be conveniently recovered in high yields
by mild alkaline hydrolysis.

Preparative HPLC of per-O-benzoylated glycosphingolipids.

    We describe here a procedure for the convenient derivati-
zation and preparative isolation of glycolipids by HPLC with UV
detection at 280 nm. Previous thin-layer chromatography (TLC),
liquid chromatography (LC) and HPLC procedures have been
encumbered by the lack of a convenient non-destructive method of
detection for the components of interest. Further, TLC isolations
are hampered by the small load capacities of each plate which
requires the streaking, scraping and elution of compounds from
multiple plates and by the ambiguities introduced by lightly
spraying of each plate with a non-destructive spray such as
methanol-water 1:1 (v/v) or primuline (15). Thick-layer TLC,
which allows large loads, frequently yields poor resolution
because the streaked sample tends to "flare" as it penetrates the
separation bed, thus causing significant overlap of bands during
migration. Larger quantities of glycolipids have been separated
by LC, with varying degrees of success, on such column packing
materials as alumina (16,17), Anasil S (18,19,20), Florisil
(21,22,23,24,25), Iatrobeads (26), silicic acid (27,28,29)
Silica gel G (30), and Unisil (31,32). All of these LC procedures
are hampered by the absence of an adequate detection system.
Although the lack of on-line detection has impeded the adaptation
of the LC procedures to HPLC, preparative HPLC of glycolipids
has been performed on silica SI 60 with post-column, off-line TLC
detection (30) and with a moving wire detector (31). The procedure
described below for per-O-benzoylation of glycolipids with benzoic
anhydride in pyridine and DMAP as catalyst avoids N-benzoylation
problem and provides a convenient method for the detection and
preparative isolation of glycolipids. The application of this
procedure for the isolation of 15 mg of glycolipids in a single
HPLC run is described.

a. Isolation of crude liver neutral glycolipids. Total lipids
were isolated from human liver by the method of Folch et al. (33).

Crude neutral glycolipids were isolated from the lipid extract essentially by the procedure of Vance and Sweeley (10). The lipids were dissolved in chloroform and placed on a silicic acid column (25 gm packing to 1 gm lipid extract) the neutral lipids and fatty acids were eluted with chloroform (20 ml/gm packing). Crude neutral glycolipids were then eluted from the column with acetone-methanol (9:1 v/v) (40 ml/gm packing). The acetone-methanol eluate was collected in 200 ml fractions, so that mono- and di-hexosyl ceramides were in greater concentration in the earlier fractions relative to the tri- and tetra-hexosyl ceramides. The acetone-methanol from each of the desired fractions was evaporated to dryness and exposed to mild alkaline methanolysis. The content of neutral glycolipids in the fractions of interest was determined by the quantitative HPLC method of Ullman and McCluer (3).

b. Per-0-benzoylation condition. Samples which had been dried under a stream of nitrogen and which contained approximately 15 mg of neutral glycolipids from liver, were transferred into a 20 mm x 150 mm screw capped culture tube and dessicated over P_2O_5 for at least three hours. A 1.5 ml portion of freshly prepared 20% benzoic anhydride in pyridine (w/v) was added to the culture tube followed by a 1.5 ml portion of 10% DMAP in pyridine (v/v). The tube was briefly flushed with nitrogen, capped tightly, and incubated at 37°C for two hours. The tube was then placed in a water bath maintained at room temperature and the pyridine was removed with a stream of nitrogen. Three ml of hexane was added to the residue and the suspension was washed four times with 1.8 ml of alkaline methanol. The alkaline methanol was prepared by the addition of 1.2 gm Na_2CO_3 to 300 ml of methanol-water 80:20 (v/v) (all of the Na_2CO_3 did not dissolve). Each time the lower phase was withdrawn and discarded. Finally, the hexane layer was washed once with 1.8 ml of methanol-water 80:20 and after removal of the lower phase the hexane was evaporated with a stream of nitrogen at room temperature. The sample was then dissolved in 5 ml of methanol and placed onto a reverse phase rapid sample preparation column (Sep Pak™, Waters Assoc.) that had been preconditioned with 30 ml of methanol. An additional wash of 5 ml of methanol was added to the column and the per-0-benzoylated glycolipids were eluted with 10 ml of methanol-acetone 9:1 in a 20 mm x 150 mm screw cap culture tube. This fraction was dried at room temperature with a stream of nitrogen and redissolved in 4% ethyl acetate in hexane (v/v) for injection.

c. HPLC. Per-0-benzoylated neutral glycolipids were applied to a 4.6 mm x 25 cm LiChrosorb SI 100, 10μ particle with a loop injector and 4% ethyl acetate in hexane pumped at 0.5 ml/min. The derivatives were then eluted isocratically with 18% ethyl acetate in hexane for 30 minutes and then with a linear gradient of 18 to 45% ethyl acetate in hexane over thirty minutes. The

flow rate was 3 ml/min and UV detection was performed at 280 nm.
Each peak was isolated and partially characterized for recovery
and purity by analytical HPLC. After the isolation was complete
the initial solvent and adsorbant conditions were regenerated
with a three minute reverse gradient.

d. Results. Excellent resolution of the major glycolipid peaks
(Fig. 2) was obtained. Because of the mono-, di-, tri-, and
tetra-hexosyl ceramides did not maintain a constant ratio in each
fraction during the isolation of the crude glycolipid fraction
from Unisil, the chromatogram in Figure 2 is not representative
of the distribution of liver glycolipids. Ethylacetate was
selected as a solvent because of the high lipid solubility of
the derivitized glycolipids because it is easily removed after
collection of the desired fractions, and because column adsorbant
activity is rapidly reestablished after gradient elution.
Detection at 280nm avoided saturation of the detector signal when
large quantities of glycolipids were applied to the column.
Larger columns and greater amounts of glycolipids could have been
used by setting the detector to a slightly higher wavelength thus
decreasing the sensitivity. Although capacity of the column was
not determined in these experiments, there was no change in
retention time or peak shape when 5, 10, or 15 mg of glycolipids
were chromatographed.
    Each sample was collected so that approximately 5% of the
leading and tailing edges of the peaks were omitted from the
collection. After a sample was collected, the isolated fraction
was dried under nitrogen, dissolved in a known volume of hexane
and an aliquot tested for purity by analytical HPLC. The sample
was then dried and the residue was subjected to mild alkaline
hydrolysis (to recover natural glycolipid), perbenzoylated in 10%
benzoyl chloride in pyridine, redissolved in hexane and again
analyzed by quantitative HPLC. The two methods for evaluation of
fraction purity were in good agreement. Usually, isolated sample
peaks were greater than 98% free of glycolipid contaminates and
it was not uncommon to isolate peaks that contained no other
glycolipids. If necessary, further purification could be obtained
by rerunning each of the isolated samples in an isocratic solvent
system with a solvent composition near the eluting composition of
the gradient. Each fraction was shown to be free of non-UV
absorbing impurities by migration of the isolated natural glyco-
lipidson silica Gel G TLC plates in chloroform-methanol water
(65:25:4) with detection by charring with 55% H_2SO_4 in water (w/w)
To determine the recovery of the neutral glycolipids a crude
glycolipid was per-O-benzoylated and 0.1% of the sample was
injected onto the analytical chromatograph to determine the amount
of each component. The remainder was injected onto the preparative
column and the total effluent, excluding the solvent front was
collected, dried under nitrogen, dissolved in initial solvent and

0.1% reinjected onto the analytical column. Two different
experiments averaged 84, 84.5, 86, and 89% recovery for mono-,
di-, tri-, and tetra- hexosyl ceramides, respectively.

After completion of the preparative run, the column was
regenerated and reused several times with little loss of
efficiency. The use of the rapid sample preparation column in
the isolation of the per-O-benzoylated derivatives was
necessitated by the large quantity of dark brown impurities in
the original lipid preparation, which also eluted from the Unisil
column with the crude glycolipid fraction. The use of a Sep-Pak^{TM}
column greatly reduced this discoloration. The isocratic portion
of the column elution was utilized to obtain better resolution
of the other, early-eluting peaks which were assumed to be
monohexosyl-containing sphingolipids. The structures of these
components are under study and presumably are similar to the "a"
and "b" components seen with the plasma glycolipids in Fig. 1.

Conclusions

Derivatives of glycosphingolipids which have large extinction
coefficients can be prepared and separated according to their
carbohydrate content by adsorbtion chromatography. This use of
modern HPLC equipment allows quantitation of less than 50 pmole
quantities of these compounds so that small amounts of body fluids
tissue or tissue culture cells can be readily analyzed for the
major nuetral glycosphingolipid components. Such components can
be further separated into molecular species by reverse phase
chromatography. The use of DMAP as a catalyst for derivatization
with benzoic acid anhydride allows the convenient preparation of
per-O-benzoyl derivatives. Parent glycosphingolipids can be
regenerated from these derivatives by treatment with mild alkali.
Thus, modern liquid chromatographic techniques with on-line
detection can be utilized for the isolation of the neutral glycosphingolipids.

Literature Cited
1. Snyder, L.R., and Kirkland, J.J., "Introduction to Modern
    Liquid Chromatography"; Wiley-Interscience: New York, N.Y.,
    1974.
2. McCluer, R.H.; Evans, J.E. Preparation and analysis of
    benzoylated cerebrosides. J. Lipid Res., 1973, 14, 611.
3. Ullman, M.D., McCluer, R.H. Quantitative analysis of plasma
    neutral glycosphingolipids by high performance liquid
    chromatography of their perbenzoyl derivatives. J. Lipid Res.,
    1977, 18, 371-377.
4. Inch, T.D.; Fletcher, H.C. N-acyl derivatives of 2-acylamino-
    2-deoxy-D-glucopyranose. J. Org. Chem., 1966, 31, 1815.
5. Sugita, M.; Iwamori, M; Evans, J.; McCluer, R.H.; Moser, H.W.;
    Dulaney, J.T.   High-performance liquid chromatography of
    ceramides: application to analysis in human tissues and
    demonstration of ceramide excess in Farber's disease. J.
    Lipid Res., 1974, 15, 223.

6. Iwamori, M.G.; Moser, H.W. Above normal urinary excretion
    of urinary ceramides in Farber's disease, and characterization
    of their components by high performance liquid chromatography.
    Clin. Chem., 1975, 21, 725.
7. Iwamori, M.; Costello, C.; and Moser, H.W. Analysis and
    quantitation of free ceramide containing nonhydroxy and
    2-hydroxy fatty acids, and phytosphingosine by high-perfor-
    mance liquid chromatography. J. Lipid Res., 1979, 20, 86.
8. Samuelsson, K. Identification and quantitative determination
    of ceramides in human plasma. Scand. J. Clin. Lab. Invest.,
    1971, 27, 371.
9. Ullman, M.D.; McCluer, R.H. Quantitative microanalysis
    of perbenzoylated glycosphingolipids by HPLC with detection
    at 230 nm. J. Lipid Res., 1978, 19, 910.
10. Vance, D.E.; Sweely, C.C. Quantitative determination of the
    neutral glycosyl ceramides in human blood. J. Lipid Res.,
    1967, 8, 621-630.
11. Fletcher, K.S.; Bremer, E.G.; Schwarting, G.A. P blood
    group regulation of glycosphingolipid levels in human
    erythrocytes. J. Biol. Chem., in press.
12. McCluer, R.H.; Gross, S.K.; Sapirstein, V.S.; Meisler, M.H.
    Testosterone effects of kidney and urinary glycolipids in the
    light eared mouse mutant. FASEB Proc., 1979, 38, 405.
13. Nanaka, G.; Kishimoto, Y. Simultaneous determination of
    picomole levels of gluco- and galactocerebroside, mono-
    galactosyl diglyceride and sulfatides by high performance
    liquid chromatography. Biochim. Biophys. Acta, 1979, 572, 423.
14. Suzuki, A.; Handa, S,; Yamakawa, T. Separation of molecular
    species of higher glycolipids by high performance liquid
    chromatography of their O-acetyl-N-p-nitrobenzoyl derivatives.
    J. Biochem., 1977, 82, 1185.
15. Skipski, V.P. Thin-layer chromatography of neutral glycosphingo-
    lipids. Methods in Enzymology, 1975, 35, 396-425.
16. Svennerholm, E.; Svennerholm, L. Isolation of blood serum
    glycolipids. Acta Chem. Scand., 1962, 16(5), 1282-1284.
17. Gray, G.M. A comparison of the glycolipids found in
    different strains of Ascites tumour cells in mice. Nature,
    1965, 207(4996), 505-507.
18. Siddiqui, B.; McCluer, R.H. Lipid components of sialosyl-
    galactosylceramide of human brain. J. Lipid Res., 1968,
    9(3), 366-370.
19. Puro, K. Isolation of bovine kidney gangliosides. Acta Chem.
    Scan., 1970, 24(1), 13-22.
20. Siddiqui, B.; Hakomori, S. A ceramide tetrasaccharide of
    human erythrocyte membrane reacting of the anti-type.
    IV. pneumococcal polysaccharide antiserum. Biochim. Biophys.
    Acta., 1973, 330, 147-155.
21. Yamakawa, I.; Irie, R.; Iwanaga, M. The chemistry of post-
    hemolytic residue of stroma of erythrocytes. IV. silicic
    acid chromatography of mammalian stroma glycolipids. J.
    Biochem. 1960, 48, 490-507.

22. Radin, N.S. Florisil chromatography. Methods Enzymol., 1969, 14, 268-272.
23. Rouser, G.; Bauman, A.J.; Kritchevsky, G.; Heller, D.; O'Brien, J. Quantitative chromatographic fractionation of complex lipid mixtures: brain lipids. J. Amer. Oil. Chem. Soc., 1961, 38, 544-555.
24. Rouser, G.; Kritchevsky, G.; Heller, D.; Lieber, E. Lipid composition of beef brain, beef liver, and the sea anemone; two approaches to quantitative fractionation of complex lipid mixtures. J. Amer. Oil. Chem. Soc., 1963, 40, 425-554.
25. Saito, T.; Hakomori, S. Quantitative isolation of total glycosphingolipids from animal cells. J. Lipid Res.,1971, 12(2) 257-259.
26. Ando, S.; Isobe, M.; Nagai, M. High performance preparative column chromatography of lipids using a new porous silica, Iatrobeads (R). Biochim. Biophys. Acta, 1976, 424, 98-105.
27. Yamakawa, T.; Yokoyama, S.; Kiso, N. Structure of main globoside of human erythrocytes. J. Biochem., 1962, 52, 228-231.
28. Svennerholm, E.; Svennerholm, L. The separation of neutral blood-serum glycolipids by thin-layer chromatography. Biochim. Biophys. Acta., 1963, 70, 432-441.
29. Martensson, E. Neutral glycolipids of human kidney. Isolation, identification, and fatty acid composition. Biochim. Biophys. Acta, 1966, 116, 296-308.
30. Viswanathan, C.V.; Hayashi, A. Ascending dry-column chromatography as an aid in the preparative isolation of glycolipids. J. Chromat., 1976, 123, 243-246.
31. Ullman, M.D.; McCluer, R.H. Isolation and quantitative analysis of neutral glycosylceramides by high performance liquid chromatography (HPLC). FASEB Proceedings, 1976.
32. Tjaden, U.R.; Krol, J.H.; Van Hoeven, R.P.; Oomer-Meulemans, E.P.M.; Emmelot, P. High pressure liquid chromatography of glycosphingolipids (with special reference to gangliosides) J. Chromat.,1977, 136, 233-243.
33. Folch, J.; Lees, M.; Sloane-Stanley, G.H. A simple method for the isolation and purification of total lipids from animal tissues. J. Biol. Chem.,1957, 266, 497-509.

RECEIVED December 10, 1979.

High Performance Liquid Chromatography of Membrane Glycolipids

Assessment of Cerebrosides on the Surface of Myelin

SHOJI YAHARA and YASUO KISHIMOTO

The John F. Kennedy Institute and Department of Neurology, Johns Hopkins University School of Medicine, Baltimore, MD 20205

JOSEPH PODUSLO[1]

Neuroimmunology Branch, National Institutes of Health, Bethesda, MD 20205

Carbohydrates occur on cells or plasma membranes primarily in the form of glycoproteins or glycolipids. Those accessible on the outer surface of the cell or membrane may have important biological functions as adhesion sites in terms of cell recognition, as receptors for hormones, toxins, or viruses, or as specific immunological determinants or antibody receptors. The presence of galactose or galactosamine as a terminal carbohydrate in these membrane surface glycolipids or glycoproteins has been determined by a procedure which utilizes the reaction with galactose oxidase (1). The enzyme converts the terminal primary alcohol group of these carbohydrates to an aldehyde. This aldehyde group can then be reduced by NaB^3H_4 to the original alcoholic group in glycoproteins or glycolipids. By this series of reactions, part of the membrane galactolipids or galactoproteins are labeled with tritium (Chart 1). Since galactose oxidase is not permeable to membrane, the identification of ^3H-labeled galactolipids or galactoproteins in extracts from cells or membranes has been considered acceptable evidence for locating these compounds on the surface of cells or membranes. This procedure has been useful for identifying a variety of carbohydrate-bearing macromolecules (in particular glycoproteins) on the surface of cell membranes (2, 3, 4).

There are, however, two major disadvantages to studying glycolipids in this manner. First, many lipids other than galactolipids are also labeled by this procedure. The exact nature of the labeling has not been elucidated, but, at least some double bonds are reduced with tritium and some ester linkage is cleaved yielding radioactive saturated lipid and radioactive alkyl alcohols, respectively. The exchange of hydrogen with tritium may also be occurring. Pretreatment of cells or membranes with non-radioactive $NaBH_4$ prior to galactose oxidase treatment helps to circumvent this problem to some extent but cannot make this procedure free from this complication.

High levels of such non-specific reduction were observed

[1]Current address: Department of Neurology, Mayo Clinic, Mayo Foundation, Rochester, Minnesota 55901

0-8412-0556-6/80/47-128-015$5.00/0

when the intact myelin sheath preparation was treated with NaB^{3}H$_4$ in the absence of galactose oxidase (5). A major radioactive peak was observed near cerebroside after separation of the lower phase lipids by TLC. Further evaluation of this material by hydrolysis showed no radioactivity in galactose, fatty acids or sphingosine. In addition, by varying the solvent system, this radioactive peak could be separated from the cerebrosides. Consequently, such non-specific reduction can easily result in erroneous interpretation of surface membrane constituents.

A second disadvantage of this procedure is that it does not have quantitative capabilities for determining surface glycolipids. It merely demonstrates whether a portion of a given glycolipid is on the surface but not the ratio of surface lipid to inaccessible lipid. Customarily, a large amount of radioactivity is used for such labeling but only a very small fraction of it is incorporated into the lipid. This was particularly the case for the nonhydroxycerebroside where the amount of labeled galactose observed after hydrolysis of this isolated glycolipid was only a minor percentage of the total label (5). Interpretation of such low levels of radioactivity may be unreliable for assessing surface membrane glycolipids, since it may, in fact, represent damage to the membrane bilayer or splitting of the lamellar.

We have recently developed a sensitive and specific method for the quantitative and qualitative determination of cerebrosides and sulfatides using high performance liquid chromatography (6, 7). In order to quantitate cerebrosides on a surface, we developed an additional new method that separately compares the amount of surface cerebrosides with the remaining cerebrosides by using high performance liquid chromatography. This method also uses galactose oxidase, but instead of reducing the aldehyde formed by NaB^{3}H$_4$, the aldehyde is converted to 2,4-dinitrophenylhydrazone followed by perbenzoylation. The product produces separate peaks from that of perbenzoylated cerebroside. Thus, the ratio of oxidized and unoxidized cerebrosides can be directly compared by high performance liquid chromatography.

In this manuscript, we will first describe the newly developed high performance liquid chromatography of cerebroside, sulfatide, and other minor galactolipids. This method allows complete analysis of a very small amount of these glycolipids in cell or membrane preparations. This will be followed by a description of our new method of determining surface galactolipids and its application to myelin cerebrosides.

Procedures

Materials. Galactose oxidase was purchased from Worthington Biochemicals (Freehold, NJ). ^{125}I was obtained from Amersham (Arlington Heights, IL). All solvents (glass-distilled) were the products of Burdick-Jackson (Muskegon, MI). Pyridine was stored

over KOH pellets and used without further purification. Benzoyl
chloride was obtained from Aldrich Chemicals (Milwaukee, WI) and
70% HClO$_4$ (double distilled from Vycor) from G. Frederick Smith
Chemical (Columbus, OH). Trypsin was obtained from Worthington,
while the trypsin inhibitor (turkey egg white), phospholipase C
(Cl. Welchii type I), catalase and phosphatidyl choline (egg yolk)
were all obtained from Sigma Chemical Co. (St. Louis, MO). Thin-
layer chromatographic plates precoated with 0.25 mm thick Silica
Gel GF were purchased from Analtech (Newark, DE). Myelin was
prepared either from brains of young Sprague-Dawley rats (Charles
River CD, Charles River Breeding, Wilmington, MA) or from spinal
cords of adult Osborn-Mendel rats (Veterinary Resources Branch at
the National Institutes of Health) according to Norton and
Poduslo (8). Standard 6-dehydrocerebrosides were prepared by
oxidizing with galactose oxidase according to Radin (9).

Instrumentation. The HPLC equipment consisted of two Model
740 solvent delivery systems combined with a Model 744 solvent
programmer, Model 714 pressure monitor and a Model 755 sample
injector (all from Spectra-Physics, Santa Clara, CA). The column
used was 25 cm x 3 mm i.d. stainless steel tube packed with either
Spherisorb silica 5 μ or Spherisorb ODS 5 μ. Detection was made
with a Schoeffel Instrument Corporation (Westwood, NJ) Model SF
770 spectromonitor. Peak areas were measured by the cut and
weight method. Radioactivity was measured by direct measurement
in a Searle Model 1185 Automatic Gamma System.

Determination of cerebrosides, sulfatides and other galactolipids in myelin by HPLC. Total lipids were extracted with
chloroform/methanol (2/1), washed according to Folch et al. (10),
and then subjected to benzoylation-desulfation as described previously (6). The total lipids were heated with 20 μl benzoyl
chloride and 100 μl pyridine and desulfated with 0.2 M HClO$_4$ in
acetonitrile (prepared by mixing 0.17 ml 70% HClO$_4$ and 10 ml
acetonitrile). With this procedure, cerebrosides were converted
to perbenzoyl derivatives while sulfatides were converted to partially benzoylated cerebroside in which the hydroxyl group at
galactose-3 is free (Chart 2). A portion of the reaction mixture
dissolved in a known volume of hexane was injected into the HPLC
equipped with Spherisorb silica 5 μ column. The column was eluted
with hexane/isopropanol (99.5/0.5, v/v) isocratically for the
first 3 min followed by gradient elution from 0.5 to 10% isopropanol in hexane in 20 min. The flow rate was maintained at 1.2
ml/min throughout. Peaks of glucocerebroside, nonhydroxycerebroside, hydroxycerebroside, monogalactosyl diglyceride, nonhydroxysulfatide, and hydroxysulfatide were separated from each other
under these conditions, and concentrations of these lipids were
determined from the peak size. Peaks for minor nonpolar galactolipids, namely cerebroside esters and 1-0-alkyl isomers of
monogalactosyl diglycerides overlap with one of the above peaks.

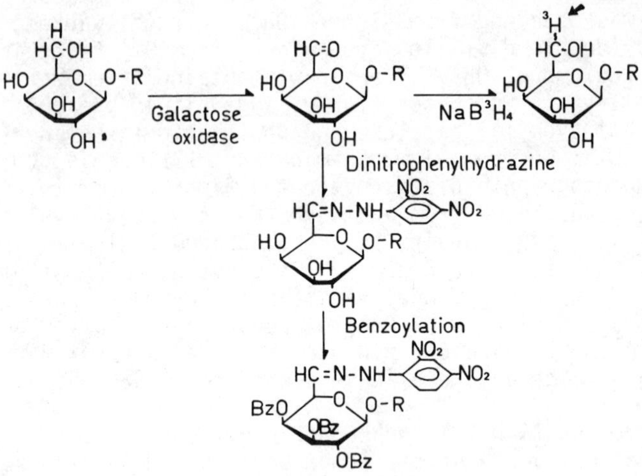
Galactose oxidase
NaB³H₄
Dinitrophenylhydrazine
Benzoylation

Chart 1

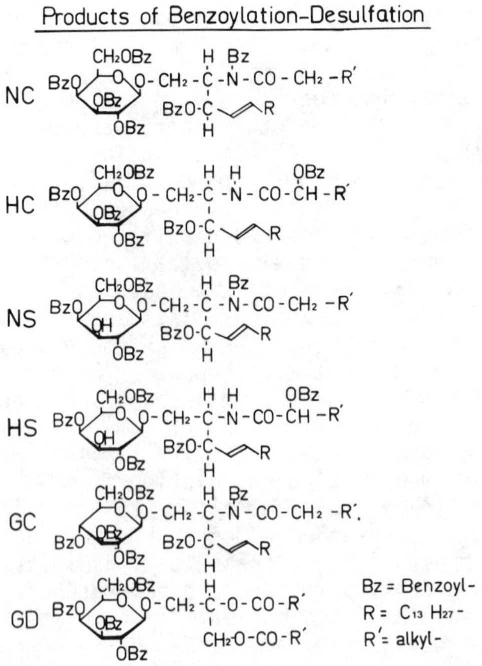
Products of Benzoylation–Desulfation
NC
HC
NS
HS
GC
GD
Bz = Benzoyl-
R = C13 H27-
R′= alkyl-

Chart 2

Eluting the column isocratically with hexane/tetrahydrofuran (90.25/9.75, v/v) provides separation of the above overlapped peaks (11).

Determination of homolog compositions of cerebrosides, sulfatides, monogalactosyl diglyceride, and its 1-0-alkyl ether isomer by reverse phase HPLC. The above benzoylation-desulfation product is placed on a TLC plate coated with Silica Gel GF. At least 1 nmol (approximately 1 µg) is required to obtain satisfactory results for each individual lipid. The plate is developed with hexane/isopropanol (98/2, v/v) once or twice depending on the relative activity of the plate. After the first development, the plate was examined under UV light. If each component is sufficiently separated, as shown in Fig. 1, a second run is not necessary. The spots were marked under the UV light allowing 1/2 height of the spot on the top and bottom of each spot (or band) so that any particular homolog was not selectively missed. The powder from the spot was scraped and mixed with 0.5 ml of 95% ethanol. The mixture was sonicated in a sonic cleaner bath for 2 min. 1.5 ml Of ether was added to the mixture and then shaken vigorously for 30 min with a W-8 Twist Action Shaker (New Brunswick Instrument, New Brunswick, NJ). The mixture is then centrifuged, and the supernatant is evaporated to dryness. The residue is dissolved in a known volume of cyclohexane, and a portion is injected to HPLC equipped with Spherisorb ODS 5 µ column.

Although spots of perbenzoylated nonhydroxycerebroside, monogalactosyl diglyceride, and hydroxycerebroside, and perbenzoylated-desulfated nonhydroxy- and hydroxysulfatide are well separated from each other, the spot of benzoylated derivative of 1-0-alkyl ether isomer of monogalactosyl diglyceride overlaps with that of benzoylated nonhydroxycerebroside. The amount of the 1-alkyl,2-acyl,3-monogalactosyl glycerol is normally so small that it will not interfere significantly with the analysis of non-hydroxycerebroside. However, if the analysis of this minor glycolipid is desired, the material eluted from the band of benzoylated nonhydroxycerebroside can be rechromatographed on another TLC system, such as the use of hexane/tetrahydrofuran on Silica Gel GF plate. These two benzoylated lipids separate well from each other under this condition. If the examination of homolog composition of monogalactosyl diglyceride and its 1-0-alkyl ether isomers is desired, a larger amount of brain sample is required, since their concentration is much smaller than cerebrosides and sulfatides.

HPLC of galactolipids from a membrane treated with galactose oxidase. The membrane treated with galactose oxidase is extracted with chloroform/methanol as described above. The extract containing up to 1 mg of total lipids is shaken with a solution of 2 mg of dinitrophenylhydrazine HCl in 100 µl pyridine for 2 h at room temperature. The solvent is evaporated to dryness under a nitrogen flow, and the residue is further dried in an evacuated desiccator for 1 hr. To the dried residue, 30 µl of benzoyl chloride and 150 µl of dry pyridine is added, and the mixture is heated at 60° for 1 h. The reaction mixture is evaporated to dryness under a nitrogen flow, and the residue is further dried in an evacuated desiccator for 30 min. The residue is dissolved in 2 ml hexane. The hexane solution is washed once with 1 ml of 3% aqueous sodium carbonate followed twice by acetonitrile/water (4/1, v/v) and then evaporated to dryness.

The residue is dissolved in a known volume of hexane and injected into the HPLC system equipped with Spherisorb Silica 5 µ column. The column was first eluted isocratically for the first 5 min with hexane/isopropanol (99.5/0.5, v/v) and then by increasing linearly the proportion of isopropanol in the next 20 min reaching the final concentration of hexane/isopropanol (96/4, v/v). The flow rate was maintained at 1.2 ml/min throughout. Two peaks due to perbenzoylated products of 2,4-dinitrophenylhydrazone of oxidation products from nonhydroxy- and hydroxycerebrosides appear after the peak of benzoylated hydroxycerebrosides under these conditions.

Treatment with galactose oxidase. Oxidation of myelin with galactose oxidase was performed as described previously for similar oxidation of rat spinal cord preparations (4). Typically, myelin containing 0.2-1.1 mg protein is incubated with 100-500 units of galactose oxidase in 1-3 ml of phosphate buffer (10-100 mM, pH 7.2-7.4) with or without catalase. After the incubation at room temperature to 30°C for the duration of 30 min to overnight, myelin is recovered by centrifugation, washed, and lyophilized. Total lipids were extracted from the dried residue and the oxidized cerebroside as well as unaltered cerebrosides were analyzed as described above. Alternatively, the incubation was stopped by the addition of 5 volumes of chloroform/methanol (2/1, v/v) and mixed. The lower layer after centrifugation of the mixture is washed and then evaporated to dryness, and the total lipids obtained were analyzed as described above.

Radioiodination of galactose oxidase. The chloramine T procedure (12) was used for the radioiodination of galactose oxidase. The enzyme, solubilized in 0.01 M sodium phosphate buffer, pH 7.4, was labeled using 1 mCi Na ^{125}I (13-17 mCi/ngI), 0.42 mM chloramine T (Eastman), and 1.14 mM sodium metabisulfite (Baker). Unreacted iodide was separated from the iodination enzyme by dialysis, and the enzyme was diluted in 0.1 M sodium phosphate buffer, pH 7.4, containing 0.001 M cupric sulfate.

Preparation of liposomes. The method for the liposome preparation was a modification of Costantino-Ceccarini, et al., (13). A mixture of 0.3 mg nonhydroxycerebroside, 0.2 mg of hydroxycerebroside, and 5 mg of egg yolk lecithin was swollen in 1 ml of a solution containing 130 mM KCl and 10 mM Tris-HCl pH 7.4 for 30 min. The tube was flashed with nitrogen and sonicated for 30 min. The sonicated mixture was centrifuged at 50,000 x$\underline{g}$ for 15 min to remove solid cerebrosides. The liposome solution obtained contained 165 µg of nonhydroxycerebroside and 110 µg of hydroxycerebroside in 1 ml when measured by high performance liquid chromatography.

Results

Myelin galactolipid analysis by HPLC. Fig. 2 and 3 show HPLC chromatograms obtained from myelin lipids on Silica column and reverse phase column, respectively. Reverse phase HPLC of monogalactosyl diglyceride and its 1-0-alkyl ether homolog was not examined but typical chromatograms of these lipids obtained from calf brain stem were presented previously (11). Myelin was obtained from 25 day-old rat brains.

Oxidation of myelin surface cerebrosides by galactose oxidase. Fig. 4 shows silica HPLC of a mixture containing benzoylated-nonhydroxy and hydroxycerebroside and benzoylated derivatives of 2,4-dinitrophenylhydrazone of oxidation products from nonhydroxy- and hydroxycerebroside. Standard curves of two 6-dehydro-derivatives were shown in Fig. 5. These standard curves demonstrate that the response of the benzoylated dinitrophenylhydrazones are linear between 0.025 nmol and 0.6 nmol. Since cerebrosides containing 5 nmol can be determined without tailing to these peaks, this method should allow the determination of as little as 0.5% of the oxidation product. The fact that each curve intersects 0 point in both the abscissa and ordinate indicates that even smaller amounts of these compounds can be detected by this technique.

We obtained unexpected findings using this method to study myelin: the oxidation by galactose oxidase of myelin-bound cerebrosides could not be detected. The oxidation did not occur either with the intact spinal cord preparation, with isolated myelin, or even with lyophilized myelin. In one experiment, lyophilized myelin containing 5 mg each of dry weight was incubated with 100, 200, and 500 units of galactose oxidase for 60 min at room temperature, and no cerebroside oxidation occurred. To examine whether the enzyme is active under the same conditions, we coated 0.1 mg each of nonhydroxy- and hydroxycerebrosides on 10 mg Celite (Analytical grade) and incubated it with 100 units of galactose oxidase for 60 min at room temperature. The result indicated that 5.6 nmol and 3.5 nmol each of nonhydroxy and hydroxycerebrosides (approximately 4.6 and 3.0% each were oxidized. Oxidation of the same cerebrosides by the same galactose oxidase in a tetrahydrofuran/water mixture as described by Radin (9) resulted in nearly complete oxidation.

To further examine the inability of galactose oxidase in oxidizing myelin-bound cerebrosides, one mg each of lyophilized

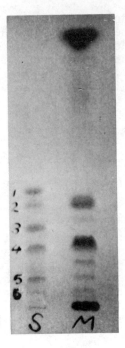

Figure 1. TLC of myelin lipids after treatment with perbenzoylation–desulfation. Line S, standard; line M, derivatized myelin lipids. Spots 1 through 6 are benzoylated–desulfated derivatives of: (1) glucocerebroside, (2) nonhydroxycerebroside, (3) monogalactosyl diglyceride, (4) hydroxycerebroside, (5) nonhydroxysulfatide, and (6) hydroxysulfatide, respectively. See text for details of TLC conditions.

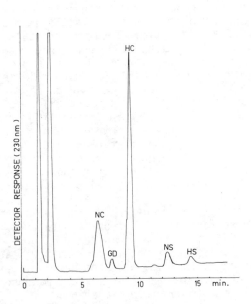

Figure 2. Silica HPLC of myelin lipids. NC, nonhydroxycerebroside; HC, hydroxycerebroside; NS, nonhydroxysulfatide; HS, hydroxysulfatide; and GD, monogalactosyl diglyceride. See text for details of TLC conditions.

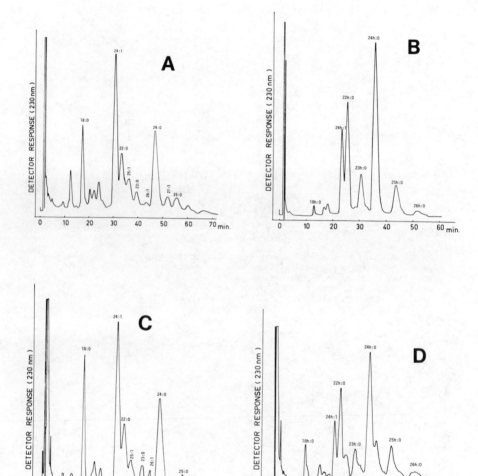

Figure 3. Reverse-phase HPLC of (A) perbenzoylated nonhydroxycerebroside, (B) hydroxycerebroside, (C) perbenzoylated–desulfated nonhydroxysulfatide, and (D) hydroxysulfatide. Each homolog peak was identified by fatty acid symbols, carbon numbers followed by number of double bonds.

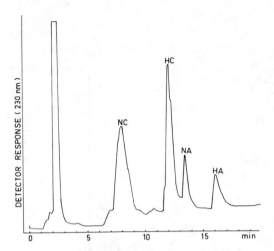

Figure 4.   Silica HPLC of perbenzoylated derivative of dinitrophenylhydrazone of
6-dehydrocerebrosides.

50 μg of nonhydroxycerebroside and 30 μg hydroxycerebroside is mixed with equal amounts of 6-dehydroderivatives of hydroxy- and nonhydroxycerebroside. These mixtures were subjected to dinitrophenylhydrazone–benzoylation as described above, and 1/20 of each reaction mixture was injected into silica–HPLC. NC, nonhydroxycerebroside; HC, hydroxycerebroside; NA, 6-dehydrononhydroxycerebroside; HA, 6-hydrohydroxycerebroside.

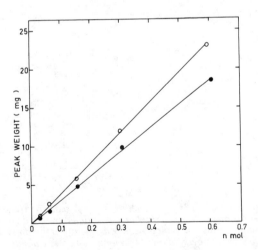

Figure 5.   Standard curve of perbenzoylated derivative of dinitrophenylhydrazone
of 6-dehydrocerebrosides as analyzed by silica HPLC. Open and closed circles:
derivatives from nonhydroxycerebroside and hydroxycerebroside, respectively.

myelin was wetted with 0.25 ml of benzene or benzene containing 167 nmols and 397 nmols each of nonhydroxy and hydroxycerebrosides, respectively. These were again lyophilized. Each dried residue was incubated with 150 units of galactose oxidase in room temperature overnight. The results indicated that the reaction product from the lyophilized myelin which was relyophilized with benzene alone showed no detectable oxidation. On the other hand, the product of the same myelin preparation but "coated" with cerebrosides showed 12.5 nmol and 8.8 nmol of nonhydroxy- and hydroxycerebroside oxidized by the enzyme reaction, as shown in Fig. 6 and Fig. 7.

One possible explanation for the lack of oxidation by galactose oxidase was thought to be steric hindrance. To investigate this possibility, lyophilized myelin containing 5.45 mg protein in 5 ml 0.1 M phosphate buffer, pH 7.4, was mixed with 0.5 ml of the same buffer solution containing 1400 units of trypsin and incubated for 1 h at 37°C. 5 mg Of trypsin inhibitor in 1 ml of the same buffer was added and the mixture was then incubated for 30 min at the same temperature. Galactose oxidase (942 units) in 1 ml of the same buffer was next added to the above mixture, and the incubation continued for 1 h at room temperature. This experiment, however, gave no evidence that oxidation by galactose oxidase occurred. In another experiment, 3.3 mg of lyophilized myelin was incubated with 3.3 mg of phospholipase C (6 units/mg) in 10 ml of buffer containing 10 mM Tris-HCl pH 7.4, 1 mM $CaCl_2$ at 37°C for 2 h (14). The reaction mixture was centrifuged at 44,000 xg for 1 h and the pellets obtained were rehomogenized in 1 ml of 10 mM pH 7.2 phosphate buffer. The homogenate was then incubated with 200 units of galactose oxidase at room temperature overnight. The incubation product did not show any detectable oxidation.

The inability of galactose oxidase to oxidize myelin-bound cerebrosides may also be due to the absorption of the enzyme by myelin. We examined this possibility by labeling galactose oxidase with [125]I. In one experiment, freshly prepared myelin containing 2.09 mg protein was incubated with 3.77 units of ^{125}I-labeled galactose oxidase containing 226,500 cpm at room temperature for various periods of time, and the mixture was centrifuged at 16,000 rpm. The radioactivity in the supernatant was counted by a γ-counter. In another experiment, the same amount of myelin was incubated under identical conditions except that 192.2 units of galactose oxidase containing the same amount of radioactivity was used. The results of these experiments, shown in Table 1, demonstrate that the galactose oxidase indeed was bound to myelin. The binding appears to be saturated within 5 min incubation. With 3.77 units of galactose oxidase used, the average of 1.67 units (44.3% of added enzyme) was bound to myelin containing 2.09 mg protein. On the other hand, when 192.2 units of the enzyme was incubated, an average of 13.5% which is 25.9 units, was bound to the same amount of myelin.

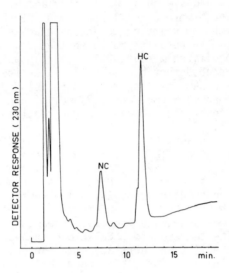

Figure 6.   Silica HPLC of product from myelin, which was treated with benzene alone and lyophilized.   See caption to Figure 4 for peak identification.

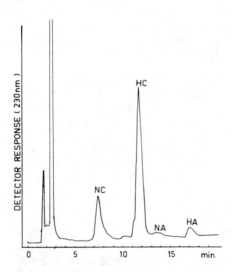

Figure 7.   Silica HPLC of product from galactose oxidase-treated myelin which were "coated" by cerebrosides.   See caption to Figure 4 for peak identification.

Table I

^{125}I-Labeled Galactose Oxidase Binding to Myelin

Incubation Time (min)	^{125}I-Galactose Oxidase Bound				
	5	10	20	30	60
			cpm		
^{125}I-Galactose Oxidase Added					
3.77 units (226,500 cpm)	102,686	97,348	98,625	101,011	102,707
192.2 units (226,500 cpm)	31,522	27,582	30,321	30,838	32,130

In a separate experiment, different amounts of ^{125}I-labeled galactose oxidase were incubated for 5 min at room temperature with fresh myelin containing 545.3 µg protein. The result, as shown in Fig. 8, indicates that the binding is saturable and the saturated amount is 4.76 units of galactose oxidase for the myelin used, or 8.73 units per mg of myelin protein. Although these experiments show that the binding of galactose oxidase to myelin occurs, they also demonstrate that the degree of absorption is too small to explain the inability of galactose oxidase to oxidize myelin-bound cerebrosides.

Oxidation of other forms of cerebrosides by galactose oxidase. Microsomes (5.49 mg protein) and cytosol (7.15 mg protein) from 25 day-old rat brain, which were prepared as described previously (6), were each incubated with 100 units of galactose oxidase overnight at room temperature. The results indicate that oxidation of cerebrosides in these brain sub-cellular fractions was not detected. We tested whether cerebrosides in artificial membrane could be oxidized by this enzyme. Liposomes were prepared as described in Procedures. 0.2 ml Of the liposomes which contained 40 µg nonhydroxycerebroside and 33 µg hydroxycerebroside was incubated with 100 units of galactose oxidase at room temperature overnight. The examination of the product as described above shows that cerebrosides in liposomes was not oxidized. Another aliquot (0.2 ml) of the liposomes was mixed with 1 ml tetrahydrofuran, 1.0 ml of 10 mM phosphate buffer, pH 7.2 containing 100 units of galactose oxidase. More than 90% of cerebrosides were oxidized by this method.

Discussion

Our new method of sphingolipid analysis using high performance liquid chromatography allows us to determine not only their quantities but also their homolog compositions in a small amount of tissue. We now feel that less than 1 mg of fresh brain or nerve tissue is sufficient for complete analysis. The application of this new method for analyzing cerebroside and sulfatide in plaques of brain from a patient with multiple sclerosis has been recently described (15).

We further extended this method to distinguish those glyco-lipids on a cell or membrane surface from those that are inaccessible to oxidation by galactose oxidase. This method is based on the impermeability of galactose oxidase. The oxidized and unreacted galactolipids can be determined separately but simultaneously by HPLC. In this method, the aldehyde, obtained by treatment with galactose oxidase, is first converted to 2,4-dinitrophenylhydrazone and then perbenzoylated before the chromatographic analysis. Although the cerebroside aldehydes could be benzoylated directly and analyzed, the peaks of the perbenzoylated aldehydes emerged slightly earlier than the corresponding perbenzoylated

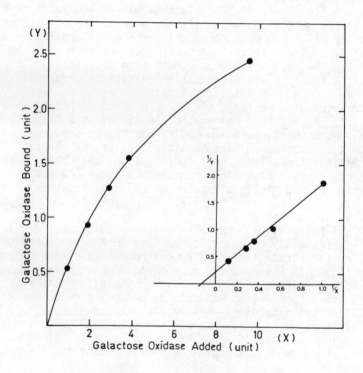

Figure 8. Binding of ^{125}I-galactose oxidase on myelin

cerebrosides on HPLC. Both aldehyde peak areas from the non-hydroxy- and hydroxycerebroside, however, are often contaminated by unidentified minor components, and this made the analysis inaccurate. Consequently, conversion to the hydrazone was performed.

Using the new procedure, attempts were made to quantitate the cerebrosides located on the surface of myelin. Myelin is composed of multilamellar bilayers of membrane of approximately 70% lipid and 30% protein (16). About 20% of the total lipid consists of cerebroside and sulfatide. Because of the lipophilic nature of the ceramide moiety and the hydrophilic nature of galactose, it has been postulated that the galactose moiety of myelin cerebrosides is facing the surface while the ceramide moiety is buried within the bilayer. Even considering the multilamellar structure of myelin, at least several percent of the cerebrosides should be present on the myelin surface. The method described in this manuscript should allow us to determine surface cerebrosides to as little as 0.5% of the total cerebrosides.

Unexpectedly, we found that myelin cerebrosides are not oxidizable by galactose oxidase, at least not in a detectable degree. We have attempted to modify the myelin structure so that galactose oxidase would have accessibility to the cerebrosides. These manipulations included lyophilization, sonication, hypotonic treatment, trypsin digestion, and phospholipase C digestion. Disruption of the myelin structure using these treatments has been reported (17). In fact, the effect of phospholipase C was obvious from the examination of lipids by thin-layer chromatography; nearly all phosphatidyl choline and ethanolamine were degraded.

This inability of galactose oxidase to oxidize cerebrosides is a direct contradiction to a recent report by Linington and Rumsby (18). In their study, cerebrosides which were not oxidized by galactose oxidase were compared with cholesterol by GLC. The cerebroside determination was made by measuring galactose after the methanolysis; oxidized cerebroside yielded 6-dehydrogalactose which was found unstable under methanolysis conditions. By measuring the loss of galactose relative to the cholesterol content, they found that approximately 36-50% of the cerebrosides in myelin were attacked by galactose oxidase.

The reason for this discrepancy between our present study and the finding of Linington and Rumsby is not clear. Our enzyme was very active. It oxidized nearly all cerebrosides when reacted in a tetrahydrofuran/buffer system. When cerebrosides were coated on celite or myelin, galactose oxidase attacked them. Two possible causes for the inability of this enzyme to oxidize cerebroside in isolated myelin were considered. The first cause may be due to the absorption of galactose oxidase by myelin by either ionic or hydrophobic interactions. If a portion of galactose oxidase is hydrophobic, it is possible that the enzyme can be incorporated into the lipid matrix. Accordingly, we labeled galactose oxidase with ^{125}I and found that such absorption was insignificant compared to the total amount of enzyme present during the incubation.

The second cause may be due to steric hinderance of neighboring components within the myelin sheath. However, we found that digestion of trypsin or phospholipase C cannot alleviate the problem. In addition, hypotonic treatment of myelin which affects the integrity of the bilayer and also causes the splitting of the lamellae at the external apposition (19, 20), did not result in the oxidation of the cerebrosides. Even cerebrosides in liposomes made from pure phosphatidyl choline could not be oxidized. Incidentally, this finding also contradicts Linington and Rumsby who reported significant oxidation of cerebrosides in liposomes made from myelin lipids.

At this time, our only alternative explanation to our findings is that galactose oxidase may not be able to oxidize the "bound" form of cerebrosides possibly because of size restrictions at the active site of the enzyme. This form includes cerebrosides in membranes, lyposomes, or micells. Matsubara and Hakomori also recently determined the proportion of lactosylceramide and globoside located in erythrocyte surface membranes (T. Matsubura and S. Hakomori, personal communication). In this study, they treated erythrocytes with galactose oxidase, reduced it by NaBD$_4$ treatment, and measured the ratio of the deuterated lipid against undeuterated lipid with mass spectrometry. They found somewhat more oxidation; 2-3% of lactosyl ceramide, and approximately 10% of globoside were labeled with deuterium. Therefore, it is likely that the longer the saccharide chain to which galactose or galactosamine is attached, the higher the rate of oxidation that can be achieved by galactose oxidase.

Although, as described above, Linington and Rumsby reported up to 50% of the oxidation of myelin cerebrosides by galactose oxidase, they also reported very little labeling of myelin cerebrosides by the galactose oxidase -- NaB^{3}H$_4$ method. They oxidized myelin (75 mg of dry weight), which presumably contained approximately 10 mg of cerebrosides in 50 mg of total lipids, with 900 units of galactose oxidase and reduced it with 5 mCi of NaB^{3}H$_4$. After 5 hrs of incubation, they obtained 2,261,450 dpm (approximately 1 μCi) of ^{3}H in cerebrosides. Although the specific activity of NaB^{3}H$_4$ used was not given, the cerebrosides labeled could be about 0.1-0.2 μg, assuming the specific activity was in the range of 5-15 Ci/mmol as reported by Poduslo et al., (4) and also assuming that this radioactivity represents specific labeling of the galactose moiety. This amount of cerebroside, therefore, represents only 0.001-0.002% of the total cerebroside.

A number of similar studies on cell surface galactolipids have been based on this galactose oxidase-NaB^{3}H$_4$ reduction procedure. However, it is now apparent that only a very small portion, less than 0.5% if any, of the cerebrosides in membranes are oxidizable by galactose oxidase. Therefore, cerebrosides and possibly other galactolipids previously identified by the surface labeling

technique apparently represent only a small portion of the total
surface galactolipids, and the results of such studies should be
interpreted with caution.

Abstract

        An HPLC method is described which determines the quantity and
elucidates the homolog composition of cerebrosides and sulfatides
in small tissue samples. Total lipids from the tissue were subjected to benzoylation-desulfation, and the product was analyzed
quantitatively by silica HPLC. Another aliquot of the product was
further fractionated by TLC. Spots of benzoylated cerebrosides
and desulfated sulfatides were analyzed by reverse phase HPLC for
the homolog compositions of these sphingolipids. Less than 1 mg
of fresh brain or nerve tissue is sufficient for complete analysis.
A new procedure has been developed which assesses the topographical distribution of cerebrosides in biological membranes. This
method involves the treatment of cells or membrane fractions with
galactose oxidase followed by extraction of the total lipids with
chloroform-methanol. The lipids were then reacted with 2,4-
dinitrophenylhydrazine HCl in pyridine, and the reaction products
were benzoylated and analyzed by silica HPLC. The cerebrosides
which are oxidized by the enzyme resulted in perbenzoylated
derivatives of 6-dehydrocerebrosides which yielded separate peaks
behind the unoxidized perbenzoylated cerebrosides. Consequently
this procedure would distinguish surface membrane cerebrosides
from the unreactive "inaccessible" cerebrosides. This technique
was applied to myelin from the central nervous system, and
unexpectedly, myelin cerebrosides were found unoxidizable by
galactose oxidase. Modifications of myelin, such as lyophilization, hypotonic treatment, trypsin digestion, and phospholipase
C digestion, were not effective in exposing myelin-bound cerebrosides. Moreover, we also found that cerebrosides bound to brain
microsomes, cytosol, or even in liposomes with lecithin were not
oxidized by the enzyme. On the other hand, cerebrosides coated
on Celite or myelin were oxidized by the enzyme. These results
suggest that cerebrosides bound in a bilayer structure may not be
available for oxidation by galactose oxidase.

Acknowledgement

        This study was supported in part by Research Grants NS-13559,
NS-13569 and HD-10891 from the National Institutes of Health,
United States Public Health Service.

Literature Cited

1. Steck, T.L., "Membrane Research"; Academic Press, New York,
    1972; pp. 71-93.

2. Steck, T.L. and Dawson, G. J. Biol. Chem. 1974, 249, 2135-2142.
3. Gahmberg, C.G. and Hakomori, S. J. Biol. Chem. 1973, 248, 4311-4317.
4. Poduslo, J.F.; Quarles, R.H. and Brady, R.O. J. Biol. Chem. 1976, 251, 153-158.
5. Poduslo, J.F. Adv. Exp. Med. Biol. 1978, 100, 189-205.
6. Nonaka, G. and Kishimoto, Y. Biochim. Biophys. Acta 1979, 572, 423-431.
7. Yahara, S.; Moser, H.W.; Kolodny, E.H. and Kishimoto, Y. J. Neurochem. in press.
8. Norton, W.T. and Poduslo, S.E. J. Neurochem. 1973, 21, 749-757.
9. Radin, N.S. Methods Enzymol. 1972, 28, 300-304.
10. Folch, J.; Lees, M. and Sloane-Stanley, G.H. J. Biol. Chem. 1957, 226, 497-509.
11. Yahara, S. and Kishimoto, Y. manuscript in preparation.
12. Greenwood, F.C.; Hunter, W.M. and Glover, J.S. Biochem. J. 1963, 89, 114-123.
13. Cestelli, A.; White, F.V. and Costantino-Ceccarini, E. Biochim. Biophys. Acta 1979, 572, 283-292.
14. Feinstein, M.B. and Felsenfeld, H. Biochemistry 1975, 14, 3041-3048.
15. Kawamura, N.; Yahara, S.; Kishimoto, Y. and Toutelotte, W.W. manuscript in preparation.
16. Norton, W.T. and Poduslo, S.E. J. Neurochem. 1973, 21, 759-773.
17. Smith, M.E. and Benjamins, J.A. "Myelin"; Plenum Press, New York, 1977, pp. 447-488.
18. Linington, C. and Rumsby, M.G. Adv. Exp. Med. Biol. 1978, 100, 263-273.
19. Finean, J.B. and Bunge, R.E. J. Mol. Biol. 1963, 1, 672-682.
20. McIntosh, T.J. and Robertson, J.D. J. Mol. Biol. 1976, 100, 213-217.

RECEIVED December 10, 1979.

Analysis of Glycosphingolipids by Field Desorption Mass Spectrometry

C. E. COSTELLO, B. W. WILSON[2], and K. BIEMANN
Massachusetts Institute of Technology, Cambridge, MA

V. N. REINHOLD
Harvard Medical School, Boston, MA 02115

Conventional electron impact or chemical ionization mass spectrometry requires that volatilization precede ionization and this is clearly a limiting factor in the analysis of many bio-chemically significant compounds. A newer ionization technique, field desorption (FD) ([1,2]) removes this requirement and makes it possible to obtain mass spectrometric information on thermally unstable or non-volatile organic compounds such as glycoconjugates and salts. This development is particularly significant for those concerned with the analysis of glycolipids and we have therefore explored the suitability of field desorption mass spectrometry (FDMS) for this class of compounds. We have evaluated experimental procedures in order to enhance the efficiency of the ionization process and to maximize the information content of spectra obtained by this technique.

In FDMS, the desorption surface is a 10 μ wire covered by a dense growth of microneedles (Figure 1) produced by slowly heating the wire in a high electric field and an atmosphere of benzonitrile ([3]). The microneedles possess much three-dimensional detail and terminate in many fine tips (Figures 2,3). The material to be analyzed is applied either by dipping the emitter into a solution of the sample or by transferring a few microliters of the solution directly to the wire by means of a syringe ([4]).

The sample-laden emitter (Figure 4) is placed directly in the ion source (Figure 5) and an electric field of about 10 kV is applied. Under these conditions field strengths approaching 10^7 to 10^8 V/cm are generated at the microneedle tip and the sample then undergoes ionization and desorption. Heating of the sample may be required. Most significant, this ionization process introduces very little excess energy into the desorbed molecules and the spectra therefore frequently consist of molecular ions showing little or no fragmentation. Other ionization-desorption processes may be observed which correspond to the addition of H^+, Na^+ or K^+ or similar cationic attachment. It is thus possible to obtain

[1]Current address: Battelle Pacific Northwest Laboratories, Richland, Washington.

0-8412-0556-6/80/47-128-035$5.00/0

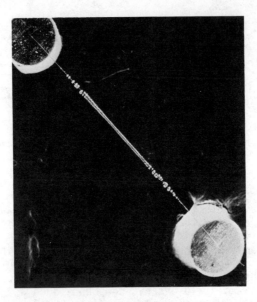

Figure 1. Electron micrograph of field desorption emitter prepared by activation of 10-μ tungsten wire in a benzonitrile atmosphere. Distance between posts is 5 mm.

Figure 2. Electron micrograph of activated emitter showing dendrite growth along 10-μ wire

Figure 3. Electron micrograph of region at dendrite tips on activated emitter

Figure 4. Electron micrograph of activated emitter to which sample has been applied by dipping emitter into a solution and allowing the solvent to air dry.

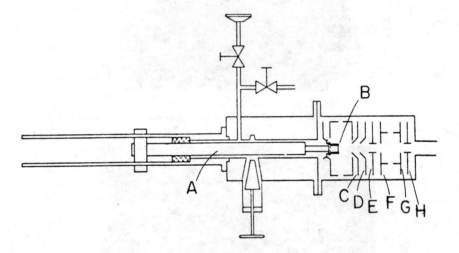

Varian MAT

*Figure 5.   Field desorption ion source showing the position of the emitter during the
analysis. A: Push rod.*

molecular weight information about nonvolatile or fragile molecules even without derivatization. By careful control of experimental conditions, it is also possible to bring about some (possibly thermal) fragmentation and thereby obtain additional structural information.

There are a number of obvious advantages to ionization by
field desorption. First, derivatization is not required (although
as we shall demonstrate below, it is sometimes advantageous).
This avoids two complications: possible lack of sample stability
during chemical manipulations and the increase in mass as a result
of derivatization, which leads to molecular weights that often
approach or even exceed the mass range of the instrument. Second,
the greater abundance of parent ions relative to fragment ions
makes possible a semi-quantitative assessment of molecular distribution in complex mixtures. Phospholipids have been studied by
this method (5,6,7) and the field desorption mass spectrum of a
glycosphingolipid, galactoceramide, has been reported (7). The
feasibility of direct analysis of polar samples without chemical
derivatization and the presence of abundant high mass ions in the
spectra make field desorption an attractive approach for mass
analysis.

We report here a study to assess the usefulness of FDMS in
the analysis of sphingolipids and glycosphingolipids, part of a
collaborative effort with the goal of developing a better understanding of the abnormal metabolism of these compounds in mammalian tissues and their implication in storage diseases. Since
benzoylation has been shown to be useful in the purification of
sphingolipids by high pressure liquid chromatography (HPLC) (8),
it also was of interest to investigate the characteristics of
these derivatives in FDMS. The field desorption mass spectra of
carbohydrates, sphingolipids and glycosphingolipids of increasingly complex structures have been obtained at different emitter
currents. In addition, permethyl, peracetyl, pertrifluoroacetyl
and heptafluorobutyryl derivatives have been prepared and the
results compared to those obtained using the underivatized
compounds.

Materials and Methods

Sphingomyelin, bovine cerebrosides, psychosine and sphingosine were obtained from Supelco, Inc. Dihydrolactocerebroside
and dihydroglucocerebroside were obtained from Miles Yeda Ltd.
Psychosine and a new analog thereof were extracted from human
brain tissue and separated by HPLC as their biphenyl carbonyl
derivatives.

Field desorption mass spectra were obtained on a Varian MAT
731 instrument (Florham Park, NJ) fitted with the combined
EI/FI/FD ion source. Emitters were prepared in the Varian apparatus according to Schulten and Beckey (3), or were pretreated
before activation by soaking in a saturated salt solution (9).

Samples were dissolved in a suitable solvent ($CHCl_3$ or 2:1 $CHCl_3/CH_3OH$) (1-10 µg/µl) and loaded by dipping the FD emitter or by adding 1-3 µl of the solution to the emitter with a micro syringe. In both cases the solvent was removed by air drying. The instrument was operated under the following conditions: Accelerating voltage, 8 or 6 KV, counter electrode voltage, 3-6 KV, ion source temperature, 100°C, emitter current increased manually to allow the recording of spectra at several temperatures. Low resolution spectra were recorded by electrical scanning at a resolution M/ΔM 1000 to 2500 depending on the molecular weight of the compound; assignments of mass were made using the instrument mass marker which had been calibrated against PFK. High resolution spectra were recorded on IONOMET photoplates at M/ΔM 5000. Some accurate mass measurements were obtained by peak matching at M/ΔM 8000.

Results and Discussion

Field desorption spectra of the sphingolipids and glycosphingolipids investigated featured intense protonated molecular ions at moderate emitter currents (19 to 22 ma). At the best anode temperature (BAT), the molecular ion clusters constituted the base peak in many of the spectra. The assignment of the $(M + H)^+$ structure to the $(M + 1)^+$ ion species was confirmed by accurate mass measurement of the sphingenine ion at m/e 300 (measured 300.2859; calculated 300.2902) (Figure 6). Attachment of a positively charged metal ion (usually Na or K) to a neutral molecule forming a positively charged complex (cationization) was observed for several of the compounds and for some, the cationized species instead of the MH^+ constituted the base peak of the spectrum. This complex may arise because salts are extracted with the sample during isolation, or may be due to association of the sample with cations present on the emitter surface when salt-saturated emitters are used (9). For example, Figure 7 shows that spectra of the biphenyl carbonyl derivatives of psychosine and a new analogue thereof are dominated by the cationized species. These materials had been purified by HPLC prior to analysis by mass spectrometry. The MH^+, $(M + Na)^+$ and $(M + K)^+$ ions in the field desorption spectrum made it apparent that these compounds differed in the degree of unsaturation, thereby answering the question of structural modification in the new compound. Cationization obviously does not prevent successful determination of molecular weights and it has even been suggested (10) to generate it deliberately to resolve ambiguities. Cationization was not observed for compounds whose exchangeable hydrogens had been replaced by derivatization.

The simplicity of FD spectra obtained at low emitter currents made possible the analysis of complex mixtures of glycolipids to obtain information about molecular weight distributions. Figure 8 shows the field desorptive mass spectrum obtained for a mixture

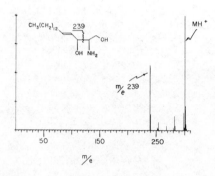

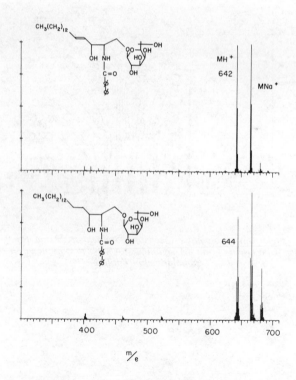

Figure 6. Field desorption mass spectrum of sphingenine, recorded at 20 ma

Figure 7. Field desorption mass spectra recorded at 22–23 ma for a biphenyl carbonyl derivative of psychosine and a biphenyl carbonyl derivative of a new compound isolated from human brain tissue. Structure indicated for the unknown was assigned on the basis of this spectrum and chemical evidence relating the unknown to psychosine. Both samples were purified by HPLC prior to FDMS.

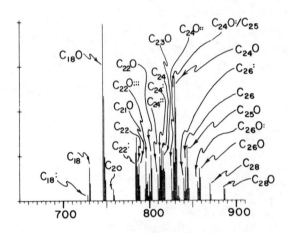

Figure 8.   Field desorption mass spectrum obtained at 22 ma for a mixture of cerebrosides from bovine brain. Assignments of MH+ are discussed in the text and summarized in Table 1.

of bovine cerebrosides. Assignment of the structures to the FD
peaks was made under the assumption that the sample consisted of
a mixture of cerebrosides varying in the nature of their side
chains and that each cerebroside gave rise to a protonated molecular ion. (For the most abundant compound, $C_{18}OH$, an ion at
m/e 785 {MH + K}$^+$ was also observed. Analogous ions are probably
also present at the same relative intensities for all other
compounds in the mixture, but the lower abundance of the components makes them less obvious.) In Table I, information regard-
ing acyl groups obtained from this single spectrum is compared
to that obtained by gas chromatographic analysis of esters of
the fatty acids obtained by methanolysis. As can be seen from
the Table, the results of the two methods are quite consistent.
Several compounds not found by GC were detected at low levels by
FDMS. (On the basis of low resolution data alone, it is not
possible to distinguish between C_n and $C_{n-1}:0$ side chains.)
Field desorption analysis therefore offers an opportunity to
survey biological extracts for abnormal distributions of these
compounds without necessitating extensive chemical workup.

At higher emitter currents, fragment ions became more significant in the spectra. A survey of compound types yielded the
spectra shown in Figures 9 and 10A+B, obtained for psychosine,
N-stearoyl dihydroglucocerebroside and N-stearoyl dihydrolactocerebroside, respectively.

The characteristic fragments, which have been observed in
this field desorption study in addition to the molecular ions,
are summarized in Scheme A. The results of Cleavages A,B,C and
F are fragments related to the aliphatic moieties of the sphingolipids, while D and E are characteristic of the polar head-
group. The non-binding orbitals of the heteroatoms in these
molecules provide sites favorable for electron removal, which
leads to ionization and subsequent fragmentation in order to
stabilize the positive charge. The strategic locations of
several heteroatoms in sphingolipid structures (the biosynthetic
consequences of conjugation) introduce points of bond lability.
The resulting fragments are important for structure determination
because they delineate the building blocks of the molecule.

Cleavage A with charge retention on the oxygen-containing
portion provides a fragment ion which makes it possible to distinguish between sphingenine and sphinganine derivatives by the
presence of an ion at m/e 239 or 241, respectively. For some
compounds, a complementary ion may be observed for charge retention on the nitrogen-containing part. Cleavage B or C with
charge retention by the amide portion of the molecule yields
important information about the length of the acyl chain attached
to the amino group by its degree of unsaturation or hydroxylation.
Cleavage D seems confined to the glycosphingolipids. The mass
difference between this ion and the molecular ion is of value in
determining the size of the carbohydrate portion of the molecule.
In the spectra of the disaccharides, Cleavage E leads to one of

TABLE 1. Fatty Acid Composition of Bovine Cerebroside

M/E MH$^+$	SIDE CHAIN	FDMS contribution	GLC contribution	SIDE CHAIN
716	$C_{16}{:}0$	ND	ND	$C_{16}{:}0$
730	$C_{18}{:}0$	3%	1%	$C_{18}{:}0$
746	$C_{18}OH$	19%	25%	$C_{18}OH$
758	$C_{20}{:}0/C_{19}OH{:}1$	<1%	<1%	$C_{20}{:}0$
774	$C_{20}OH$	<1%	<1%	$C_{20}OH$
784	$C_{22}{:}1$	5%	ND	$C_{22}{:}1$
786	$C_{22}{:}0/C_{21}OH{:}1$	3%	2%	$C_{22}{:}0$
788	$C_{21}OH$	3%	ND	$C_{21}OH$
796	$C_{22}{:}OH{:}3$	3%	ND	—
798	$C_{23}{:}1/C_{22}OH{:}2$	2%	ND	$C_{23}{:}1$
800	$C_{23}{:}0/C_{22}OH{:}1$	2%	2%	$C_{23}{:}0$
802	$C_{22}OH$	4%	8%	$C_{22}OH$
810	$C_{24}{:}2$	2%	ND	—
812	$C_{24}{:}1$	4%	2%	$C_{24}{:}1$
814	$C_{24}{:}0/C_{23}OH{:}1$	6%	7%	$C_{24}{:}0$
816	$C_{23}OH$	5%	6%	$C_{23}OH$
826	$C_{25}{:}1$	2%	4%	$C_{25}{:}1$
828	$C_{25}{:}0/C_{24}OH{:}1$	6%	8%	$C_{25}{:}0/C_{24}OH{:}1$
830	$C_{24}OH$	13%	29%	$C_{24}OH$
840	$C_{26}{:}1$	1%	ND	$C_{26}{:}1$
842	$C_{26}{:}0/C_{25}OH{:}1$	3%	1%	$C_{26}{:}0$
844	$C_{25}OH$	6%	ND	$C_{25}OH$
856	$C_{27}{:}0/C_{26}OH{:}1$	2%	3%	$C_{27}{:}0/C_{26}OH{:}1$
858	$C_{26}OH$	3%	2%	$C_{26}OH$
870	$C_{28}{:}0/C_{27}OH{:}1$	<1%	ND	$C_{28}{:}0/C_{27}OH{:}1$
886	$C_{28}OH$	<1%	ND	$C_{28}OH$

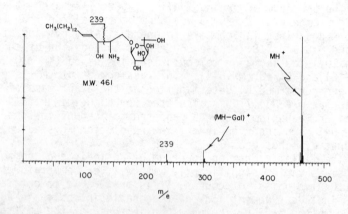

Figure 9. Field desorption mass spectrum of psychosine, recorded at 21 ma

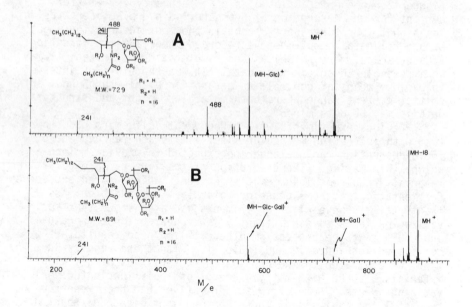

*Figure 10. Field desorption mass spectra recorded at 23–24 ma for samples of:
(A) N-stearoyl dihydroglucocerebroside and (B) N-stearoyl dihydrolactocerebroside.
(Both samples contain small amounts of the lower homolog N-palmitoyl (n = 14).*

the most abundant ions in the spectrum, which, along with the
molecular ion, allows identification of the carbohydrate
structure.

The spectrum of sphingomyelin (Figure 11) taken at higher
emitter temperatures (25-28 ma) shows a transition from the
pattern described in Scheme A to one dominated by ions at m/e
834 and 916 and 918. These correspond to addition of choline
(mass 104) to the major molecular species $C_{18}:0$ (MW 730), $C_{24}:1$
(MW 812) and $C_{24}:0$ (MW 814). This assignment of $(M + choline)^+$
has been confirmed by determination of the exact mass of the m/e
832 ion as m/e 834.7106 (calculated for $C_{46}H_{97}O_7N_3P$: 834.7060).
A similar mechanism has been proposed to account for the
$(M + 104)^+$ ion observed in the low resolution FD spectra of
phosphatidyl cholines (7).

The field desorption spectrum of an individual compound in
this class varies significantly with emitter temperatures. For
structural studies, of course, it is preferable to obtain spectra
which include fragment ions of reasonable intensity in addition
to the molecular ion. However, an increase of the emitter current
to produce fragmentation often leads to rapid depletion of the
sample and loss of signal. These effects must be balanced to
obtain maximum information. The line marked "underivatized" in
Figure 12 represents a plot of signal intensity vs. emitter
current for the desorption of N-stearoyl dihydrolactocerebroside.
The high temperature spectra contain the most information about
structural detail, but must be recorded under adverse conditions
(faster scans, more noise). When sample size is limited, it may
be impossible to obtain such data. It was therefore of interest
to explore means of controlling desorption evenly throughout an
emitter current range where both molecular weight and structural
details may be determined.

Previous experiments with carbohydrate oligomers had shown
that with increasing molecular size, limits are reached where
field desorption can no longer be expected to provide molecular
weight information. The desorption of the larger, more polar
materials required higher emitter currents which caused excessive
fragmentation at the expense of molecular ions and high mass
fragments. These earlier experiments had also shown that derivatization leads to significantly lower desorption temperatures
and could make the process smoother (11). These results are
summarized in Figure 13 and Table II. With these results in
mind, we evaluated the desorption characteristics of glycolipid
materials blocked with methyl, acetyl and polyfluorinated deri-
vatizing agents. Some of these derivatives had been reported to
be useful for analysis of these compounds by electron impact or
chemical ionization mass spectrometry (12,13,14), although for
the larger molecules, more elaborate derivatization schemes have
proved necessary (15-20).

All of the glycosphingolipid derivatives investigated
desorbed at emitter currents lower than those observed for the

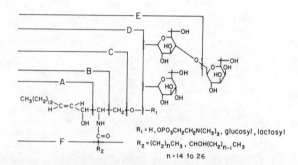

Scheme A.   Important fragmentations observed in field desorption mass spectra of sphingolipids and glycosphingolipids

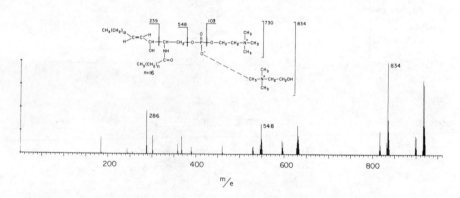

Figure 11.   Field desorption mass spectrum of sphingomyelin obtained at high emitter current (28 ma) and therefore dominated by peaks that correspond to transfer of choline (mass 104) to the three major molecular species present: n = 16, MW 730; n = 22.1, MW 812; and n = 22, MW 814.   The (M + choline) adducts are observed at m/e 834, 916, and 918, respectively.   For the higher MW compounds, the fragment at m/e 548 when n = 16 occurs at m/e 630 and 632.

48   CELL SURFACE GLYCOLIPIDS

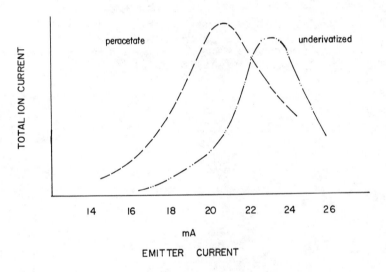

Figure 12.   Plot of total ion current vs. emitter current for underivatized (---) and peracetylated dihydrolactocerebroside (-···-)

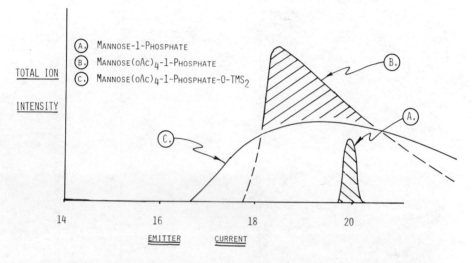

TOTAL ION

INTENSITY

Figure 13. *Plot of total ion current vs. emitter current for: (A) mannose-1-phosphate, (B) mannose (OAc)₄-1-phosphate, and (C) mannose (OAc)₄-1-phosphate-O-TMS₂ (11).*

TABLE II

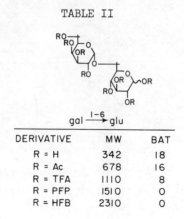

gal $\xrightarrow{1-6}$ glu

DERIVATIVE	MW	BAT
R = H	342	18
R = Ac	678	16
R = TFA	1110	8
R = PFP	1510	0
R = HFB	2310	0

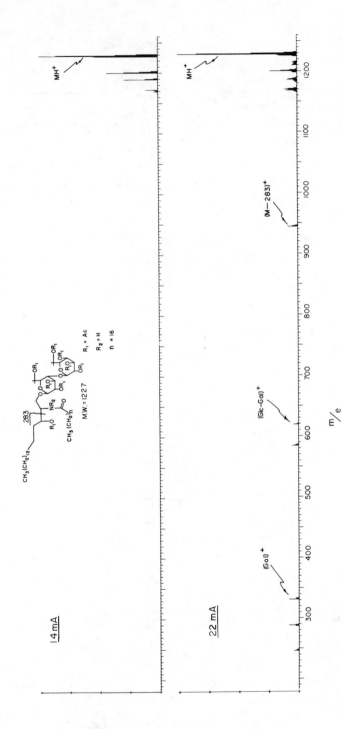

Figure 14. Field desorption mass spectra of peracetyl derivative of N-stearoyl dihydrolactocerebro-side recorded at 14 and 22 mA

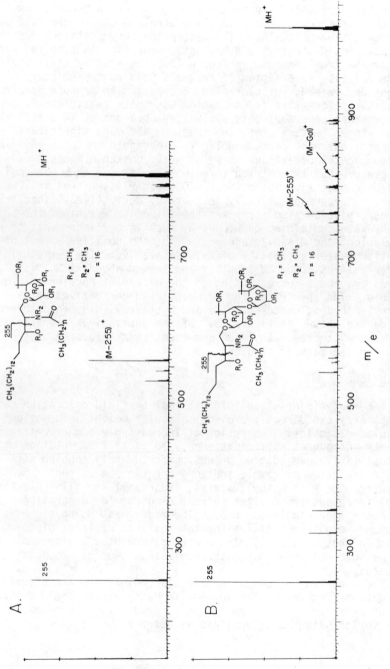

Figure 15. Field desorption mass spectra recorded at 18 ma for permethylated derivatives of: (A) N-stearoyl dihydroglucocerebroside and (B) N-stearoyl dihydrolactocerebroside.

free compounds, but not all of the spectra included structurally
diagnostic fragments. In general, the desorption process was
made much smoother and more easily controlled by the decrease in
polarity of the molecules. The desorption curve obtained for the
peracetylated N-stearoyl dihydrolactocerebroside is compared with
that of the free compound in Figure 12. The spectra still varied
in the relative intensities of fragment ions as the emitter
current was raised, but controlled desorption to produce spectra
exhibiting informative fragmentation was quite feasible. Figure
14 shows the spectra of this derivative obtained at 14 and 22 ma.
At the higher temperatures, the spectrum had many significant
fragments, but none corresponding to Cleavage A in Scheme A. The
trifluoroacetyl derivative showed a weak, unstable cluster of
ions centered at the MH$^+$ ion (m/e 1682). Its spectrum was dominated by consecutive losses of CF_3CO (mass 97), so that it
appears to be an analytically less useful derivative. Since tri-
fluoroacetylation also increases the molecular weight of the
carbohydrate-containing compounds by much larger increments than
acetylation, the disadvantages outweigh the positive aspect of
slightly greater volatility of this derivative. The permethylated derivatives desorbed smoothly and showed abundant MH$^+$ ions as
well as fragments corresponding to the different portions of the
molecule. The spectra of the permethylated derivatives of N-
stearoyl dihydroglucocerebroside and N-stearoyl dihydrolactocerebroside are shown in Figure 15. Of the derivatives prepared in
this survey, the permethylated compounds seem to be the most
useful for structural studies.

Conclusions

     The analysis of glycolipid material by field desorption mass
spectrometry can directly provide molecular weight information
on isolated samples. Desorption at increasingly higher emitter
currents introduces sufficient energy to cause fragmentation at
the most labile conjugating groups, unfortunately coupled with
a rapid depletion of sample and more erratic ion current. Sample
derivatization with methyl, acetyl, fluoroacyl and silyl blocking
groups decreases the emitter current required for desorption and
produces a more stable ion beam. The added mass using higher
molecular weight derivatives for the analysis of large glycolipids (globosides, etc.) may impose instrumental limitations in
mass measurement even though the ionization process proceeds
adequately.
     The opportunity to obtain molecular weight and structural
information from very polar involatile materials by field desorption as described in this preliminary survey offers many advantages for the study of glycolipid structures.

Acknowledgements

The authors are grateful to R.H. McCluer for providing the biphenyl carbonyl derivatives of psychosine and its new analogue, and for the gas chromatographic analysis of the methanolysis products of bovine brain cerebrosides, and to Y. Kishimoto for furnishing some of the lipids used in these studies and to both of them for many helpful discussions. The electron micrographs were kindly obtained by J.D. Geller of JEOL, USA, Inc. This work was supported by a grant from the NIH Division of Research Resources (RR00317).

Literature Cited

1. Beckey, H.D. Int. J. Mass Spectrom. Ion Physics, 1969, 2, 500.

2. Beckey, H.D. "Principles of Field Ionization and Field Desorption Mass Spectrometry"; Pergamon Press, Oxford, 1977.

3. Schulten, H.R.; Beckey, H.D. Org. Mass Spectrom., 1972, 6, 885.

4. Olson, K.L.; Cook, J.C. (Jr.); Rinehart, K.L. (Jr.). Biomed. Mass Spectrom., 1974, 1, 358.

5. Wood, G.W.; Lau, P.Y. Biomed. Mass Spectrom., 1974, 1, 354.

6. Wood, G.W.; Lau, P.Y.; Rao, G.H. Biomed. Mass Spectrom, 1976, 3, 172.

7. Wood, G.W.; Lau, P.Y.; Morrow, G.; Rao, G.H.; Schmidt, D.E. (Jr.); Teubner, J. Chem. Phys. Lipids, 1977, 18, 316.

8. McCluer, R.H.; Evans, J.E. J. Lipid Res., 1973, 14, 611.

9. Kambara, H. Personal communication.

10. Röllgen, F.W.; Schulten, H.R. Org. Mass Spectrom., 1975, 10, 660.

11. Reinhold, V.N. Paper FAMOB9, 27th ASMS Conference, Seattle, Wa., 1979.

12. Anderson, B.A.; Karlsson, K.A.; Pascher, I.; Samuelsson, B.E.; Sten, G.O. Chem. Phys. Lipids, 1972, 19, 89.

13. Markey, S.P.; Wenger, D.A. Chem. Phys. Lipids, 1974, 12, 183.

14.   Ledeen, R.W.; Kundu, S.K.; Price, H.C.; Fong, J.W. Chem.
      Phys. Lipids, 1974, 13, 429.

15.   Karlsson, K.A. FEBS Letters, 1973, 32, 317.

16.   Karlsson, K.A.; Pascher, I.; Pimlott, W.; Samuelsson, B.E.
      Biomed. Mass Spectrom., 1974, 1, 49.

17.   Karlsson, K.A. Biochemistry, 1974, 13, 3642.

18.   Karlsson, K.A.; Pascher, I.; Samuelsson, B.E. Chem. Phys.
      Lipids, 1974, 12, 271.

19.   Holm, M.; Pascher, I.; Samuelsson, B.E. Biomed. Mass
      Spectrom. 1977, 4, 77.

20.   Oshima, M.; Ariga, T.; Murata, T. Chem. Phys. Lipids, 1977,
      19, 289.

RECEIVED December 10, 1979.

High Resolution Proton NMR Spectra of Blood-Group Active Glycosphingolipids in DMSO-d_6

J. DABROWSKI—Max-Planck-Institut für Medizinische Forschung, Heidelberg, Jahnstrasse 29

H. EGGE—Institut für Physiologische Chemie, Universitat Bonn, Nussallee 11

P. HANFLAND and S. KUHN—Institut für Experimentell Hamatologie und Bluttransfusionswesen, Universitat Bonn, Annaberger Weg

It has been shown recently that the application of high resolution proton magnetic resonance spectroscopy is a useful technique for the structural elucidation of glycosphingolipids. Martin-Lomas and Chapman (1) and the group of Karlsson (2) investigated acetylated and methylated and reduced derivatives of glycosphingolipids. This group succeeded in analyzing non-derivatized cerebrosides (3) and higher glycosphingolipids by high resolution NMR spectroscopy. In order to avoid solubility problems, like formation of micelles and aggregates in aqueous solution, these products have been successfully measured in DMSO-d_6. In simpler glycosphingolipids such as glucosylceramide, galactosylceramide and lactosylceramide the signals of all protons that are linked to carbons bearing negatively charged substituents could be assigned. In this paper the results obtained with some of the more complex glycosphingolipids are presented.

In the oligosaccharide moiety of these glycosphingolipids, well analyzed by conventional methods, the linkages of the sugar components glucose, galactose, N-acetyl-glucosamine and N-acetylgalactosamine exhibit a number of variations with respect to sequence, anomeric configuration and site of attachment. The observed changes of the H^1 and H^2 resonances of the sugar rings, resulting from the interaction of the neighboring sugar units, can be condensed to a number of rules. These rules greatly facilitated the complete structural elucidation of a hitherto unknown ceramidedecasaccharide isolated from rabbit erythrocyte membranes (4).

Methods

The spectra were obtained at 338° K on the Bruker HX-360 spectrometer equipped with a Bruker 2000 computer with 32 K memory capacity. The operating frequency was 360 MHz and the spec-

0-8412-0556-6/80/47-128-055$5.00/0

tra widths amounted to 3.3 KHz. The deuterium-exchanged samples
were dissolved in DMSO-d_6 containing 2% D_2O. The sample
concentration amounted to 0.2%. The free induction decays were
mutiplied by a resolution enhancement function (Lorentzian-to-
Gaussian transformation). The chemical shifts of the H^2 protons of all sugar units were determined by spin-decoupling difference spectroscopy (SDDS).

Results and Discussion

From the data presented in Tables 1 and 2, two sets of
rules can be deduced concerning the influence of substitution on
the chemical shifts of the H^1 and H^2 protons of the sugar
rings. Considering one sugar in an oligosaccharide chain, the
shifts of H^1 and H^2 of this sugar will be influenced a) by
the sugar units attached to C_3, C_4 or C_6 and b) by the
aglycon or sugar moiety to which it is glycosidically linked itself.

a) The H^1 signal of a β-linked sugar (coupling constant
   $J_{1.2}$ 8 Hz) is shifted about 0.05 - 0.07 ppm to lower
   field after substitution with another sugar unit. Further
   extension of the oligosaccharide chain to the nonreducing
   end has no apparent effect on the position of this signal.
   The only exception to this rule is observed in the spectrum
   of the Forssman hapten (V) (Fig. 5, Table 1). This downfield shift, though regularly observed in all compounds so
   far analyzed, is neither specific with regard to the type
   nor the site of attachment of the substituting sugar, like
   position 3 or 4 in the series presented here.

   The H^1 signal of an α-linked galactose (coupling constant
   $J_{1.2}$ 4 Hz) is shifted downfield by only 0.02 ppm after
   substitution by another sugar. Again, further extension of
   the sugar chain towards the non-reducing end has no effect
   on the position of the signal, exactly as in the case of the
   β-linked sugars.

b) In contrast to the relations just described, the chemical
   shifts of the anomeric protons are influenced in a rather
   specific way by the type of sugar or aglycon to which the
   sugar under observation is glycosidically linked. The δH^1
   values also differ depending on the point of attachment to
   another sugar unit. Thus, the H^1 signal of a non-terminal
   galactose linked β-glycosidically to position 4 of glucose
   appears at 4.27 ppm whereas that of a galactose β-glycosidically linked to position 4 of N-acetylglucosamine appears at
   4.30 ppm. Thus, two distinctly resolved H^1 signals can be
   observed for the two β 1-4 linked galactose residues in compound (VI) (Table 1, Fig. 6). In the same way, terminal

Table 1:   H^1 and J$_{1,2}$ Coupling Constants of Glycosphingolipids in DMSO-d$_6$.

GalNAc 1 $\xrightarrow{\alpha}$ 3GalNAc 1 $\xrightarrow{\beta}$ 3Gal 1 $\xrightarrow{\beta}$ 4Gal 1 $\xrightarrow{\beta}$ 4 Glc 1 $\xrightarrow{\beta}$ 1Cer

	GalNAc	3GalNAc	3Gal	4Gal	4 Glc	1Cer
I δ (J$_{1,2}$)					4.10 (7.7)	
II δ (J$_{1,2}$)				4.23 (7.3)	4.17 (7.7)	
III δ (J$_{1,2}$)			4.81 (4.0)	4.28 (7.7)	4.17 (8.1)	
IV δ (J$_{1,2}$)		4.54 (8.1)	4.83 (3.6)	4.28 (7.7)	4.17 (8.1)	
V δ (J$_{1,2}$)	4.74 (3.6)	4.59 (8.5)	4.83 (3.6)	4.29 (7.7)	4.21 (7.7)	

Gal 1 $\xrightarrow{\alpha}$ 3Gal 1 $\xrightarrow{\beta}$ 4GlcNac 1 $\xrightarrow{\beta}$ 3Gal 1 $\xrightarrow{\beta}$ 4Glc 1 $\xrightarrow{\beta}$ 1Cer

	Gal	3Gal	4GlcNac	3Gal	4Glc	1Cer
VI δ (J$_{1,2}$)	4.85 (3.8)	4.30 (7.3)	4.70 (8.4)	4.27 (8.0)	4.17 (7.8)	

Gal 1 $\xrightarrow{\alpha}$ 3Gal 1 $\xrightarrow{\beta}$ 4GlcNAc β1

VII δ (J$_{1,2}$)   4.84 (3.6)   4.30 (7.4)   4.42 (8.1) 6   Gal 1→4GlcNAc 1 $\xrightarrow{\beta}$ 3Gal 1→4Glc 1→Cer

3

Gal 1 $\xrightarrow{\alpha}$ 3Gal 1 $\xrightarrow{\beta}$ 4GlcNAc β1

δ (J$_{1,2}$)   4.84 (3.6)   4.30 (7.3)   4.67 (8.5)   4.30 (7.3)   4.67 (8.5)   4.27   4.17 (7.7)

Table 2:   Chemical Shifts of H^2 of Glycosphingolipids in DMSO-d_6

$$\text{Gal } 1 \xrightarrow{\alpha} 4\text{Gal } 1 \xrightarrow{\beta} 4 \text{ Glc } 1 \xrightarrow{\beta} 1 \text{ Cer}$$

I			2.98
II		3.33	3.06
III	3.66	3.34	3.05

$$\text{Gal } 1 \xrightarrow{\alpha} 3 \text{ Gal } 1 \xrightarrow{\beta} 4 \text{ GlcNAc } 1 \xrightarrow{\beta} 3\text{Gal } 1 \xrightarrow{\beta} 4 \text{ Glc } 1 \xrightarrow{\beta} 1 \text{ Cer}$$

IV	3.59	3.42	3.44	3.42	3.04

VII $\text{Gal } 1 \xrightarrow{\alpha} 3 \text{ Gal } 1 \xrightarrow{\beta} 4 \text{ GlcNAc } 1$

      3.59   3.42   3.45 $\searrow^{\beta} 6$

                                              $\text{Gal } 1 \xrightarrow{\beta} 4\text{GlcNAc } 1 \xrightarrow{\beta} 3\text{Gal } 1 \xrightarrow{\beta} 4\text{Glc } 1 \xrightarrow{\beta} 1\text{Cer}$

                                              3
                                              $\nearrow 3.45$ 3.44   3.46   3.05
                                              β

      $\text{Gal } 1 \xrightarrow{\alpha} 3 \text{ Gal } 1 \xrightarrow{\beta} 4 \text{ GlcNAc } 1$

      3.59   3.42   3.44

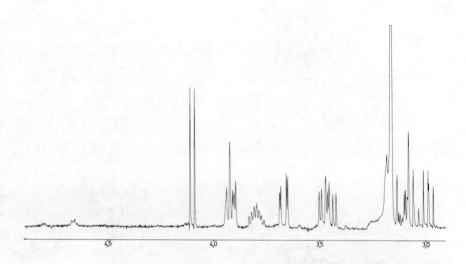

Figure 1. 360-MHz proton NMR spectrum of glucosylceramide in DMSO-d_6. For other conditions, see Methods.

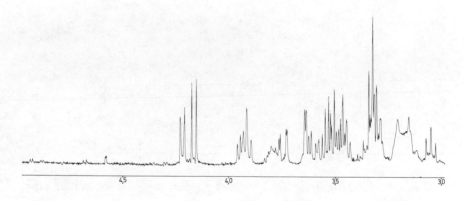

Figure 2. 360-MHz proton NMR spectrum of lactosylceramide in DMSO-d_6

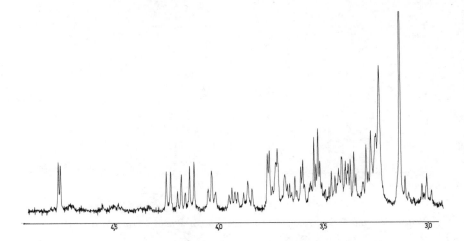

Figure 3.   360-MHz proton NMR spectrum of globotriaosylceramide in DMSO-d$_6$

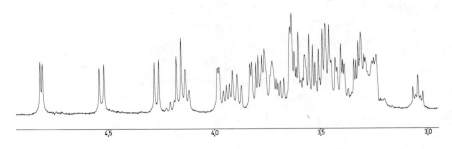

Figure 4.   360-MHz proton NMR spectrum of globotetraosylceramide in DMSO-d$_6$

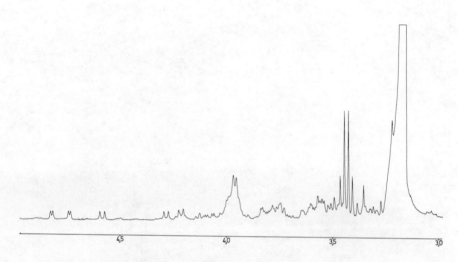

Figure 5. *360-MHz proton NMR spectrum of Forrsman hapten in DMSO-d$_6$*

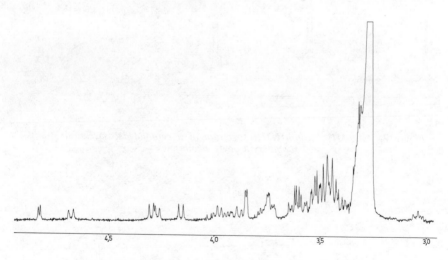

Figure 6. *360-MHz proton NMR spectrum of IV^3 Gal-β-neolactotetraosylceramide in DMSO-d$_6$*

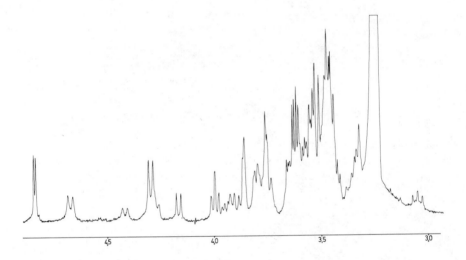

*Figure 7. 360-MHz proton NMR spectrum of a ceramidedecasaccharide from
rabbit erythrocyte membranes*

galactose residues linked α-glycosidically to position 3 or 4 of another galactose unit can be distinguished by the H^1 signals at 4.85 and 4.81 ppm, respectively (Table 1, Figs. 3 and 6). Hence, specific features of B-blood-group active glycosphingolipids can in principle also be deduced from the NMR spectra.
The chemical shifts of the H^2 protons are strongly influenced by the substitution in position 3 or 4. Thus, the sites of attachment deduced by H^1 shifts, as outlined above, can be confirmed independently by the H^2 values. These values - all determined by the SDDS method - are shown in Table 2. The sequences Gal-β-1-4-gal-β- (Fig. 3) and Gal-β-1-3-Gal-β- (Fig. 6) can be clearly distinguished by a large downfield shift of δH^2 from 3.34 to 3.42 ppm in the α-1-3 substituted galactose. The same shift can be observed in compound VI with the sequence GlcNAc-β-1-3-Gal- (Figs. 3 and 6, Table 2).
The rules exemplified above greatly facilitated the structural elucidation of a ceramidedecasaccharide isolated from rabbit erythrocytes (Fig. 7, Tables 1 and 2). The resonances at 4.17 ppm (1 proton) and 4.27 ppm (1 proton) in the spectrum of (VII) (Fig. 7) clearly correspond to the H^1 signals attributed to the sequence Gal-β-1-4-Glc-β-1-1-ceramide. The signal at 4.84 ppm, with the intensity of two protons, apparently belongs to two terminal galactoses linked α-1-3 to a galactose, as in compound (VI). The doublet at 4.30 ppm (3 protons) corresponds to three H^1 protons of galactose residues linked β-1-4 to N-acetylglucosamine. The indicated branching point is supported by the results of SDDS. After irradiation at 4.30 ppm two distinctly different spectra were obtained for the respective H^2 protons at 3.42 ppm and 3.45 ppm. Whilst overlapping resonances precluded an assignment of these signals on the basis of integrals a clear decision can be reached by comparison with the H^2 resonances of compound (VI). Since in both β-1-4 linked galactose residues the H^2 signal appears at 3.42 ppm, the signal at 3.45 obviously has to be attributed to the doubly substituted galactose. Two of the glucosamine residues exhibit a common H^1 signal at 4.67 ppm (2 protons). This shows that they form part of the sequence GlcNAc-β-1-3 Gal. On the other hand the signal at 4.42 ppm (1 proton) belongs to the glucosamine linked β-1-6 to the galactose at the branching point. Analogous upfield shifts for H^1 resonances have been reported for 1-6 linked mannose (5) and glucose (6) derivatives.
The proposed structure is in full agreement with the results obtained by mass spectrometry, immunodiffusion and analysis of partially methylated alditolacetates (4).

References

1. Martin-Lomas M.; Chapman D. Chem. Phys. Lipids 10, 152 (1973).
2. Falk, K. E.; Karlsson, K. A.; Samuelsson, B. E. Arch. Biochem. Biophys. 192, 164 (1979); ibid. 191, 177 (1979); ibid. 192, 191 (1979).
3. Dabrowski, J.; Egge, H.; Hanfland P. Chem. Phys. Lipids, submitted.
4. Hanfland, P.; Egge, H.; Dabrowski, J.; Roelke, D., in Glycoconjugates, R. Schauer, P. Boer, E. Buddeke, M. F. Kramer, J. F. G. Vliegenthart, A. Wiegandt eds., J. Thieme Verlag Stuttgart 520 (1979).
5. Dorland, L.; Haverkamp, J.; Vliegenthart, J. F. G.; Strecker, G.; Michalski, J. -C.; Fournet, B.; Spik, G.; Montreuil, J. Eur. J. Biochem. 87, 323 (1978).
6. De Bruyn, A.; van Beeumen, J.; Anteunis A.; Verhegge, G. Bull. Soc. Chim. Belg. 84, 799 (1975); A. De Bruyn (1979) Dissertation, Rijksuniversiteit Gent, Belgium.

RECEIVED December 10, 1979.

Glycophosphoceramides from Plants

ROGER A. LAINE, THOMAS C.-Y. HSIEH[1], and ROBERT L. LESTER

Department of Biochemistry, University of Kentucky College of Medicine, Lexington, KY 40536

Glycophosphoceramides contain a phosphodiester linkage between the carbohydrate moiety and the ceramide. They occur in plants and fungi (1,2,3) and have not been reported in animals. These negatively charged as well as ubiquitous glycophosphoceramides in plants may be analogous to, and rival in complexity, the sialic acid-containing glycosphingolipids in animal cell membranes, which have not been reported to occur in plants.

Carter and his co-workers (1) reported the preparation of major phytosphingosine-containing glycolipids from soybean, corn, flaxseed, peanut, sunflower seed, cotton seed, and wheat phospholipids. These materials were obtained by an alkaline saponification procedure (1 N KOH at 37°C for 24 h) which was designed to hydrolyze the esters of glycerol-containing lipids. They reported that these materials, comprising about 5% of the total crude phospholipids, were obtained as white amorphous powders of similar composition, optical activity, and solubility properties from various plant sources, and were named "phytoglycolipids". Composition analyses of these substances indicated the presence of phytosphingosine, fatty acids, phosphate, inositol, glucosamine, hexuronic acid, galactose, arabinose, and mannose. A preparation of oligosaccharides from corn phytoglycolipids (4) was obtained by barium hydroxide treatment, which presumably would not hydrolyse the glycosidic linkages of the oligosaccharide chain. The first indication of the heterogeneity of Carter's oligosaccharide preparation was provided by paper chromatography (4). They reported that all efforts to obtain separate discrete spots from the sample failed. However, partial fractionation was achieved on carbon-Celite columns eluted with increasing concentrations of aqueous ethanol. Further separation was obtained by anion ex-

[1]Current address: R & D, Brown & Williamson Tobacco Corp., 1600 W. Hill St., Louisville, KY 40232

0-8412-0556-6/80/47-128-065$5.00/0

change chromatography. These workers concluded that the oligosaccharide mixture obtained by alkaline hydrolysis of the "purified" corn phytoglycolipids had the following approximate
distribution:

Fraction		%
A	GlcNH$_2$-GlcUA-Inositol	9
B	Tetrasaccharide	41
C	Pentasaccharide	10
D	Hexasaccharide	10
E	Heptasaccharide	14
F	Octa- and higher oligosaccharides	8

The complete structure of a tetrasaccharide, the major oligosaccharide of corn phytoglycolipids, was reported by Carter,
et al. (5). The N-acetylated carboxyl-reduced tetrasaccharide
was oxidized by periodate and the products were reduced with
sodium borohydride and then hydrolyzed with acid. Isolation of
D-arabitol as one of the polyol products showed that in the tetrasaccharide the inositol was 2,6-disubstituted. Proton magnetic
resonance studies on the derived glycosylinositol and N-acetylated
carboxyl-reduced trisaccharide suggested that glucuronic acid moiety was attached to the C-6 position of inositol. The mannose,
therefore, according to these workers was attached to the C-2
position of inositol in the tetrasaccharide (5). All -α anomeric
configurations were also deduced from proton magnetic resonance
spectra. Although the intact phytoglycolipid preparation was a
mixture of members with varying carbohydrate chain lengths, they
carried out another periodate experiment on this mixture. The
major polyol isolated was a tetritol fraction which was shown by
paper chromatography to be a mixture of erythritol-threitol (8:1).
D-arabitol was isolated (tetritol:pentitol, approximately 11:1)
and a small amount of hexitol was also isolated. The weight of
evidence suggested to these workers that in mild acid hydrolysis
(reflux in 2 N formic acid for 3 h) of phosphorylated oligosaccharide (from corn and flax), inostiol-1-phosphate was detected as
the major product. Thus these workers proposed the complete
structure of the major member of phytoglycolipids from corn seeds
as follows.

$$\begin{array}{l} \text{Man}(\alpha 1{\rightarrow}2) \\ \hspace{3.5cm} \diagdown \\ \hspace{4cm} \text{Myoinositol-1-0-phosphoceramide} \\ \text{GlcNH}_2(\alpha 1{\rightarrow}4)\text{GlcUA}(\alpha 1{\rightarrow}6) \diagup \end{array}$$

No further work was reported on characterization of the more complex members in this series of phytoglycolipids from plants.
Wagner, et al. (6) reported to have isolated from peanuts a
phytoglycolipid-like material for which a tentative structure was
proposed as follows:

Cer-phosphate-Inos(?-4)GlcUA(αl→3)GlcNH$_2$(1→?)(Gal, Ara, Man).

Carter and Koob (70 isolated a phytoglycolipid fraction from bean leaves (Phaseolus vulgaris). They extracted these glycophosphoceramides by refluxing in hot 70% ethanol (0.1 N in HCl) for 20 min. This acidic extraction procedure may have caused partial breakdown of these complex compounds. Wagner, et al. (8) reported isolation of a glycophosphoceramide similar to phytoglycolipids from the green alga Scenedesmus obliquus, but the only carbohydrates detected were glucose and glucuronic acid. This was also the first indication that algae synthesize sphingosine or sphingolipids. Carter and Kisic (9) reported partial characterization of another class of glycophosphoceramides from crude inositol lipids of plant seeds, which was related to corn phytoglycolipids but contained no amino sugars.
    Kaul and Lester (3) developed a mild extraction procedure to obtain a crude concentrate of glycophosphoceramides from fresh mature tobacco leaves. Thin layer chromatography of this concentrate indicated the presence of a dozen or more polar lipids containing inositol, phosphate, and carbohydrate. Two of the major members were purified by chromatography on porous silica gel beads and partially characterized as GlcNac-GlcUA-Inositol-phosphoceramide (termed PSL-I) and GlcNH$_2$-GlcUA-Inositol-phosphoceramide (termed PSL-II). Although not fully characterized, the other members in the concentrate were reported to be inositol-containing glycophosphoceramides with a higher carbohydrate content (10). The reported amount of glycophosphoceramide concentrate (about 100 μmol per Kg fresh weight) was of comparable magnitude to the estimate of phytoglycolipids present in the crude extract from bean leaves (0.1% of dry weight) (7) with leaf moisture taken into consideration. The proposed structure of PSL-I and PSL-II, as well as the properties of the other glycophosphoceramides in the tobacco leaf concentrate indicated their close similarity to the phytoglycolipids studied by Carter and his group. Kaul and Lester (3) reported that the trisaccharide-containing PSL-I and PSL-II constituted a total of approximately 40% of the tobacco glycophosphoceramides, in contrast to the report by Carter, et al. (4) that trisaccharides constituted only about 9% of corn phytoglycolipid.

GlcNAC(αl→4)GlcUA(αl→2)-Myoinositol-1-0-phosphoceramide

PSL-I

Kaul and Lester (10) reported the preparation of six novel
glycophosphoceramide fractions from the above crude concentrate
from tobacco leaves. The crude concentrate was first resolved
into two groups by column chromatography on diethylaminoethyl-
cellulose. The first group contained no acetyl residues, whereas
the second group contained one N-acetyl per phosphorus. Three
lipid fractions from each group were further resolved by chromatography on Porasil columns. The chemical composition and the percent of the total P in the crude concentrate of these lipid fractions obtained are as follows:

PSL-IA:	PSL-I-$(Ara)_2(Gal)_2$	0.96%
PSL-IB:	PSL-I-$(Ara)_3(Gal)_2$	0.27%
PSL-IC:	PSL-I-$(Ara)_4(Gal)_2$	1.56%
PSL-IIA:	PSL-II-$(Ara)_3Gal$	0.75%
PSL-IIB:	PSL-II-$(Ara)_{2-3}(Gal)_2$	3.90%
PSL-IIC:	PSL-II-$(Ara)_2(Gal)_2Man$	0.85%

Apparently, these glycophosphoceramide fractions were related to
but much less abundant than the major members PSL-I and PSL-II in
the concentrate.

PSL-I: The Major Glycophosphoceramide from Tobacco Leaves

For characterization of PSL-I, the major glycophosphoceramide
previously isolated from the tobacco glycophosphoceramide concentrate by Kaul and Lester (3), the carboxyl-reduced (11) trisaccharide moiety was first obtained by alkaline degradation of the
carboxyl-reduced PSL-I, followed by alkaline phosphatase treatment
on the resulting trisaccharide and phospho-trisaccharide mixture
(12). Methylation linkage analyses (13,14,15,16) were performed
on the trisaccharide by combined gas chromatography/mass spectrometry in both electron-impact and chemical ionization modes (17,
18) and the data (Figure 1) suggested a partial structure
GlcNAcp(1→4)Glcp(1→?)Inos for the carboxyl-reduced PSL-I trisaccharide (12). Carbohydrate composition and CrO_3 oxidation products for anomeric configuration on the trisaccharide were analyzed by gas chromatography (19,20,21). The data suggested the
structure GlcNAcp(α1→4)Glcp(α1→?)Inos for the PSL-I carboxyl-
reduced trisaccharide. Periodate oxidation experiments to determine the linkage between glucuronic acid and myoinositol were
carried out on the intact PSL-I (12). The phospho-alcohol product
from myoinositol was separated from other products by anion
exchange chromatography and the final derivative examined by
chemical ionization mode of gas chromatography/mass spectrometry
was shown to be erythritol, indicating that the glucuronic acid
was attached to the C-2 position of the myoinositol ring (Figures
2,3a,3b). This completed the characterization of PSL-I as

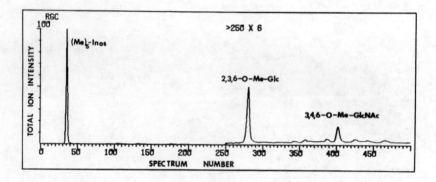

PSL-I Carboxyl-Reduced Trisaccharide:

GlcNAcp(1→4)Glcp(1→?)Inos

Figure 1. Methylation linkage analysis of PSL-I by GC/MS: total ion chromatogram of partially methylated alditol and myoinositol acetates (PMAA) from PSL-I carboxyl-reduced trisaccharide by gas chromatography/mass spectrometry in electron-impact mode.

Peaks identified: penta-O-methyl-mono-O-acetylmyoinositol derived from mono-linked myoinositol, 2,3,6-tri-O-methyl-1,4,5-tri-O-acetylglucitol derived from a 4-linked glucose, and 3,4,6-tri-O-methyl-1,5,di-O-acetyl-2-acetamido-2-N-methylglucitol derived from a terminal N-acetylglucosamine. The PMAA sample was chromatographed on a 1.5 m × 2 mm ID column packed with 3% OV-210 in a Finnigan automated GC/MS model 3300/6110. Temperature program: 150° to 215°C at 6°C/min.

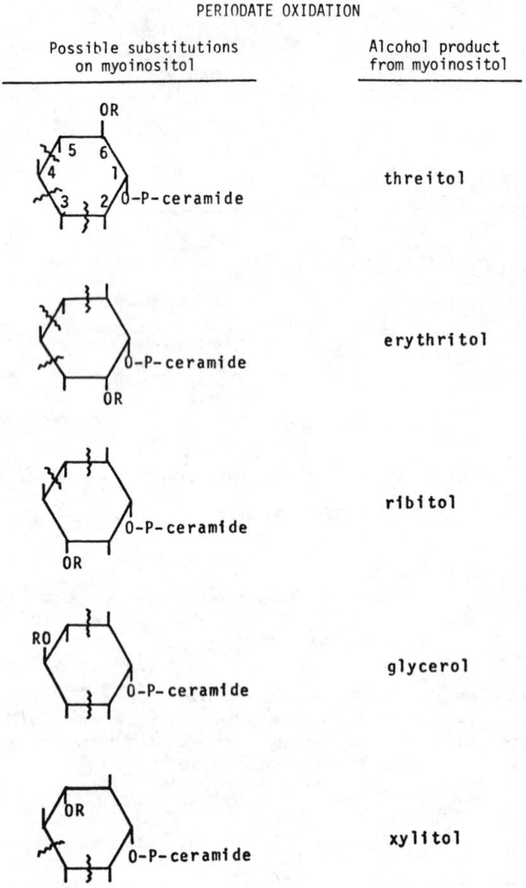

PERIODATE OXIDATION

Possible substitutions on myoinositol	Alcohol product from myoinositol

threitol

erythritol

ribitol

glycerol

xylitol

Biochemistry

Figure 2.   Possible substitutions on myoinositol by glucuronic acid. Shown are the bonds susceptible to periodate oxidation (wavy lines) and the predicted corresponding final myoinositol-derived alcohol products after periodate oxidation, followed by NaBD₄ reduction, hydrolysis, anion exchange chromatography and dephosphorylation (12).

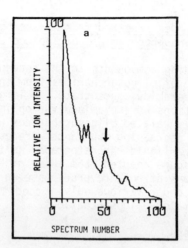

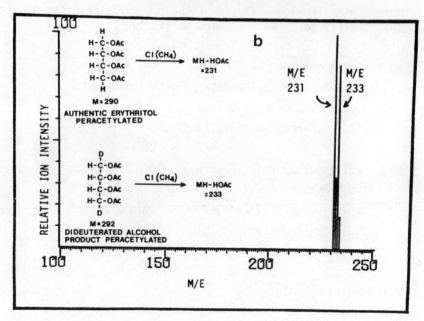

Figure 3. Chemical ionization (methane) GC/MS of the acetylated final product derived from periodate oxidation of the myoinositol ring in PSL-I. (a): Total ion chromatogram of co-injected mixture of the unknown dideuterated alcohol product and the authentic erythritol. (b): Chemical ionization spectrum of peak indicated by an arrow in (a). Inset diagrams depict the fragmentation.

GlcNAcp(α1$\rightarrow$4)GlcUAp(α1$\rightarrow$2)Inos-1-0-phosphoceramide (Figure 4)(12).

Major Oligosaccharides Prepared from the Carboxyl-Reduced Concentrate of Glycophosphoceramides of Tobacco Leaves

For the remaining components in the concentrate, Hsieh (22) prepared a mixture of oligosaccharides from the carboxyl-reduced (23) glycophosphoceramide concentrate. A large number of chromatographic conditions were examined for optimal fractionation. A series of closely related oligosaccharides with increasing complexity and in decreasing abundance were observed on reverse-phase high pressure liquid chromatography as the peracetylated derivatives [procedure adapted from those of Wells and Lester (24)]. Combinations of both reverse-phase and normal-phase columns were used under various solvent conditions to achieve isolation of the major oligosaccharides.

Methylation Analyses

Methylation linkage analysis of the partially methylated alditol acetates gave the following derivatives:

Major trisaccharide:

    3,4,6-tri-0-methyl-2-deoxy-2-methylaminoglucitol

    2,3,6-tri-0-methylglucitol

    1,3,4,5,6-penta-0-methylinositol

Major tetrasaccharide: (Figure 5)

    2,3,4,6-tetra-0-methylgalactitol

    3,6-di-0-methyl-2-deoxy-2-methylaminoglucitol

    2,3,6-tri-0-methylglucitol

    1,3,4,5,6-penta-0-methylinositol

Minor tetrasaccharide:

    2,3,4,6-tetra-0-methylmannitol

    3,4,6-tri-0-methyl-2-deoxy-2-methylaminoglucitol

    2,3,6-tri-0-methylglucitol

    tetra-0-methylinositol

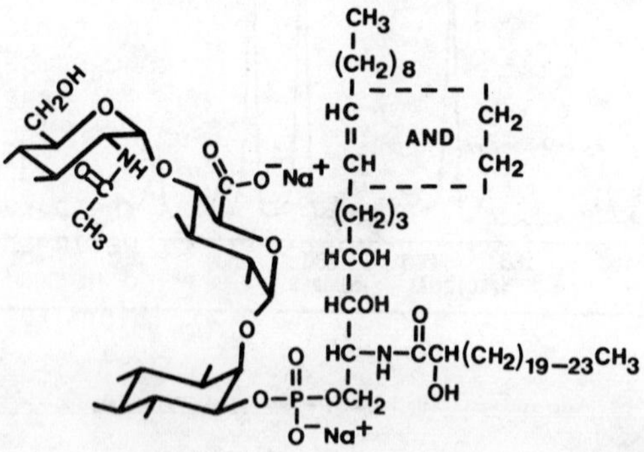

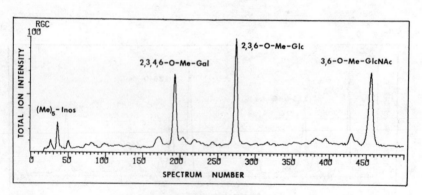

Biochemistry

Figure 4. Proposed structure of PSL-I: GlcNAcp($\alpha1\rightarrow4$)GlcUAp($\alpha1\rightarrow2$ myoinositol-1-O-phosphoceramide (12)

Figure 5. Methylation linkage analysis of the major tetrasaccharide from tobacco glycophosphoceramide concentrate

Total ion chromatogram: penta-O-methyl-mono-O-acetylmyoinositol derived from mono-linked myoinositol, 2,3,4,6-tetra-O-methyl-1,5-di-O-acetylgalactitol derived from a terminal galactose, 2,3,6-tri-O-methyl-1,4,5-tri-O-acetylglucitol derived from a 4-linked glucitol, and 3,6-di-O-methyl-1,4,5-tri-O-acetyl-2-acetamido-2-N-methylglucitol from a 4-linked N-acetylglucosamine.

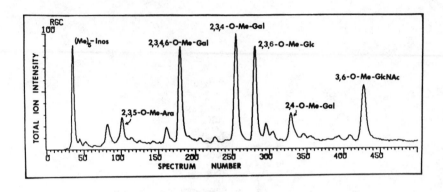

Major component: Galp(1→6)Galp(1→4)GlcNAcp(1→4)Glcp(1→2)Inos

Minor component:
Galp(1→6)
Araf(1→3) ⟩Galp(1→4)GlcNAcp(1→4)Glcp(1→2)Inos

Figure 6. Preliminary methylation linkage analysis of the major pentasaccharide from tobacco glycophosphoceramide concentrate

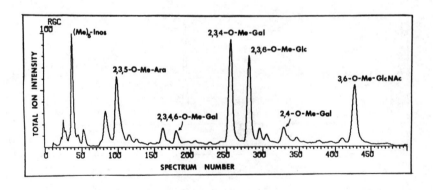

Major component: Araf(1→6)Galp(1→4)GlcNAcp(1→4)Glcp(1→2)Inos

Minor component:
Galp(1→6)
Araf(1→3) ⟩Galp(1→4)GlcNAcp(1→4)Glcp(1→2)Inos

Figure 7. Preliminary methylation linkage analysis of the minor pentasaccharide from tobacco glycophosphoceramide concentrate

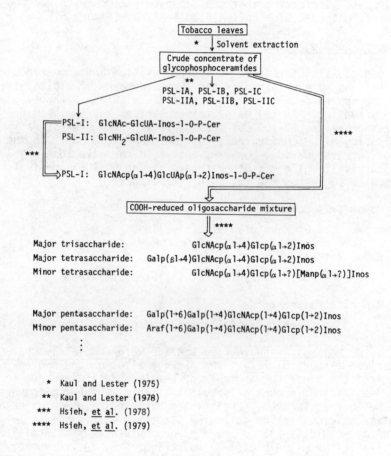

Major trisaccharide: GlcNAcp(α1→4)Glcp(α1→2)Inos
Major tetrasaccharide: Galp(β1→4)GlcNAcp(α1→4)Glcp(α1→2)Inos
Minor tetrasaccharide: GlcNAcp(α1→4)Glcp(α1→?)[Manp(α1→?)]Inos

Major pentasaccharide: Galp(1→6)Galp(1→4)GlcNAcp(1→4)Glcp(1→2)Inos
Minor pentasaccharide: Araf(1→6)Galp(1→4)GlcNAcp(1→4)Glcp(1→2)Inos

　*　 Kaul and Lester (1975)
　**　 Kaul and Lester (1978)
　***　 Hsieh, et al. (1978)
　****　 Hsieh, et al. (1979)

Figure 8.   Summary of structural characterization of glycophosphoceramides from tobacco leaves

Sequence Analysis

The carbohydrate sequence of the major tetrasaccharide was determined by examining the nitrous acid deamination products (25) as permethylated disaccharides by chemical ionization mode of gas chromatography/mass spectrometry. The products were identified as hexosyl-2,5-anhydromannitol and hexosyl-myoinositol, indicating that the major tetrasaccharide had the sequence Galp(1→4)GlcNAcp(1→4)Glcp(1→2)Inos (Figure 5).

Anomeric Configuration

Additional information on the composition and anomeric configurations were obtained by gas chromatography of alditol acetates prepared from the oligosaccharides with and without CrO_3 oxidation. In the major trisaccharide, and in the minor tetrasaccharide, 80-100% of the sugars survived CrO_3 oxidation indicating all α configuration of the anomeric bonds. In the major tetrasaccharide, however, the yield for galactose was 29% survival, while the other sugars showed 80-100% survival. This data suggested the following structures:

Major trisaccharide:

GlcNacp(α1→4)Glcp(α1→2)Inos

Major tetrasaccharide:

Galp(β1→4)GlcNAcp(α1→4)Glcp(α1→2)Inos

Minor tetrasaccharide

GlcNAcp(α1→4)Glcp(α1→?)[Man(α1→?)]Inos

Thus, the major tri-and tetrasaccharide were completely characterized (Figure 8) (22). The linkage sites on the myoinositol of the minor tetrasaccharide remain undetermined due to the insufficient amount of sample available. Higher oligomers are being fractionated. Preliminary data indicate that a major pentasaccharide has the following structure Galp(1→6)Galp(1→4)GlcNAcp(1→4)Glcp(1→2)Inos and a minor pentasaccharide Araf(1→6)Galp(1→4)GlcNAcp(1→4)Glcp(1→2)Inos (Figures 6, 7). A summary of the results is shown in Figure 8.

Acknowledgements

This investigation was supported in part by Research Grant PCM7609314 from the National Science Foundation , Project KTRB-053 from the Tobacco and Health Research Institute, University of Kentucky, and Grant IROIGM23902 from the National Institutes of Health.

Abstract

Chemical structures of certain glycophosphoceramides from tobacco leaves were studied. The structures which have been characterized to date are as follows:

(1) major glycophosphoceramides PSL-I:
GlcNAcp(α1→4)GlcUAp(α1→2)Inos-1-0-P-Cer

(2) the oligosaccharides isolated from the glycophosphoceramide concentrate after carboxyl-reduction:

(a) major trisaccharide:
GlcNAcp(α1→4)Glcp(α1→2)Inos

(b) major tetrasaccharide:
Galp(β1→4)GlcNAcp(α1→4)Glcp(α1→2)Inos

(c) minor tetrasaccharide:
GlcNAcp(α1→4)Glcp(α1→?) [Manp(α1→?)]Inos

(d) major pentasaccharide:
Galp(1→6)Galp(1→4)GlcNAcp(1→4)Glcp(1→2)Inos

(e) minor pentasaccharide:
Araf(1→6)Galp(1→4)GlcNAcp(1→4)Glcp(1→2)Inos

Literature Cited

1. Carter, H.E., Celmer, W.D., Galanos, D.S., Gigg, R.H., Lands, W.E.M., Law, J.H., Mueller, K.L., Nakayama, T., Tomizawa, H.H., and Weber, E. J. Am. Oil. Chem. Soc., 1958, 35, 335.
2. Lester, R.L., Smith, S.W., Wells, G.B., Rees, D.C., and Angus, W.W. J. Biol. Chem., 1974, 249, 3388.
3. Kaul, K., and Lester, R.L. Plant Physiol., 1975, 55, 120.
4. Carter, H.E., Brooks, S., Gigg, R.H., Strobach, D.R., and Suami, T. J. Biol. Chem., 1964, 239, 743.
5. Carter, H.E., Strobach, D.R., and Hawthorne, J.N. Biochemistry, 1969, 8, 383.
6. Wagner, H., Zofcsik, W., and Heng, I. Z. Naturforsch, 1969, 24, 922.
7. Carter, H.E., and Koob, J.L. J. Lipid Res., 1969, 10, 363.
8. Wagner, H., Pohl, P., and Munzing, A. Z. Naturforsch, 1969, 24, 360.
9. Carter, H.E., and Kisic, A. J. Lipid Res., 1969, 10, 356.
10. Kaul, K., and Lester, R.L. Biochemistry, 1978, 17, 3569.
11. Taylor, R.L., Shively, J.E., Conrad, H.E., and Cifonelli, J.A. Biochemistry, 1973, 12, 3633.
12. Hsieh, T.C.-Y., Kaul, K., Laine, R.A., and Lester, R.L. Biochemistry, 1978, 17, 3575.
13. Björndal, H., Lindberg, B., and Svensson, S. Carbohyd. Res., 5, 433.

14. Björndal, H., Lindberg, B., Pilotti, A., and Svensson, S.
     Carbohydrate Res., 1970, 15, 339.
15. Hakomori, S. J. Biochem. (Tokyo), 1964, 55, 205.
16. Stellner, K., Saito, H., and Hakomori, S. Arch. Biochem.
     Biophys., 1973, 155, 464.
17. Hancock, R.A., Marshall, K., and Weigel, H. Carbohyd. Res.,
     1976, 49, 351.
18. Laine, R.A., Hodges, L.C., and Cary, A.M. J. Supramol.
     Struct., 1977, 5, Suppl. 1, 31.
19. Hoffman, J., Lindberg, B., and Svensson, S. Acta Chem.
     Scand., 1972, 26, 661.
20. Laine, R.A., and Renkonen, O. J. Lipid Res., 1975, 16, 102.
21. Albersheim, P., Nevins, D.J., English, P.D., and Karr, A.
     Carbohyd. Res., 1967, 5, 340.
22. Hsieh, T.C.-Y., Ph.D. dissertation: "Chemical Characteriza-
     tion of Glycophosphosphingolipids from Tobacco"; University
     of Kentucky: Lexington, Kentucky, 1979.
23. Taylor, R.L., Shively, J.E., and Conrad, H.E. Methods in
     Carbohyd. Chem., 1976, 7, 149.
24. Wells, G.B., and Lester, R.L. Anal. Biochem., 1979, 97, (in
     press).
25. Bayard, B., and Roux, D. FEBS Lett., 1975, 55, 206.

RECEIVED December 10, 1979.

Glycolipids of Rat Small Intestine with Special Reference to Epithelial Cells in Relation to Differentiation

M. E. BREIMER, G. C. HANSSON, K.-A. KARLSSON, and H. LEFFLER

Department of Medical Biochemistry, University of Göteborg, Göteborg, Sweden

Saccharides may be structurally very complex. In addition to the variation in type and sequence of monomers as for peptide, the heterocyclic carbohydrate monomer may vary in ring size, the glycosidic bond may have both different positions and configurations, and there is often branching of the saccharide chains. A great variability may also mean a rich biochemical language (provided there is specificity of expression) and this is one of the reasons why cell surface carbohydrates are being considered in biological recognition (1, 2).

The membrane-bound carbohydrates exist as glycoproteins and glycolipids. Although the functional importance of these substances is far from proven they appear to be essential parts in phenomena such as cellular adhesion, control of differentiation and cell growth, and the binding by cells of enzymes, hormones and toxins.

One system that we consider of great interest for the study of cell surface glycolipids is the small intestine. Firstly, the epithelial cells lining the intestine exist in a great number on the enlarged surface area and each cell has in itself a large cell surface involved in transport processes and recognition phenomena. Secondly, these cells, arranged as a single columnar layer on the basement membrane, are rapidly renewed (1-3 days) and undergo a successive maturation on their way from the crypt depth to the villus tip (3). Thirdly, these cells are possible to prepare by a gentle washing technique (4), the oldest, less strongly adhered cells (villus tip) being obtained in the first, and the youngest, cells (crypt) obtained in the final fractions. Lastly, the concentration of complex glycolipids is high in relation to protein (see 5), which may be explained by a large amount of surface membrane in relation to other membranes.

Our study was divided into two different parts and applied on two separate strains of rat, which were shown to differ in blood groups. In the first stage, following improvement and adaptation of methods, glycolipids were prepared and characterized from pooled whole small intestine of the black and white strain. In the second stage, the knowledge of the general glycolipid

0-8412-0556-6/80/47-128-079$6.50/0

composition allowed a characterization on a smaller scale of
epithelial cells and non-epithelial residue, and a comparison of
the two strains. The different components of tissue are visualized
in Fig. 1.

The experience obtained has now been used for a similar investigation on human material (in preparation).

Methods

The animals used were from inbred strains of white, and black
and white rat. The preparation of epithelial cells in separate
stages of differentiation was modified from the technique of
Weiser (4). The completeness of removal of epithelial cells from
non-epithelial residue was checked by conventional microscopy.
The preparation of total glycosphingolipids free of contaminants
has been improved to an important extent but is based on
conventional steps such as chloroform–methanol extraction, mild
alkaline degradation, dialysis, acetylation and chromatography
on DEAE-cellulose and silicic acid. Thin-layer chromatography was
done on HPTLC plates with silica gel 60 (Merck). Conditions for
mass spectrometry (6,7) and NMR spectroscopy (8, 9, 10) have been
described. Gas chromatography after degradation of native or permethylated glycolipids was done according to standardized techniques (11) except that the analysis was performed on capillary
columns.

Non-Epithelial Tissue

The non-acid pattern of the residue after exhaustive washing
and removal of epithelial cells from small intestine is shown in
Fig. 2, for the black and white (B_s) and white (W_s) strain,
respectively. The two samples look identical with a major component corresponding to four sugars. Most of the glycolipids have
been isolated and characterized. To present an overview the total
glycolipids of white rat were subjected to a novel application
of mass spectrometry (7) after permethylation and reduction with
LiAlH$_4$. Figs. 3 and 4 show some of the results. The mixture of
glycolipid derivatives is introduced into the ion source and
successively heated (5°C/min) as shown on the scale below the
curves. Scans (each scan producing a mass spectrum such as that
of Fig. 7) were taken each 38 sec, and the change in relative
intensity of selected ions for separate glycolipids was reproduced as curves along the temperature and scan scales. In this
case the ions selected contained the complete saccharide and the
fatty acid as shown in the explaining formulas (usually relatively abundant ions, which is demonstrated for the A active glycolipids in Figs. 7 and 8). Curves corresponding to specific ions
for nine separate glycolipids are reproduced. Two series of
compounds are shown, one without (Fig. 3) and the other with
hexosamine (Fig. 4). The curve in Fig. 3 for m/e 516 (monohexosyl-

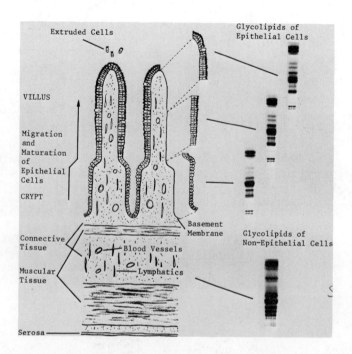

Figure 1. *Acid and non-acid glycosphingolipids were prepared and characterized from different compartments of rat small intestine: non-epithelial residue, total epithelial cells, and epithelial cells of different maturity (crypt, intermediate, and villus fractions).*

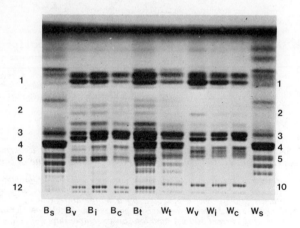

Figure 2. *Thin-layer chromatogram of non-acid glycolipids of small intestine of black and white (B) and white (W) rat*

The following samples were applied: 40 μg of total glycolipids (t); glycolipids corresponding to 4 mg protein of non-epithelial residue (s); glycolipids corresponding to 2 mg protein for epithelial cells of villus (v), intermediate (i), and crypt (c) fractions. Figures in the margins indicate number of sugars. Anisaldehyde was used for the detection, and the solvent was chloroform–methanol–water 60:35:8 (by volume).

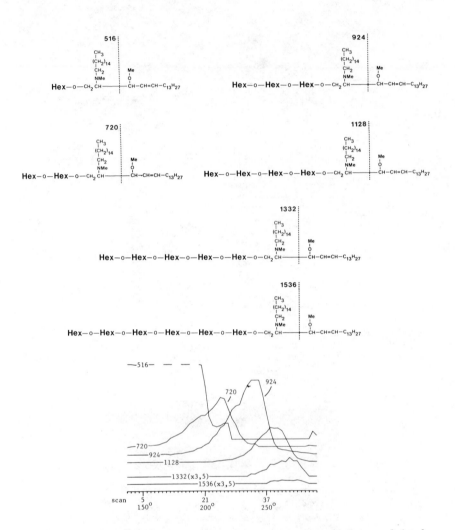

Figure 3. Selected ion monitoring from mass spectrometry of a permethylated-reduced mixture of non-acid glycolipids from non-epithelial residue of the white rat

The curves reproduced correspond to relative abundance of saccharide plus fatty acid ions (see formulas) of glycolipids lacking hexosamine as a function of evaporation temperature. A total of 200 μg was evaporated by a temperature rise of 5°C/min, and spectra were recorded each 38 sec. The electron energy was 34 eV, acceleration voltage 4 kV, trap current 500 μA, and ion source temperature 280°C.

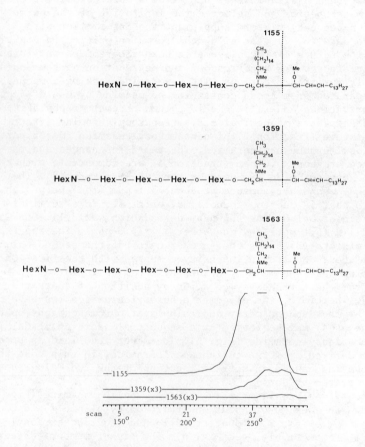

Figure 4.   Selected ion monitoring of saccharide plus fatty acid ions of hexosamine-containing glycolipids from the same experiment as for Figure 3

ceramide) appears at lower temperature while those for higher
members (Figs. 3 and 4) come up later, in some cases indicating
a complete separation of glycolipid species. The relative intensities of the separate bands on the chromatogram (Fig. 2) are
not directly comparable with the ion curves as the relative
abundance of ions decreases rapidly with ion mass.

The space available does not allow a more detailed presentation. Mass spectra and selected ion monitoring of the permethylated (non-reduced) mixture supplement the information with
sequence data (6, 7) that allow the formulas written in Figs. 3
and 4. The nature of the oligomeric hexosylceramides was further
substantiated by NMR spectroscopy and degradation of some pure or
partially purified fractions. The hexosamine-lacking compounds
were separated from those containing hexosamine by use of acetylated derivatives and silicic acid column chromatography. Fig. 5
shows NMR spectra of derivatized tetrahexosylceramide (fraction A)
and a mixture (fraction B) of pentahexosylceramide (major part)
and fucosyltetrahexosylceramide. The two β-resonances and the α-
resonance at about 5.0 ppm (fraction A) are comparable with those
of trihexosylceramide of human erythrocyte membrane (8). Therefore, the second α-resonance at about 5.1 ppm (the sharp signal
close to that is due to ethanol) may originate in a terminal Gal
(the ratio of Gal:Glc as shown by degradation was 3:1). One Gal
is bond 1→3 (5.1 ppm) and the other 1→4 (5.0 ppm). The spectrum
of the mixture (B) shows the same signals but the second Galα
has now about doubled in intensity compared with the first Galα,
suggesting that the major pentahexosylceramide is formally derived from the tetrahexosylceramide by addition of another
Galα1→3. Therefore, the oligomeric hexosylceramides (we have detected by mass spectrometry and thin-layer chromatography up to
eight hexoses) may be formed by a sequential addition of Galα1→3
to globotriaosylceramide (Fig. 6). The minor fucolipid is probably also derived from the tetrahexosylceramide, in this case the
fucose having caused an upfield location of the two Galα resonances (indicated by dots).

The tetraglycosylceramide with terminal hexosamine (Fig. 4)
was shown to consist of about one third of cytolipin K and two
thirds of cytolipin R (globotetraosyl- and isoglobotetraosyl-
ceramide, respectively, see Fig. 6). The higher members detected
in this series (Fig. 4) are probably formed by an elongation of
globotriaosylceramide as for the first series and a termination
by GalNAcβ1→3 (Fig. 6). Of particular interest was the identification of a blood group B active hexaglycosylceramide based on
galactosamine (Fig. 6). The glycolipids detected in non-epithelial
tissue are summarized in Fig. 6. The ganglioside composition will
be commented on below.

Epithelial Cells

The glycolipid pattern of epithelial cells is distinctly

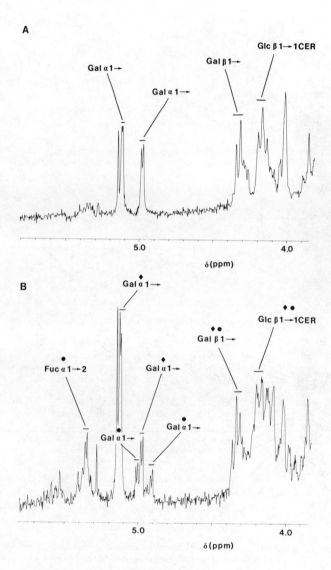

Figure 5.   NMR spectra of two permethylated-reduced glycolipid samples (A and B) lacking hexosamine and isolated from whole intestine of black and white rat; 2 mg in 0.5 mL chloroform and 2300 pulses at 40°C (sample A), and 1 mg in 0.5 mL chloroform and 5300 pulses at 40°C (sample B).

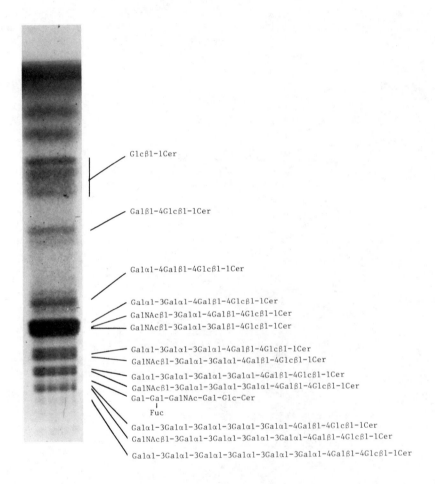

Glcβ1-1Cer

Galβ1-4Glcβ1-1Cer

Galα1-4Galβ1-4Glcβ1-1Cer

Galα1-3Galα1-4Galβ1-4Glcβ1-1Cer
GalNAcβ1-3Galα1-4Galβ1-4Glcβ1-1Cer
GalNAcβ1-3Galα1-3Galβ1-4Glcβ1-1Cer

Galα1-3Galα1-3Galα1-4Galβ1-4Glcβ1-1Cer
GalNAcβ1-3Galα1-3Galα1-4Galβ1-4Glcβ1-1Cer

Galα1-3Galα1-3Galα1-3Galα1-4Galβ1-4Glcβ1-1Cer
GalNAcβ1-3Galα1-3Galα1-3Galα1-4Galβ1-4Glcβ1-1Cer
Gal-Gal-GalNAc-Gal-Glc-Cer
                 |
                Fuc

Galα1-3Galα1-3Galα1-3Galα1-3Galα1-4Galβ1-4Glcβ1-1Cer
GalNAcβ1-3Galα1-3Galα1-3Galα1-3Galα1-4Galβ1-4Glcβ1-1Cer

Galα1-3Galα1-3Galα1-3Galα1-3Galα1-3Galα1-4Galβ1-4Glcβ1-1Cer

Figure 6.   Thin-layer pattern with deduced chemical formulas of non-acid glyco-
lipids of white rat non-epithelial residue (cf. Figure 2).

different from that of non-epithelial tissue (Fig. 2). Bands
corresponding to one and three sugars are dominating. In addition,
there are a number of compounds that have been prepared from
pooled whole intestines of the black and white strain. Two series
of fucolipids were identified, one with blood group H and one
with blood group A determinants. Mass spectra of permethylated-
reduced derivatives of two of the A-glycolipids are shown in
Figs. 7 and 8, respectively. In addition, a 6-sugar A active
compound was characterized, thus completing a series with 4, 6
and 12 sugars.

Concerning the 12 sugar compound the mass spectra of the
permethylated derivative (not shown) and of the permethylated-
reduced derivative (Fig.8) are remarkable in that they together
afford a conclusion on the type, number and sequence of sugars
including branching of the chain, in addition to ceramide
structure (to be published). The saccharide plus fatty acid peaks
in the interval m/e 2835-2977 (Fig. 8) are evidence for five
hexoses, five hexosamines, two fucoses and a varying fatty acid,
mainly from 16:0 (m/e 2835) to 24:0 (m/e 2947) nonhydroxy fatty
acid, but also 24:0 hydroxy acid (m/e 2977). In the spectrum of
the non-reduced derivative (not shown) m/e 396 showed that the
dominating base is phytosphingosine. According to the relative
intensity of the series of peaks at m/e 2835-2977 the major
molecular species contained phytosphingosine and 20:0 nonhydroxy
fatty acid. Evidence for the sequence and branching point was
obtained by the absence or presence of several ions. Some primary
and secondary (loss of methanol, mass 32) ions with a successive
increase in the number of sugars from the non-reducing end are
shown up to nine sugars (m/e 1915). The absence of sequence ions
between m/e 871 and 1915 speaks against a linear sequence
between these two fragmentation points. (There were analogous
ions obtained from the non-reduced derivative). The absence of
ions for smaller saccharides with two fucoses (in spectra of both
derivatives) is evidence for fucose location in separate chains.
Finally, there is a number of rearrangement ions containing the
fatty acid and an increasing part of the saccharide from the
ceramide end (some of them indicated below the formula). These
ions have taken up one or two hydrogens depending on the location
of the branch (see peaks at m/e 614, 818, 1049, 1239, 1470, 1848,
2093, 2324).

Therefore, the evidence obtained from the two derivatives is
conclusive concerning the sequence of the 12-sugar glycolipid.
This substance represents the largest biomolecule structurally
determined by mass spectrometry thus far.

Compared with the B-active glycolipid found in non-epithelial tissue, the fucolipids in epithelial cells were based on
glucosamine instead of galactosamine (see Fig. 9). The H active
fucolipids of black and white rat had three and five sugars,
respectively. The glycolipids found in epithelial cells of the
two strains are summarized in Fig. 10.

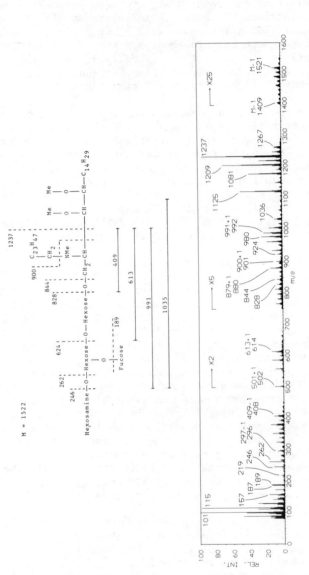

Figure 7. Mass spectrum of permethylated-reduced derivative (60 μg) of a blood-group A active tetraglycosylceramide. Electron energy was 44 eV, acceleration voltage 4 kV, trap current 100 μA, ion source temperature 290°C, and probe temperature 215°C.

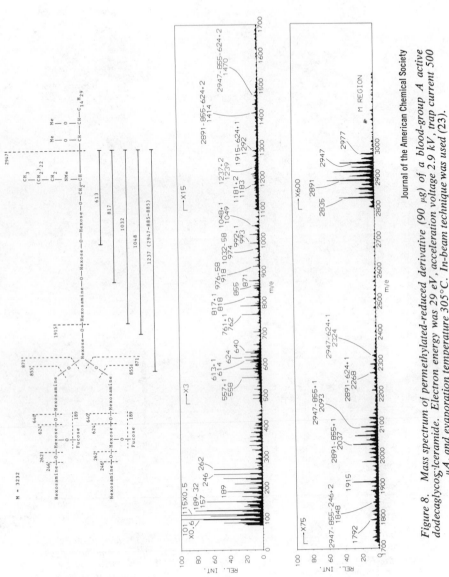

Figure 8. Mass spectrum of permethylated-reduced derivative (90 μg) of a blood-group A active dodecaglycosylceramide. Electron energy was 29 eV, acceleration voltage 2.9 kV, trap current 500 μA, and evaporation temperature 305°C. In-beam technique was used (23).

Journal of the American Chemical Society

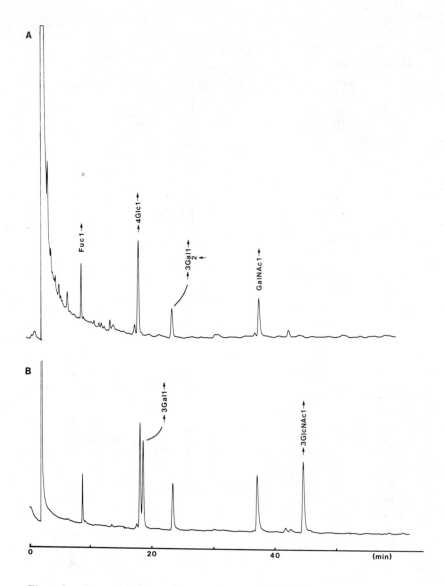

Figure 9.   Open tubular gas chromatogram of partially methylated alditol acetates obtained from blood-group A active tetraglycosylceramide (A) and hexaglycosylceramide (B), respectively. Stationary phase was OV-1, and carrier gas was N₂. Column temperature was kept at 175°C for 14 min, then raised 1°C/min. The designation above the peaks indicate actual binding positions.

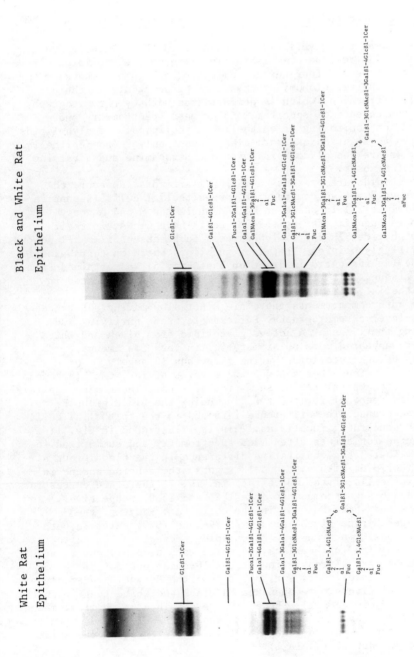

Figure 10. Thin-layer pattern with deduced chemical formulas of non-acid glyco-lipids of epithelial cells of the two rat strains (cf. Figure 2)

Differences between the two compartments and between strains

As shown, the glycolipid patterns of epithelial cells and non-epithelial residue are distinctly different. Glycolipids with one, two, three and four hexoses exist in both compartments. Concerning the two globosides these are present only in the non-epithelial fraction, which is demonstrated both by chromatography and by the absence of specific ions at mass spectrometry and selected ion monitoring of epithelial glycolipids. The glycolipids with five to eight hexoses are also present only in non-epithelial tissue, as are the glycolipids with one hexosamine and a varying number of hexoses.

Fucolipids are present in both compartments. However, the blood group B active compound of non-epithelial cells (absent in epithelial cells) is based on GalNAc while the H and A active substances specific for epithelial cells have GlcNAc in their core saccharide. In fact, GlcNAc seems to be absent from all non-epithelial glycolipids. The minor fucolipid based on tetrahexosyl-ceramide (as indicated in fraction B of Fig. 5) was obtained from pooled tissue. This glycolipid has been shown to be located in the epithelial cells.

In both compartments there are minor slow-moving substances on thin-layer chromatography. For example, when purifying and enriching the 12-sugar A active glycolipid from black and white rat there appeared more polar material in very low amounts, probably being glycolipids having more than 12 sugars.

The difference between the two strains of rat, the black and white and the white strain, seems rather clear. The non-epithelial tissue is identical for the two, including the blood group B active substance. The difference is found in the epithelial cells and only concerning fucolipids. This is illustrated in Fig. 11 by selected ion curves after mass spectrometry and summarized in Fig. 10. There is a qualitative difference in the blood group A type glycolipids with 4, 6 and 12 sugars, these being absent in epithelial cells of the white rat. In Fig. 11 there are curves for the 4- and 6-sugar compounds (m/e 1125 and 1560, respectively) in the black and white but not in the white rat. However, the 3- and 5-sugar H-type glycolipids (m/e 894 and 1329) exist in both samples. The 10-sugar H-type glycolipid, present in the white rat does not show up in the black and white rat, probably due to a complete GalNAcα glycosylation of the 10-sugar but not of the 3- and 5-sugar glycolipids. For some reason the H-type 3-sugar glyco-lipid is relatively more abundant in black and white than in white rat (Figs. 2, 10 and 11). These results obtained by chemical means were confirmed by immunology, which showed the black and white rat glycolipids to be blood group A active, while those of the white rat were non-active (Table I).

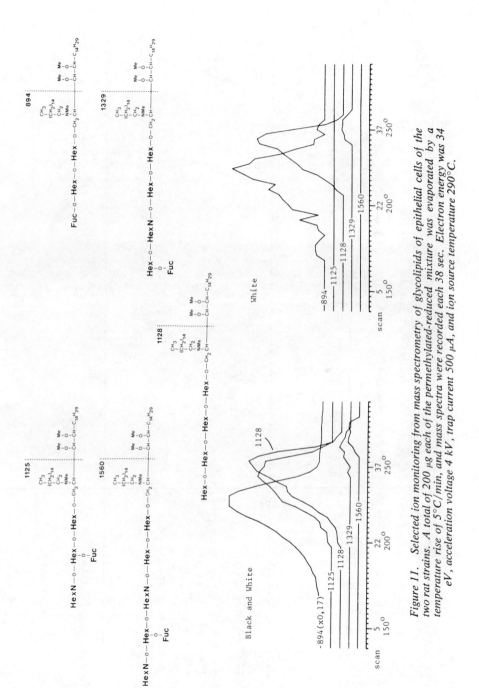

Figure 11. Selected ion monitoring from mass spectrometry of glycolipids of epithelial cells of the two rat strains. A total of 200 µg each of the permethylated-reduced mixture was evaporated by a temperature rise of 5°C/min, and mass spectra were recorded each 38 sec. Electron energy was 34 eV, acceleration voltage 4 kV, trap current 500 µA, and ion source temperature 290°C.

Table I. Some Characteristics of Different Compartments of Rat Small Intestine. The Blood Group Activities Concern Glycolipid Fractions.

| | White Rat (8 rats) | | Black and White Rat (7 rats) | |
	Epithelial Cells	Non-Epithelial Residue	Epithelial Cells	Non-Epithelial Residue
Total protein (g)	1.14	1.19	1.32	1.26
Blood group A activity *	-	-	4+	-
Blood group H activity **	2+	ND	ND	ND
Blood group B activity *	-	+	-	+
Non-acid glycolipids (mg)	8.2	10.9	14.6	8.2

* Antisera dilution 1:1
** Antisera dilution 1:0
ND means: not determined

Gangliosides

The situation for gangliosides is also complex, with a number
of separate species. However, two of these are quite dominating,
and are hematoside with N-acetyl and N-glycoloyl substitution,
respectively. Fig. 12 shows that the N-acetyl type exists in non-
epithelial while the N-glycoloyl type is mostly present in epithelial cells.

Epithelial Cells of Different Location and Maturity

Epithelial cells of small intestine were prepared in a fract-
ional way (4), the older, less adherent villus tip cells being
washed out by EDTA-containing phosphate buffer first, while mitotic crypt cells appeared in the final fractions. The enzyme
characteristics of the series of fractions obtained (Fig. 13)
followed conventional criteria for differentiated (villus) and
less differentiated (crypt) cells (3, 4). The thymidine kinase
activity decreased from crypt to villus while the activity of
alkaline phosphatase increased (Fig. 13).
The cells obtained were pooled in three fractions, a
villus (v), an intermediate (i), and a crypt (c) fraction. The
patterns of glycolipids of these are shown in Fig. 2 (non-acid)
and Fig. 12 (acid). The only significant differences between the
three locals concern the three major glycolipids and are a successive increase of monoglycosylceramide (Fig. 2) and hematoside
(Fig. 12) from crypt to villus, but a successive decrease of trihexosylceramide (Fig. 2). These facts have been noticed before
(12, 13). Other differences exist but we have not yet resolved
and quantitated all minor bands to allow comments on this. There
is also a change in relative intensity of the two bands of each
of mono- and trihexosylceramide (Fig. 2). The slower-moving band
is increasing towards the villus. Analogous changes are also
apparent for minor glycolipids. The reason for the two bands is a
heterogeneity in the ceramide portion, mainly concerning 2-hyd-
roxylation of the fatty acid. As the base is almost exclusively
phytosphingosine a change in the mass spectral fragments for
ceramide indicated by the formula of Fig. 14 should reflect the
fatty acid change. Monitoring of these ions through the temperature interval shown should give the composition of all glycolipids present. However, as mono- and trihexosylceramides dominate
the two major peaks indicated at about 190°C and 225°C mainly
reflect these two glycolipids, respectively. One should also bear
in mind that the relative proportion of these two substances
changes between the two fractions (see Fig. 2, fractions B_v and B_c).
With this knowledge one may interprete from the curves of Fig. 14
a relative lengthening of the fatty acid and an increased hydroxylation from crypt to villus. The relative increase in chain
length is shown by m/e 722 (24:0 hydroxy) compared with m/e 666
(20:0 hydroxy) and m/e 610 (16:0 hydroxy) in the two fractions,

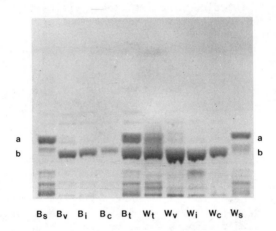

Figure 12. Thin-layer chromatogram of gangliosides of small intestine of black and white (B) and white (W) rat

The fractions and amounts were analogous to those of Figure 2, except for the total fractions (t), where 20 µg glycolipid were used. Bands for N-acetyl (a) and N-glycoloyl (b) type of hematoside are indicated. Resorcinol was used for the detection, and the solvent was methyl acetate–2-propanol–CaCl₂ (8 mg/mL)–NH₃ (5M) 45:35:15:10 (by volume).

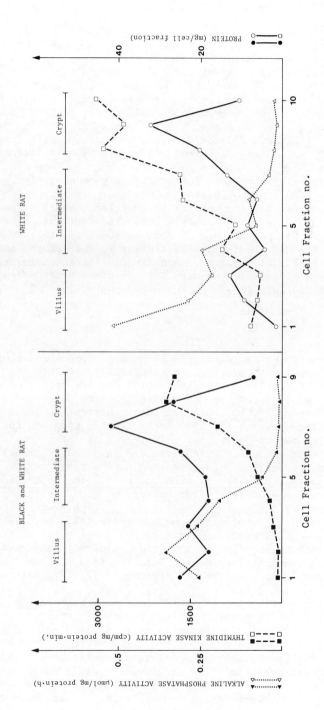

Figure 13. Protein and enzyme curves for epithelial cells fractionated from small intestine of the two rat strains. Experimental conditions were modified from Weiser (4). Pooled fractions used for glycolipid analysis are defined at the top. Data were from single rats.

but also by m/e 636 (20:0) which is quite dominating in the trihexosylceramide peak of the crypt fraction (225°C) while m/e 692
(24:0) is the most abundant ion of the villus fraction. The
change in hydroxylation is not clear from the curves of Fig.14
without an integration. However, from earlier experience of the
behaviour of molecular species of glycolipids on thin-layer
chromatography (14) and knowledge of major fatty acids present
(Fig. 14) one may conclude that the two bands (Fig. 2) are mainly composed of 20, 22, 23 and 24 carbon nonhydroxy acids (upper
band) and 20, 22, 23 and 24 carbon hydroxy fatty acids (lower
band).

The change in fatty acid composition may be shown for separate major or minor glycolipids in the mixture by selecting
fragments specific for the species in question, namely saccharide
plus fatty acid ions which are relatively abundant (see spectra
of Figs. 7 and 8). One example of this is shown for tetrahexosyl-
ceramide in Fig. 15. The change is similar to that of the total
glycolipids (Fig. 14). However, an analogous retrieval for the
4-sugar A-type glycolipid (compare Fig. 7) did not demonstrate
that clear difference in chain length between villus and crypt
cells.

Discussion

Small intestine is relatively rich in glycosphingolipids
(Table I). Compared to myelin, a metabolically stable poly-
membrane structure (15), also with a high content of glycolipid
(one sugar), small intestine has a pattern often dominated by
complex fucolipids (5, 16). Of particular interest is the finding
in this and recent works (5, 16) of the localization of these more
complex substances to epithelial cells which are structurally
complex and asymmetrical. These cells are involved in important
transport and recognition processes and have a rapid turnover (3).
This situation has provided us with an interesting object for the
study of structure-function relationships of glycosphingolipids.
Although there is strong evidence for one particular ganglioside
being the specific receptor for cholera toxin (1), there is at
present no good idea about a physiological function of a glycolipid. A possible exception is sulfatide, the only substance with
a rather consequent stoichiometric relation to a surface membrane
function, in this case Na^+ and K^+ transport (17, 18). Although the
postulated role (selection of K^+ ions, 17) is due to the sulfate
group, the sugar part carrying this group may be specifically required close to the membrane matrix.

In our initial studies reported here of glycolipids of rat
small intestine, preparative and structural methods were adapted
to characterize epithelial and non-epithelial tissue and epithelial cells of different location and level of differentiation. The
two compartments were distinctly different with core saccharides
with GalNAc being restricted to non-epithelial cells while those

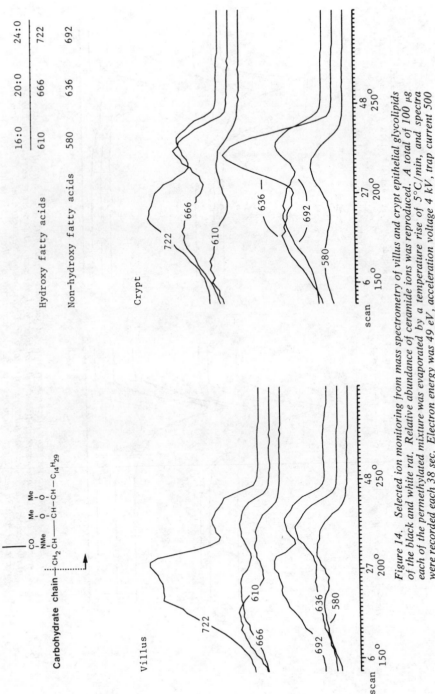

Figure 14. Selected ion monitoring from mass spectrometry of villus and crypt epithelial glycolipids of the black and white rat. Relative abundance of ceramide ions was reproduced. A total of 100 µg each of the permethylated mixture was evaporated by a temperature rise of 5°C/min, and spectra were recorded each 38 sec. Electron energy was 49 eV, acceleration voltage 4 kV, trap current 500 µA, and ion source temperature 290°C.

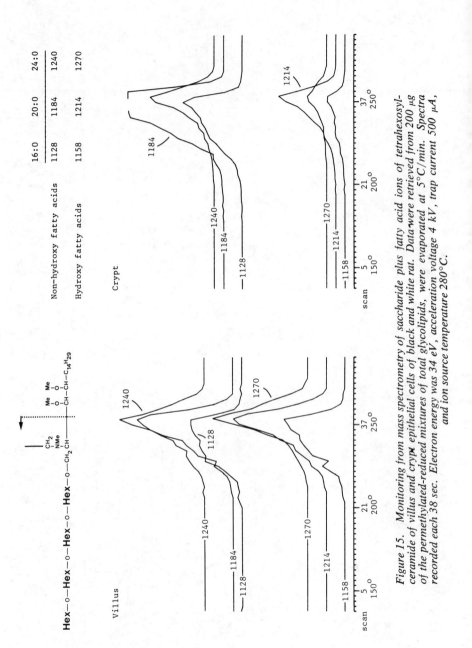

Figure 15. Monitoring from mass spectrometry of saccharide plus fatty acid ions of tetrahexosyl-ceramide of villus and crypt epithelial cells of black and white rat. Data were retrieved from 200 μg of the permethylated-reduced mixtures of total glycolipids, were evaporated at 5°C/min. Spectra recorded each 38 sec. Electron energy was 34 eV, acceleration voltage 4 kV, trap current 500 μA, and ion source temperature 280°C.

with GlcNAc were confined to epithelial cells. An unusual blood group B active hexaglycosylceramide based on GalNAc and restricted to non-epithelial cells may be identical with a glycolipid detected in rat macrophages and granuloma (19). All other fucolipids were found in epithelial cells and based on GlcNAc or lacking hexosamine. Two series of fucolipids were found in the black and white strain, one H active with 3, 5 and 10 sugars, and one A active with 4, 6 and 12 sugars. The fucolipids with 3 and 4 sugars are novel species and based simply on lactosylceramide, demonstrating that the simple derivatives of reducing lactose found in milk (20) have counterparts in membrane glycolipids. In large intestine of rat we have detected difucosyl substances which are absent from small intestine (unpublished). Further work may show if these also are analogous to the simple lactose saccharides in milk (20).

It will be interesting to see if the novel series of glycolipids in non-epithelial tissue, probably formed by a sequential addition of Galα, have specific immunological properties or can bind certain lectins. Apparently, the two strains of rat both have these glycolipids but differ in epithelial cells being blood group A positive or negative. The difference between the two strains may be explained by the absence of an α-N-acetylgalactosaminyltransferase in the white strain. Of interest is the lack of A activity in red cells and red cell glycolipids of the black and white rat, which is strongly A positive in intestinal glycolipids. Both strains had, however, B activity both in their intact red cells and in red cell glycolipids. Whether this B activity is based on the same glycolipid as found in non-epithelial tissue remains to be shown. We have preliminary evidence that this glycolipid is a major component of the complex glycolipids of rat liver. According to Table I, the epithelial cells of the black and white strain were richer in glycolipids, and according to Figs. 10 and 11 the same strain contained more fucolipid expressed as the H-type 3-sugar glycolipid.

In view of current discussions on a possible role of cell surface saccharides in control of growth and differentiation (1, 2), the changes found in epithelial cells undergoing a successive maturation from crypt to villus tip are, as a first impression, surprisingly small. An increase in monoglycosylceramide and hematoside and a decrease in trihexosylceramide, the three major glycolipid components, was found. Also, the ceramide of these glycolipids undergoes a successive change from crypt to villus with a chain lengthening and a 2-hydroxylation of the fatty acid. Concerning the more complex fucolipids, these are present already in the crypt cells (see Fig. 2 for 10- and 12-sugar compounds) indicating "a need" for these surface saccharides already in crypt cells. An extension of the saccharide chains parallel to the process of maturation (1, 2) was therefore not found.

One should, however, bear in mind the extreme complexity of the epithelial cell being highly asymmetric with a surface

membrane (where glycolipids are supposed to be located) divided
mainly into a brush border, facing the intestinal lumen, and a
basolateral membrane, being in contact with other epithelial cells
and the basal membrane. So far we have only studied whole cells
and not yet resolved minor components for a precise quantitation.
A subcellular fractionation into separate type of surface membrane
and glycolipid analysis may reveal interesting both qualitative
and quantitative differences. In fact, Lewis and coworkers (21)
have shown that preparations of brush border and basolateral
membranes of guinea-pig small intestine had different glycolipid
patterns. The glycerolipids of the two regions were fairly similar
but tri- and tetraglycosylceramides were more concentrated in the
basolateral membranes, whereas mono- and diglycosylceramides and
sulfatide were enriched in the brush border membranes.
    For human (16) and dog small intestine (5, 22) it has been
shown that globoside and the Forssman hapten, respectively, are
located in non-epithelial cells, while fucolipids are present in
epithelial cells. This is similar to the findings of this paper.
Also, glycolipids of epithelial cells (5, 16, 22) had a more
hydroxylated ceramide (phytosphingosine and 2-hydroxy fatty acid)
than non-epithelial cells (sphingosine and nonhydroxy fatty acid).
An analogous situation was found for rat small intestine, although the differences were not that clearcut, as nonhydroxy acids
were also present in epithelial cells and phytosphingosine was
also present to some extent in non-epithelial cells. The extent of
2-hydroxylation increased from crypt to villus tip (Figs. 2 and
14). The meaning of these differences in ceramide hydroxylation
(from one to three hydroxy groups) is not known. A model has,
however, been proposed, with a system of laterally oriented hydrogen bonds along the membrane at this level of ceramide in the
membrane matrix (17). The epithelial cells of intestine, especially those of the villus, are exposed to an intestinal content of
highly varying composition (both hydrophilic and hydrophobic) and
may need a more tight and stable surface membrane produced by an
increased hydroxylation of ceramide.
    As already mentioned the epithelial cells of small intestine
are involved in a number of enlarged transport processes and also
in biological recognition. Surprisingly, the acid glycolipid
fraction of rat small intestine lacked the animal sulfatide (ceramide galactose-3-sulfate), which is a major component of human
intestine (16) and also of small intestine of several animals
(cat, guinea-pig, hen and rabbit, unpublished). As for the rat,
this lipid was absent in small intestine of mouse and cod fish
(unpublished). The lack of sulfatide is unexpected in view of
the postulated role of this lipid as a K^+ receptor in Na^+ and K^+
transport (17, 18) and the dominance of Na^+ transport in small
intestine as a primary drive for the transport of a number of
other molecules. However, in these cases the molecule may be replaced by the glycerol-based sulfatide, which is removed in the

standard procedure of mild alkaline degradation.
The recognition processes of interest in relation to cell
surface saccharides and intestinal epithelial cells are of at
least two kinds. One is the exposure of primarily the brush
border membrane for a number of foreign molecules and micro-
organisms (or products of these) in the intestinal contents. A
role for carbohydrate in the binding of bacteria in the mechanism
of infection in epithelia has been postulated (1). The second
kind of recognition is the association of autologous cells with
each other, which should take place in the alteral membranes,
and the attachment of the cells to the basal membrane during
movement from crypt to villus tip. As a first step in the study
of small intestine the present work has defined to some extent
the difference concerning cell surface glycolipids between epithelial and non-epithelial cells and between whole epithelial
cells of different maturity. As a next step it would be relevant
to investigate the composition of separate types of surface
membranes. Also, the large intestine of the same strains of rat,
with a somewhat separate profiel of functions, may profile
supplementary information.

Acknowledgement

The work was supported by a grant from the Swedish Medical
Research Council (No. 3967).

References

1.   Hughes, C.L.; Sharon, N. Trends Biochem. Sci. 1978, 3, N 275.
2.   Marchesi, V.T.; Ginsburg, V.; Robbins, P.W.; Fox, C.F.; Eds.
     "Cell Surface Carbohydrates and Biological Recognition";
     Alan R. Liss, Inc.; New York, 1978.
3.   Lipkin, M. Physiol. Rev. 1973, 53, 891.
4.   Weiser, M.M. J.Biol. Chem. 1973, 248, 2536.
5.   McKibbin, J.M. J. Lipid Res. 1978, 19, 131.
6.   Karlsson, K.-A. In Witting, L.A., Ed. "Glycolipid
     Methodology"; American Oil Chemists' Society: Champaign,
     Illinois, 1976; p. 97.
7.   Breimer, M.E.; Hansson, G.C.; Karlsson, K.-A.; Leffler, H.:
     Pimlott, W.; Samuelsson, B.E. Biomed. Mass Spectrom. 1979,
     6, 231.
8.   Falk, K.-E.; Karlsson, K.-A.; Samuelsson, B.E. Arch. Biochem.
     Biophys. 1979, 192, 164.
9.   Falk, K.-E.; Karlsson, K.-A.; Samuelsson, B.E. Arch. Biochem.
     Biophys. 1979, 192, 177.
10. Falk, K.-E.; Karlsson, K.-A.; Samuelsson, B.E. Arch. Biochem.
     Biophys. 1979, 192, 191.

11.   Laine, R.A.; Stellner, K.; Hakomori, S.-i. In Korn, E.D.,
       Ed. "Methods in Membrane Biology"; Plenum Press: New York,
       1974; Vol. 2, p. 205.
12.   Bouhours, J.-F.; Glickman, R.M. Biochim. Biophys. Acta 1976,
       441, 123.
13.   Glickman, R.M.; Bouhours, J.-F. Biochim. Biophys. Acta 1976,
       424, 17.
14.   Karlsson, K.-A.; Samuelsson, B.E.; Steen, G.O. Biochim.
       Biophys. Acta 1973, 306, 317.
15.   Morgan, I.G.; Gombos, G.; Tettamanti, G. In Horowitz, M.I.;
       Pigman, W.; Eds. "The Glycoconjugates"; Academic Press:
       New York, 1977, Vol. I, p. 351.
16.   Falk, K.-E.; Karlsson, K.-A.; Leffler, H.; Samuelsson, B.E.
       FEBS Lett. 1979, 101, 273.
17.   Karlsson, K.-A. In Abrahamsson, S.; Pascher, I.; Eds.
       "Structure of Biological Membranes"; Plenum Press: New York,
       1977, p. 245.
18.   Hansson, G.C.; Heilbronn, E.; Karlsson, K.-A.; Samuelsson,
       B.E. J. Lipid Res. 1979, 20, 509.
19.   Hanada, E.; Handa, S.; Konno, K.; Yamakawa, T. J. Biochem.
       1978, 83, 85.
20.   Kobata, A. In Horowitz, M.I.; Pigman, W.; Eds. "The Glyco-
       conjugates"; Academic Press: New York, 1977, Vol. I, p. 423.
21.   Michell, R.H.; Coleman, R.; Lewis, B.A. Biochem. Soc. Trans.
       1976, 4, 1017.
22.   Smith, E.L.; McKibbin, J.M.; Karlsson, K.-A.; Pascher, I.;
       Samuelsson, B.E. Biochim. Biophys. Acta 1975, 388, 171.
23.   Dell, A.; Williams, D.H.; Morris, H.R.; Smith, G.A.;
       Feeney, J.; Roberts, G.C.K. J. Am. Chem. Soc. 1975, 97,
       2497.

RECEIVED December 10, 1979.

Galactoglycerolipids of Mammalian Testis, Spermatozoa, and Nervous Tissue

ROBERT K. MURRAY, RAJAGOPOLIAN NARASIMHAN, MARK LEVINE, and LES PINTERIC
Department of Biochemistry, University of Toronto, Toronto, Ontario, Canada M5S 1A8

MARGARET SHIRLEY, CLIFFORD LINGWOOD, and HARRY SCHACTER
Department of Biochemistry, Hospital for Sick Children, Research Institute, Toronto, Ontario, Canada M5G 1X8

The two major classes of glycolipids present in mammalian cells are glycosphingolipids and glycoglycerolipids (1). It is with certain members of the latter class that this article is concerned. Glycoglycerolipids are well established constituents of plant and bacterial cells (2,3,4). Galactosyl- and digalactosyl- diacylglycerols are the major glycoglycerolipids found in plant cells, although trigalactosyldiacylglycerol, 6-0-acylgalactosyl- diacylglycerol and sulfoquinovosyldiacylglycerol have also been described (4). In bacteria, mono- and di- glycosyldiacylglycerols occur most frequently, with the latter generally being the major species. Glucose, galactose and mannose are the usual sugars present in these compounds. Certain of these lipids also contain u-ronic acids. Halobacterium cutirubrum contains a glycolipid with galactose-sulfate, mannose and glucose linked to a phytanyl di-ether glyceride (5). Acyl substitutions on the sugar residues of diglycosylglycerolipids have also been described, as have phospho-glycoglycerolipids (4). The presence of glycoglycerolipids in mam-malian tissues, specifically nervous tissue, has been known since 1963 (6). Most of the mammalian glycoglycerolipids have been found to contain galactose as their sole sugar; however, the pres-ence in gastric juice and saliva of a novel series of glucoglycer-olipids has been described recently (7,8,9). Of the galactoglycer-olipids, galactosyl- and digalactosyl- diacylglycerols have re-ceived especial attention. An analog of galactosyldiacylglycerol, galactosylalkylacylglycerol, was also found in brain (10) shortly after the initial report of the presence of the diacyl compound in that organ (6). Interest in mammalian galactoglycerolipids accel-erated when it was discovered that the sulfated derivative of the lipid described by Norton and Brotz (10) was the major glycolipid of rat (11) and boar (12) testis. This sulfated galactolipid was partially characterized in a number of studies (e.g. 13 and 14) and has subsequently been fully characterized as 1-0-alkyl-2-0-acyl-3-0-β-D-(3'-sulfo)-galactopyranosyl-sn-glycerol (15). A var-iety of topics emerging from the study of this particular glyco-lipid have been reviewed previously (16). The present article will

0-8412-0556-6/80/47-128-105$5.25/0
© 1980 American Chemical Society

concentrate primarily on features concerning this and several
closely related galactoglycerolipids that have arisen since the
above mentioned review was written in mid 1975. Many aspects of
the biochemistry of the various sulfate-containing glycolipids
found in mammalian tissues have recently been reviewed by Sweeley
and Siddiqui (1), Dulaney and Moser (17) and Farooqui (18).

Nomenclature, Classification and Tissue Distribution

Both 1-0-alkyl-2-0-acyl-3-0-β-D-(3'-sulfo)-galactopyranosyl-
sn-glycerol and its non-sulfated species are major glycolipids of
the testis and spermatozoa of a number of higher animals, including humans (16). Despite the previous usage of names such as
seminolipid (12), sulfoglycerogalactolipid (19) and sulfogalacto-
glycerolipid (20) to describe the sulfated species, it now appears that sulfogalactosylalkylacylglycerol is the most chemically
informative trivial name with which to refer to this compound.
This arises from the fact that a closely related compound, almost
certainly 1-0-acyl-2-0-acyl-3-0-β-D-(3'-sulfo)-galactopyranosyl-
sn-glycerol, has been isolated from brain (21,22). The term sulfo-
galactoglycerolipid would not distinguish between these two compounds, particularly when referring to an organ such as rat brain,
in which they co-exist (22,23). Hence, it is more precise to refer to the ether-containing lipid as sulfogalactosylalkylacylglycerol (SGG) and to the diacyl-containing lipid as sulfogalactosyldiacylglycerol (22). The non-sulfated species of these two lipids
will be referred to as galactosylalkylacylglycerol (GG) and galactosyldiacylglycerol respectively.
     A classification of mammalian galactoglycerolipids is given
below.

Table I. Classification of Mammalian Galactoglycerolipids
     Diacyl Sub-Class   Alkylacyl Sub-Class

(A) Galactosyldiacylglycerol   (D) Galactosylalkylacyl-
(B) Sulfogalactosyldiacylglycerol   glycerol (GG)
(C) Digalactosyldiacylglycerol   (E) Sulfogalactosylalkylacyl-
                                           glycerol (SGG)
                                      (F) Digalactosylalkylacyl-
                                           glycerol

     Several features of this classification merit comment. Six
lipids have been included in the Table, but the identification of
two of them [(C) and (F)] is not firmly established. Lipid (C)
was tentatively identified in human brain (24); extracts of rat
brain appear to be able to catalyse its formation when incubated
under appropriate conditions (25) (this is discussed in more detail subsequently). Lipid (F) was detected in human testis and
sperm (26), and exhibited chromatographic and other properties
corresponding to what would be expected from a digalactosyl-

containing alkylacylglycerol. The systematic names for lipids (A)
and (C) are 1-0-acyl-2-0-acyl-3-0-β-D-galactopyranosyl-sn-glycer-
ol and 1-0-acyl-2-0-acyl-3-0-[α-D-galactopyranosyl(1→?)β-D-galac-
topyranosyl]-sn-glycerol; these lipids are still widely referred
to as monogalactosyl and digalactosyl diglyceride respectively.
Systematic names for lipids (B), (D) and (E) were indicated a-
bove. It is premature to assign a systematic name to lipid (F);
it will be of interest to determine whether the anomeric natures
of the two galactosyl residues are similar to those in lipid (C).
     With regard to their distribution in mammalian tissues, com-
pounds (A), (B) and (C) have been detected only in nervous tissue,
compounds (D) and (E) in both nervous tissue and testis and spermatozoa, and compound (F) only in human testis and spermatozoa.
However, preliminary evidence has been obtained (M. Levine and
R.K. Murray, unpublished observations), suggesting that small a-
mounts of compounds (A) and (B) may be present in dog testis a-
long with larger amounts of compounds (D) and (E).

Extraction of Galactoglycerolipids

     We have found the method of Suzuki (27) to be satisfactory
for extracting these lipids from testis, sperm and brain. A moderate loss of sulfate-containing galactoglycerolipids into the
upper phase of the Folch extract employed in this method occurs.
Using the method of column chromatography on silicic acid developed by Vance and Sweeley (28), the galactoglycerolipids shown in
Table I are all eluted by acetone subsequent to initial elution
of the column by chloroform. After evaporation of the acetone,
individual glycolipids can be purified by preparative thin layer
chromatography. If the glycolipid composition of the tissue under
study is complex (cf. human testis (26)), fractionation of these
lipids by chromatography using DEAE-cellulose (29) is useful.

Chemical Characterization of Galactosylalkylacylglycerols

     Table II lists the main procedures that have been used to
quantitate the amounts of these lipids present in testis, sperm
and brain and to determine their chemical structures. Reference
to some of the techniques applied to the characterization of sulfogalactosyldiacylglycerol are also included.
     One technique that we have found useful in permitting an
initial distinction between glycosphingolipids, galactosylalkylacylglycerols and galactosyldiacylglycerols is the use of brief
hydrolysis in mild alkali (cf. 20). This can be applied to either
the total glycolipid extract or to purified glycolipids. Typical
results of this procedure using a member of each of the above
three classes of glycolipids are shown in Figure 1. It should be
apparent that the use of this treatment to remove alkali-labile
contaminating lipids (e.g. phospholipids) from a glycolipid extract is unwise, until after a preliminary analysis has been

Table II. Procedures Used to Quantitate and Characterize
Galactoglycerolipids

Procedure	Compound Studied	Reference
General Analyses:		
Determination of sugar,	Rat testis SGG	(11)
glyceryl ethers and fatty	Boar testis SGG	(12)
acids by GLC		
Similar analyses by GLC-MS	Human testis SGG	(15)
	Rat brain SGG	(23)
Elemental analysis	Boar testis SGG	(12)
Quantitation by HPLC	Rat testis SGG	(30)
Analyses of the Sulfate Moiety:		
Detection of [^{35}S] sulfate	Rat testis SGG	(11)
Benzidine method	Rat testis SGG	(11)
Sodium rhodizinate method	Boar testis SGG	(12)
Estimation of lipid-bound sulfate	Rat testis SGG	(11)
IR spectroscopy	Boar testis SGG	(12)
Removal of sulfate by hydro- lysis in mild acid	Rat testis SGG	(11)
Removal of sulfate by solvo- lysis in dioxane	Rat testis SGG	(19)
Removal of sulfate by aryl- sulfatase A	Rat testis SGG	(31)
Elution in salts fraction during DEAE-cellulose chromato- graphy	Rat brain SGG	(20)
Permethylation *	Boar testis SGG	(12)
Determination of attachment to galactose by resistance to treatment with periodate	Rat brain sulfo- galactosyldiacyl- glycerol	(21)
Analyses of the Galactose Moieties:		
Determination of anomeric link- age by IR and NMR	Boar testis SGG	(12)
Determination of anomeric link- age by use of β-galactosidase	Rat brain SGG	(23)
Estimation of amount using galactose dehydrogenase	Guinea pig testis SGG	(13)
Estimation of amount using fluorimetry	Human testis SGG	(15)
Analyses of Glyceryl Ethers:		
Determination of isomers by TLC	Rat testis SGG	(14)
Stereochemical analysis by optical rotatory dispersion	Human testis SGG	(15)

Table II (continued)

Procedure	Compound Studied	Reference
Susceptibility to de-acylation by mild alkali	Guinea pig testis SGG	(13)
	Rat brain SGG	(20)
Other Analysis:		
Isolation of galactosyl-glycerol	Rat brain sulfo-galactosyldiacyl glycerol	(21)
Use of Spray Reagents to Exclude Other Constituents:		
Ninhydrin (free amino group)	Boar testis SGG	(12)
Benzidine (sphingosine)	Rat testis SGG	(11)
2,4-dinitrophenylhydrazine (plasmalogenic linkage)	Rat testis SGG	(11)
Bial's orcinol reagent (sialic acid)	Boar testis SGG	(12)
Acid molybdate (phosphate)	Rat testis SGG	(11)

*This procedure naturally also yields information on the nature of the galactose moieties.
The methods referred to in this Table have been used to quantitate and characterize the testicular SGG and GG species and also the corresponding lipids and sulfogalactosyldiacylglycerol from brain. References to studies performed prior to 1972 that characterized the galactoglycerolipids of nervous tissue have not been included.
Abbreviations: GLC, gas-liquid chromatography; MS, mass spectrometry; HPLC, high performance liquid chromatography; IR, infra-red; DEAE, diethylaminoethyl; NMR, nuclear magnetic resonance; TLC, thin layer chromatography.

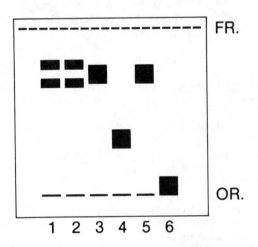

Figure 1.   Schematic of the effects of brief treatment with mild alkali on the thin-layer chromatographic migrations of three types of glycolipids.   1, 2: control and alkali-treated neutral glycosphingolipid; 3, 4: control and alkali-treated galactosyl-alkylacylglycerol; 5, 6: control and alkali-treated galactosyldiacylglycerol.   OR: origin; FR: solvent front.

The neutral glycosphingolipid is represented as a characteristic double band. We have not observed galactoglycerolipids to migrate as double bands on thin layer chromatography. Glucosyl- and lactosyl-ceramides exhibit the behavior (i.e., lack of effect of mild alkali on their migrations) of the compound shown in channels 1 and 2. SGG and GG behave in the same way as the compound represented in channels 3 and 4; the slower migrating product in channel 4 in the case of these two compounds would correspond to lyso-SGG and lyso-GG, respectively. Both galactosyl- and digalactosyldiacylgycerols show the behavior of the compound in channels 5 and 6; the product migrating at the origin in channel 6 in the case of these two compounds would correspond to galactosylglycerol and digalactosyl-glycerol, respectively. Cerebroside esters are one type of glycosphingolipid whose migration would be affected by the above treatment. Conversely, if galactosyldialkyglycerols exist in mammalian cells, their chromatographic migrations would not be affected by treatment with mild alkali.

performed on control and alkali-treated samples to determine if
the chromatographic migrations of any of the glycolipids present
are affected by it.

The initial characterization studies of the SGG derived from
rat (11) and boar (12) testis revealed the presence of approximately stoichiometric amounts of sulfate, galactose, fatty acid
and glyceryl ether. Using NMR spectroscopy, the study of Ishizuka
et al. (12) also suggested the β nature of the galactosidic linkage to glycerol and the position of the acyl chain on carbon 2 of
glycerol. In addition, analyses of the products of permethylation
indicated that the sulfate was attached to the 3' position of galactose. Measurement of the optical rotatory dispersion of the
glyceryl ether moiety (15) established the definitive structure
of SGG.

Perhaps the most remarkable feature of the SGG derived from
testis is its extremely restricted alkyl and acyl composition.
In the case of the SGG of rat (11,14), boar (12), guinea pig (13)
and human (15,26) testis, over 80% of the alkyl and acyl composition is comprised of saturated 16 carbon moieties [glyceryl-1-
hexadecyl ether (chimyl alcohol) and hexadecanoic acid (palmitic
acid) respectively]. The SGG present in rat brain appears to have
a less restricted alkyl and acyl composition (20).

Biosynthesis of Testicular and Other Galactoglycerolipids

A number of studies (11,13,14,32) have shown that [^{35}S]
sulfate is incorporated in vivo into testicular SGG. With regard
to the mechanism involved, both Knapp et al. (19) and Handa et al.
(32) have demonstrated formation of this lipid in vitro from GG
by transfer of sulfate from 3'-phosphoadenosine-5'-phosphosulfate
(PAPS), in analogy with the pathway of biosynthesis of sulfogalactosylceramide from galactosylceramide (reviewed in 17). Other
glycolipids with a terminal β-galactosyl residue (galactosyl- and
lactosyl- ceramides and galactosyldiacylglycerol) were found to
be sulfated by the enzyme preparations employed, whereas compounds
with a terminal α-galactosyl residue (galactosylgalactosylgluco-
sylceramide and digalactosyldiacylglycerol) were not. Both of
these studies suggested that primarily one sulfotransferase was
involved in the sulfation of the various glycolipid substrates;
however, this issue is not settled conclusively. The sulfotransferase activity in rat testis (19) was markedly enriched in a
Golgi apparatus fraction of that organ, confirming the involvement of that organelle in both sulfation processes (33) and in
the biosynthesis of glycolipids (34,35).

Well before the above studies were performed, the biosynthesis of galactosyldiacylglycerol in rat brain had been examined by
Wenger et al. (36). These workers found a β-galactosyl transferase
activity capable of catalysing the following reaction:

1,2-diacylglycerol + UDP-gal → Galactosyldiacylglycerol + UDP

Wenger et al. (25) later described the presence in rat brain of an
α-galactosyl transferase activity, that used the product of the
above reaction as substrate and UDP-gal as donor to catalyse the
formation of a second lipid, tentatively assigned the structure
of digalactosyldiacylglycerol (see earlier discussion). Subsequently, Flynn et al. (21) demonstrated the presence in rat brain
of a sulfotransferase activity capable of sulfating galactosyldiacylglycerol. The properties of this enzyme activity were described in more detail by Subba Rao et al. (37). Interestingly, significant differences were observed between the formation of sulfogalactosyldiacylglycerol and sulfogalactosylceramide, when catal-
ysed by the enzyme preparation used. The data did not necessarily
lead to the conclusion that two sulfotransferases were present,
but they did indicate how certain factors (e.g. ATP and Mg^{2+} concentrations) could control the relative amounts of these two lipids that were synthesized.

        In analogy with the reaction shown for formation of galacto-
syldiacylglycerol in brain, Levine et al. (38) have examined the
ability of rat testicular extracts to catalyse the following reaction:

        1,2-alkylacylglycerol + UDP-gal → GG + UDP

So far, although a variety of conditions of incubation have been
investigated, convincing evidence for the occurrence of this
reaction in rat testis has not been obtained. The significance of
a negative finding of this nature is limited, as it may only reflect a failure to select appropriate conditions. Alternatively,
the putative galactosyl transferase may be extremely labile or
present in very low activity. However, it is also possible that
another pathway, using a different acceptor molecule and/or a
different galactosyl donor, may be involved.
        The utilization of galactose for the in vivo biosynthesis of
GG and SGG by rat testis has also been examined (38,39). [^{14}C]-
galactose was injected into the testes of adult rats and the specific activities of the galactosyl moieties of these two lipids determined at various time intervals. Labelled galactose appeared
in GG by 10 minutes, the peak specific activity occurring by 2 h
after injection, and declining thereafter relatively rapidly. In
contrast, the appearance of radioactive galactose in the SGG was
much slower (detectable by 1 h), its peak specific activity occurring by 72 h after injection. Moreover, the specific activity of
the SGG subsequently declined very slowly over the following 2
weeks. These results are consistent with the hypothesis that GG
is the precursor of SGG in vivo; however, they do not prove this,
nor do they indicate from which precursor GG itself is synthesized.
Thus, although it appears reasonable to assume that sulfation is
the final step in the biosynthesis of testicular SGG, little is
known of the earlier steps.
        The pathway of biosynthesis of the glyceryl ether backbone of

the testicular SGG also remains unexplored; in view of the highly
restricted alkyl and acyl composition of the lipid, it would be
of interest to determine the substrate specificities of the enzymes involved in formation and transfer of these moieties.

Catabolism of SGG

Yamato et al. (31) purified arylsulfatase A from boar testis.
The specific activities of the enzyme preparation towards three
substrates - 4-nitrocatechol sulfate, SGG and sulfogalactosylceramide - increased at almost the same rate through the various
purification steps employed. The optimal pH for action on both of
the glycolipid substrates was 4.5. The activity of the enzyme was
somewhat greater using the sphingolipid as substrate, as compared
with SGG. A variety of procedures indicated that the two glycolipids were both substrates for the enzyme. It was suggested that
SGG may be the physiological substrate for arylsulfatase A in
testis. Essentially similar results were obtained by Fluharty et
al. (40), who examined the action of the same enzyme, but purified
from human urine, on rat testicular SGG and on sulfogalactosylceramide. Neither SGG nor classical sulfatide was a substrate for
arylsulfatase B. Again, these workers concluded that SGG appears
to be a physiological substrate for arylsulfatase A. Fluharty et
al. also pointed out that arylsulfatase A has been found in rabbit
sperm acrosomes, in which it was suggested that it might be involved in the penetration of spermatozoa through the investments
of the ovum (41).

An interesting extension of the above work was performed by
Yamaguchi et al. (42). They compared the activities towards nitro-
catechol sulfate, SGG and sulfogalactosylceramide of enzyme extracts from normal human brain and from two cases of a late in-
fantile form of metachromatic leukodystrophy (MLD). The activities
towards all three substrates were markedly deficient (1-5% of control activities) in the extracts from the diseased brains. The
authors concluded that the enzyme deficiency in the type of MLD
studied was due to a single sulfatase, catalysing the degradation
of all three substrates used. It has so far not proven possible
to determine whether SGG accumulates in the testes of adults with
late developing forms of MLD. Nor has it been established whether
SGG can accumulate in human brain in this condition; indeed, two
studies have failed to demonstrate its presence in that organ
(20,23). However, it is possible that the lipid could have a very
restricted location in human brain.

Reiter et al. (43) have shown that a second enzyme can also
act to degrade SGG. They found that secondary lysosomes from rat
liver contained not only arylsulfatase A, but also a lipase activity that could act to de-acylate SGG. Under the conditions used,
more product was formed by the action of the lipase on SGG than by
the action of arylsulfatase A. These workers also found that the
latter enzyme could use the lyso-SGG as a substrate. It would be

of interest to study the activity of the lipase on GG and also on
the diacyl-containing galactoglycerolipids. At the present time,
the relative physiological significance of the two pathways of
degradation of SGG has not been established. However, the lipase
activity has not so far been reported to be present in testis.
Further steps in the catabolism of SGG in testis - e.g. removal
of the galactosyl residue and degradation of the glyceryl ether
moieties - have apparently not yet been examined. It has been
shown that a β-galactosidase (E.C. 3.2.1.23) from Charonia lampas
is capable of removing the galactosyl residue from both the GG
and galactosyldiacylglycerol species derived from rat brain (23).

Appearance of Sulfatides During Testicular Development

     One approach towards determining the particular cell stage
at which phenotypic products (in the present case, certain specific glycolipids) of differentiation appear in testis is to remove
that organ from animals of known age and to correlate the appearance of the compound(s) under study with the appearance of a
particular cell type as determined by histologic examination. The
time of appearance in the testis of the various cell types involved in spermatogenesis has been particularly well established in
the case of the rat by Clermont and Perey (44). Using this approach , Kornblatt et al. (14) found that primary spermatocytes
appeared to be the earliest spermatogenic cells to contain high
levels of the SGG. Examination of the levels of SGG in the testes
of immature rats, hypophysectomized rats and normal and sterile
mice indicated that the majority of the SGG was located in the
germinal (spermatogenic) cells (as opposed to non-germinal cells,
such as Sertoli and Leydig cells) of the testis.
     Another finding that reinforces the probable germ cell loca-
tion of the SGG was made by Suzuki et al. (30). These workers fed
adult rats a diet deficient in vitamin A for 46 days. This resulted in a decline of SGG to 13% of its level in the testes of
appropriate control animals. Total lipid, phospholipid and DNA
(expressed appropriately) were only slightly reduced. Histologic
examination showed that the testes were aspermatogenic. Vitamin
A is obviously necessary for the maintenance of germ cell maturation; it would be of great interest to determine if it plays any
specific role in the biosynthesis of SGG.
     A dramatic increase (approx. 50-fold) of the activity of the
sulfotransferase involved in the biosynthesis of the SGG also
occurred when spermatocytes first began to appear in rat testis
(19); the rise in the activity of this enzyme preceded by several
days a marked rise in the amount of the SGG. Studies on pre-pubertal human testis (which is temporarily blocked in spermatogenesis
at a stage prior to the appearance of primary spermatocytes) have
shown that neither SGG nor GG is present (15,26). Similarly, the
testis of the pre-pubertal fowl also lacks sulfogalactosylceramide,
the sulfatide found in mature fowl testis (26). All of these find-

ings are consistent with the hypothesis that sulfatides are synthesized in the testis of a variety of species when early spermatocytes appear in that organ.

Letts et al. (45) have attempted to answer the question of
which cell population in rat testis synthesizes the SGG by using
methods that allowed fractionation of different testicular cell
types. Their results indicated that sulfation of SGG occurred at a
cell stage prior to the late (pachytene and diplotene) spermatocyte
stage. Letts et al. (45) also assayed the amount of radioactive
SGG in extracts of testis and epididymis at increasing times after
the injection of [^{35}S] sulfate into the testes of adult rats. The
epididymis showed no radioactive SGG for 4 weeks following injection,
but exhibited a dramatic appearance of the [^{35}S]-labelled compound
at 5 weeks. From previous studies on the kinetics of spermatogenesis in rats, it was possible for these workers to conclude that
sulfate incorporation into SGG must occur prior to the spermatid
stage. These workers also noted that the level of labelled SGG in
testis decreased steadily with time after injection of label.
Kornblatt (46) has made similar observations to the above. With
respect to the last point, she found that the rate of the decrease
of [^{35}S]-labelled SGG in testis coincided exactly with the rate of
decrease of [^{3}H]thymidine-labelled DNA levels in testis. This indicates that the loss of lipid was due to cell death and that
there was minimal turnover of SGG in surviving cells. To summarize,
the results from both of these studies suggest that the SGG is sulfated at the early primary spermatocyte stage. The sulfolipid then
appears to undergo little or no turnover in the germinal cells
during spermatogenesis and eventually appears in the spermatozoa.
This is an intriguing finding which implies that the lipid appears
in testis at a cell stage well before the spermatozoon and persists
in a metabolically stable form throughout all the complex cell mod-
elling processes that precede and accompany the appearance of the
highly specialized sperm cell. It should be noted, however, that
the above studies with [^{35}S] sulfate do not exclude the possibility
that other moieties of the SGG - e.g. the acyl group - could exhibit turnover.

Suzuki et al. (13) showed that boar spermatozoa possessed little or no capacity to incorporate [^{35}S] sulfate into SGG.
Narasimhan et al.(39) have confirmed the very limited, if not negligible, capacity of sperm to synthesize SGG by incubating bovine
spermatozoa with labelled glycerol and galactose. No radioactivity
was detected in the SGG following incubation with these compounds.
Radioactivity from these compounds was, however, found to be incorporated into SGG when they were injected into the testes of
mature rats.

Also relevant to the appearance of SGG during testicular differentiation were the results of a study performed by Ishizuka and
Yamakawa (47). These workers analysed the glycolipid composition
of three human seminoma (testicular) tumors. Unlike the control
human testicular tissue, no SGG or GG was detected in the tumors.

As many malignant tumors resemble fetal tissue in their biochemical composition, this result is consistent with the observed absence of SGG from immature human testis (15,26). Another tentative interpretation of this finding is that seminoma cells derive from a cell stage prior to that of the primary spermatocyte, thus accounting for their inability to synthesize SGG.

Subcellular Location of SGG in Testis and Spermatozoa

Both Letts et al. (45) and Kornblatt (46), using subcellular fractionation techniques, have obtained evidence indicating that at least some of the SGG in testis is present in the plasma membrane fraction of germinal cells. Further studies of this subject are in progress in the laboratories of these workers. Levine et al. (38) have isolated head and tail fractions of bovine spermatozoa following mild treatment of these cells with pronase; the SGG was found to be distributed in both fractions. This latter result is consistent with a location of the SGG in the plasma membrane, as this structure is continuous around the spermatozoon. It is apparent that treatment with arylsulfatase A might yield information on the exposure of the sulfate group of the SGG on the surface of these cells and could also provide a useful tool for studying the effects on spermatozoal function of modifying the structure of the lipid. However, preliminary attempts to use the arylsulfatase A of pig testis (31) to desulfate the SGG of intact bovine spermatozoa have not been successful (M. Levine and R.K. Murray, unpublished observations), despite the fact that the enzyme preparation was very active when isolated SGG was used as a substrate. The production of an antiserum to SGG (48,49) may permit the application of immunocytochemical methods to determine both its cellular and subcellular locations.

Attempted Labelling of Galactoglycerolipids Using Galactose Oxidase

The studies of Gahmberg and Hakomori (50) and Steck and Dawson (51) demonstrated the ability of galactose oxidase, in conjunction with NaB^3H_4, to label at least certain galactose and N-acetylgalactosamine residues of cell surface glycoproteins and glycolipids. In anticipation of employing this method to determine whether the galactose moieties of the SGG and GG of testicular cells and spermatozoa are exposed on the surface of these cells, Lingwood (52) has used this method to attempt to label several purified galactoglycerolipids in vitro. Using conditions that resulted in extensive labelling of galactosylceramide, GG was found to label to a maximum of 10% of the radioactivity incorporated into the sphingolipid. In addition, very low labelling of SGG, galactosyldiacylglycerol and sulfogalactosylceramide was also observed, in comparison with galactosylceramide. The labelling of the latter compound was not inhibited in the presence of GG, SGG or sulfogalactosylceramide.

The chemical explanation for the poor labelling of the galactoglycerolipids has not been elucidated. However, it does not appear to be due to decomposition of the borohydride or to degradation of these lipids during the labelling procedure. Possibly, some subtle difference in the physical states of the galactoglycerolipids as compared with the galactosphingolipids is involved. It is also of interest that sulfation of the galactose residue inhibits the action of galactose oxidase. In view of these findings, Lingwood (52) points out than an inability to be labelled by the galactose oxidase procedure does not necessarily mean that a galactolipid is absent from the cell surface. These results suggest that the galactose oxidase technique - at least as presently employed - is unlikely to be useful in determining the possible surface location of galactoglycerolipids in testicular cells and spermatozoa.

Antiserum to Testicular SGG

The pioneering studies of Rapport and his colleagues (53) clearly demonstrated the antigenicity of various glycolipids. Subsequent work has shown that antisera to glycolipids may be used to determine their cellular and subcellular locations (54). Antibodies to SGG and GG could thus prove useful in investigating the cellular and/or subcellular location of these lipids in testes and spermatozoa. The production of antibodies (complement-fixing) to sulfogalactosylceramide has been reported previously (55,56). Lingwood et al. (48,49) have thus attempted to produce antibodies to SGG in rabbits. The animals were injected by the intravenous route with liposomes containing SGG. Antibodies to SGG were detected by a complement fixation assay. Control sera showed no anti-SGG activity, but did show low antibody activity to GG, sulfogalactosylceramide and galactosylceramide. All of the anti-SGG activity was located in the IgG fraction. Anti-SGG was purified by adsorption to and elution from cholesterol particles coated with SGG. The eluted anti-SGG reacted with SGG but not with sulfogalactosylceramide or galactosylceramide; a low titer towards GG remained. These studies demonstrate the feasibility of preparing antibodies to SGG. It remains to be seen if these antibodies will prove useful for immunohistochemical approaches towards determining the location of SGG in testis and spermatozoa.

Sulfogalactolipids of the Testis of Various Species

The glycolipids of the testis of a number of animals have been analysed to determine whether SGG is a universal constituent of testicular tissue of all chordates. The results of these studies are summarized in Table III.

At least four points concerning these results merit comment: (1) SGG has been detected in the testes of all of the limited number of mammals so far examined

Table III. Distribution of Sulfogalactolipids in the Testis (or
Sperm) of Various Chordata

Animal	Class	SGG	SGC	SGGC	References
Human	Mammalia	+	+	+	(15,26)
Rat	"	+	–	–	(11,14)
Mouse	"	+	–	–	(11,14)
Guinea Pig	"	+	–	–	(11,13)
Rabbit	"	+	–	–	(11)
Boar	"	+	–	–	(12)
Duck	Aves	–	+	–	(26)
Fowl	"	–	+	–	(26)
Salmon (milt)	Osteichthyes	–	–	+	(26)
Trout	"	–	–	+	(26)
Puffer fish	"	–	+	–	(57)
Skate fish	Chondrichthyes	–	+	–	(26)
Green Monkey	Mammalia	+	+	–	M. Levine (unpublished observations)
Dog	"	+	+	–	"
Bull (sperm)	"	+	–	–	"
Opossum	"	+	+	–	"
Turtle	Reptilia	–	+	+	"
Bull frog	Amphibia	–	+	–	"

The presence or absence of each of the three sulfogalactolipids
studied is indicated by + or – respectively. It is possible that
trace amounts of one or other of the three glycolipids listed
may be present in certain of the testicular tissues marked as –,
as in most cases the estimates were based on visual examinations
of appropriately stained thin layer chromatograms.
Abbreviations: SGC, sulfogalactosylceramide; SGGC, sulfogalact-
osylglucosylceramide.

(2) SGG has not been detected in the testes of the limited numbers of birds, fish, reptiles and amphibians analysed
(3) In the testes of these latter classes of animals that lack
SGG, two other sulfogalactolipids were found to be the major
glycolipids - i.e. sulfogalactosyl- and sulfogalactosylglucosyl-
ceramides
(4) The two sphingosine-containing sulfatides are also found in
the testes of certain mammals - for instance, human testis contains both of them, in addition to SGG.

These observations indicate that it should be revealing - in
terms of increasing understanding of the mechanisms that operate
to regulate the sulfatide profile of a tissue - to compare the
capacities of testicular extracts from one or more animals of
each of the classes listed in Table III to synthesize the various
constituent parts of the above three lipids. As partially discussed earlier, the biosynthesis of these sulfatides can be considered to occur in 3 stages: (1) assembly of the lipid moieties -
i.e. ceramide and possibly 1-0-alkyl-2-0-acyl-sn-glycerol
(2) glycosylation and (3) sulfation. The specificity of the sulfation reaction appears to be relatively low, as the sulfotransferase involved in the biosynthesis of SGG will sulfate a number
of lipids with a terminal β-galactosyl residue, including GG,
galactosyl- and galactosylglucosyl- ceramides (19,32). As human
testis contains each of these three lipids (15,26), this can explain, at least in part, why it exhibits all three sulfatides.
It thus seems more likely that the varied sulfatide profiles displayed by the testis of the animals listed in Table III will be
explained by differing potentials, among species, of that organ
to synthesize the lipid moieties, and by the specificities for
both the lipid acceptors and the sugar donors of the glycosyl
transferases involved in the second stage of sulfatide biosynthesis (cf.32).

Two other points arising from this line of work also deserve
brief discussion. Firstly, it is relevant to mention that sulfo-
quinovosyl diglyceride has been reported to be the major glycolipid of the spermatozoa of sea urchins (58). It will thus be of
interest to extend studies of the comparative biochemistry of
testicular glycolipids to lower classes of animals as well as to
further members of the classes listed in Table III. Secondly, it
is apparent that, whatever the precise phylogenetic distribution
of glycolipids in testis may turn out to be, the results to date
strongly support the hypothesis that sulfatides play an important
role in testicular and/or spermatozoal function in chordates.

Galactoglycerolipids of the Nervous System

As this is a relatively large subject area, it will only be
touched upon insofar as it relates to work performed by the
authors. Galactosyldiacylglycerol was reported to be a constituent of brain in 1963 (6); subsequently, the same lipid derived

120   CELL SURFACE GLYCOLIPIDS

from bovine spinal cord was thoroughly characterized (59). Also
in 1963, GG was detected in rat brain (10); a later study confirmed the presence of GG in the brains of several other species
(60). Both of these glycolipids are also present in peripheral
nerves (cf. 61). A compound corresponding in its properties to digalactosyldiacylglycerol was also found to be a constituent of
human brain (24). Pathways for the biosynthesis of both galacto-
syl- and digalactosyl- diacylglycerols by extracts of rat brain
have been referred to earlier. The presence of GG in rat brain
suggested to Levine et al. (20) that SGG might also be located in
that organ. These workers did indeed find small amounts of SGG
(approx. one-fifteenth the amount of sulfogalactosylceramide) in
adult rat brain. They also detected small amounts of the same lipid in rabbit brain, but not in a portion of the frontal lobes of
human brain. In addition, evidence was obtained in their study
suggesting that a lesser amount of sulfogalactosyldiacylglycerol
might also be present in rat brain. However, this point was clearly established by Flynn et al. (21) and Pieringer et al. (22),
who provided unequivocal evidence, including the isolation of
sulfogalactosylglycerol, for the presence of that compound in rat
brain. These workers found larger amounts of the diacyl- than of
the alkylacyl- containing galactoglycerolipid; however, in contrast to Levine et al. (20), they used immature (approx. 22 day
old) animals. Ishizuka et al. (23) confirmed that both lipids
were present in rat brain and they developed appropriate methodology, including analyses by gas-liquid chromatography-mass spectrometry, for thoroughly characterizing them. They also showed
that the diacyl-containing lipid was the predominant compound in
the brains of rats of age up to 19 days, but thereafter the alkylacyl type predominated, consituting 85% of the sum total of these
two lipids by 68 days of age. SGG was detected in cod brain, but
neither sulfolipid was detected in normal human brain nor in the
brain of a case with metachromatic leukodystrophy.

     The studies of Levine et al. (20) revealed that the turnover
of the SGG in rat brain was similar to that of sulfogalactosylceramide. This suggested that the SGG like the classical sulfatide (62), might be located predominantly in myelin. Pieringer et
al. (22) demonstrated that the diacyl form of the sulfogalactoglycerolipids present in rat brain had a faster turnover than
that of the alkylacyl form. Because previous studies (63,64) had
indicated that the galactosyldiacylglycerol of rat brain was an
excellent marker metabolite for myelination, these workers also
studied the possible association of the two sulfogalactoglycerolipids (i.e. the diacyl and the alkylacyl species) with myelination. Support for the association of these two compounds with
myelination was found by showing that they were absent from rat
brain before 10 days of age and that they accumulated in that organ between 10 and 25 days of age (the period of maximum myelination). Further support was derived from the finding that the
sulfotransferase involved in the biosynthesis of the diacyl-con-

taining galactolipid (and presumably, but not conclusively established, also of the alkylacyl species) increased maximally in activity during the same time period. In addition, the amounts of the sulfogalactoglycerolipids and the activity of the sulfotransferase were greatly decreased in the brains of non-myelinating jimpy mice. Ishizuka et al. (23) also found that synthesis of rat brain SGG was most active around 18 days of age.

Conclusion

It is evident that knowledge of the galactoglycerolipids has grown in recent years. Instead of being recognized solely as quantitatively relatively minor glycolipids of nervous tissue, members of the alkylacyl sub-class are now also seen to constitute major glycolipids of mammalian testis and spermatozoa. Nevertheless, their tissue distribution is extremely restricted in comparison with that of the glycosphingolipids. It will be of interest to determine whether additional galactoglycerolipids occur in mammalian cells and also if other types of glycoglycerolipids exist. In this respect, as mentioned earlier, the Slomianys (7,8,9) have provided evidence that a novel series of glyceryl ether-containing glucoglycerolipids may exist in gastric juice, saliva and perhaps other secretions. However, their results have not as yet received independent confirmation (cf. 65,66).

With respect to function, one wonders if the common location of GG and SGG in the brain and testis of certain species reflects some physiological entity that both of these organs share - for instance, a blood barrier. However, the apparent absence of SGG from human brain (20,23) does not support this conjecture. Similarly, the sharing of GG and SGG by these two "sequestered" organs raises thoughts as to whether this could be of immunological significance in some situations. Yet another line of speculation is whether the presence of relatively large amounts of glyceryl ether-containing galactoglycerolipids in testis may somehow be related to the fact that the testes of most mammals are confined in a scrotum maintained at a temperature lower than the rest of the body. These surmises reflect the humbling fact that there is as yet very little understanding of the functions of the various non-sulfated and sulfated galactolipids present in mammalian cells. A ray of hope for this area is provided by the hypothesis of Karlsson and his colleagues (67,68) that sulfogalactosylceramide may act as a cofactor site for the activity of $Na^+K^+ATPase$. The possibility that SGG could be involved in such a site in spermatozoa has been raised (16,69).

The most useful function of this review will be if it stimulates further research in this area. For this reason, it seems appropriate to conclude by posing a number of fairly obvious - but nevertheless basic - questions, that will hopefully be answered in future investigations. What physical differences exist between alkylacyl and diacyl galactoglycerolipids, between galactoglycerolipids and galactosphingolipids and between sulfated and non-sulfated galactolipids (assuming in all cases that the pairs of lipids mentioned differ only with respect to the specified moieties)? Assuming that physical differences do exist, what are their functional implications? What are the details of the pathway of biosynthesis of the testicular galactoglycerolipids and what factors (e.g. genetic, hormonal, enzyme specificity etc.) control the expression of this pathway during testicular differentiation? Can this pathway be interfered with by pharmacological agents (e.g. analogs of glyceryl ethers), and if so, what effects could that have on testicular and possibly nervous system function? What are the precise cellular and/or subcellular locations of the galactoglycerolipids in testicular cells and spermatozoa? Finally, what is the function of the SGG in mature spermatozoa - is it involved in ion transport, in motility, in sperm-ovum interactions or is it merely a passenger molecule, having fulfilled its function at some earlier stage of the life history of these beautifully specialized cells?

Acknowledgements

We thank Ms. B. Palmer for her excellent technical assistance in a number of the studies whose results are summarized above. The authors are grateful for support from the Ford Foundation, from N.I.H. (Grant No. RO-1HD07889 from the National Institute of Child Health and Human Development) and from the Medical Research Council of Canada. The patience and care displayed by Ms. Stephanie Amos during the typing of this manuscript is warmly acknowledged.

Literature Cited

1. Sweeley, C.C.; Siddiqui, B. in "The Glycoconjugates" (Horowitz, M.; Pigman, W., eds.), Academic Press, New York, N.Y., 1976, vol. 1, pp.459-540.
2. Carter, H.E.; Johnson, P.; Weber, E.J. Annu. Rev. Biochem., 1965, 34, 109-142.
3. Kates, M. Advan. Lipid Res., 1970, 8, 225-265.
4. Sastry, P.S. Advan. Lipid Res., 1974, 12, 251-340.
5. Kates, M.; Palameta, B.; Perry, M.P.; Adams, G.A. Biochim. Biophys. Acta, 1967, 137, 213-216.
6. Steim, J.M.; Benson, A.A. Fed. Proc., 1963, 22, 299 (abstr. no. 830).
7. Slomiany, B.L.; Slomiany, A.; Glass, G.B.J. Eur. J. Biochem., 1977, 78, 33-39.
8. Slomiany, B.L.; Slomiany, A.; Glass, G.B.J. Biochemistry, 1977, 16, 3954-3958.
9. Slomiany, B.L.; Slomiany, A. Biochem. Biophys. Res. Commun. 1977, 79, 61-66.

10. Norton, W.T.; Brotz, M. Biochem. Biophys. Res. Commun., 1963, 12, 198-203.
11. Kornblatt, M.J.; Schachter, H.; Murray, R.K. Biochem. Biophys. Res. Commun., 1972, 48, 1489-1494.
12. Ishizuka, I.; Suzuki, M.; Yamakawa, T. J. Biochem., 1973, 73, 77-87.
13. Suzuki, A.; Ishizuka, I.; Ueta, N.; Yamakawa, T. Japan. J. Exp. Med., 1973, 43, 435-442.
14. Kornblatt, M.J.; Knapp, A.; Levine, M.; Schachter, H.; Murray, R.K. Canad. J. Biochem., 1974, 52, 689-697.
15. Ueno, K.; Ishizuka, I.; Yamakawa, T. Biochim. Biophys. Acta, 1977, 487, 61-73.
16. Murray, R.K.; Levine, M.; Kornblatt, M.J. in "Glycolipid Methodology" (Witting, L.A., ed.), Amer. Oil. Chem. Soc., Champaign, Ill., 1976, pp.305-327.
17. Dulaney, J.T.; Moser, H.W. in "The Metabolic Basis of Inherited Disease" (Stanbury, J.B.; Wyngaarden, J.B.; Fredrickson, D.S., eds.), McGraw-Hill, New York, N.Y., 1978, 4th edition, pp.770-809.
18. Farooqui, A.A. Int. J. Biochem., 1978, 9, 709-716.
19. Knapp, A.; Kornblatt, M.J.; Schachter, H.; Murray, R.K. Biochem. Biophys. Res. Commun., 1973, 55, 179-186.
20. Levine, M.; Kornblatt, M.J.; Murray, R.K. Canad. J. Biochem., 1975, 53, 679-689.
21. Flynn, T.J.; Desmukh, D.S.; Subba Rao, G; Pieringer, R.A. Biochem. Biophys. Res. Commun., 1975, 65, 122-128.
22. Pieringer, J.; Subba Rao, G.; Mandel, P.; Pieringer, R.A. Biochem. J., 1977, 166, 421-428.
23. Ishizuka, I.; Inomata, M.; Ueno, K.; Yamakawa, T. J. Biol. Chem., 1978, 253, 898-907.
24. Rouser, G.; Kritchevsky, G.; Simon, G.; Nelson, G.J. Lipids, 1967, 2, 37-40.
25. Wenger, D.A.; Subba Rao, K.; Pieringer, R.A. J. Biol. Chem. 1970, 245, 2513-2519.
26. Levine, M.; Bain, J.; Narasimhan, R.; Palmer, B.; Yates, A.J.; Murray, R.K. Biochim. Biophys. Acta, 1976, 441, 134-145.
27. Suzuki, K. J. Neurochem., 1965, 12, 629-638.
28. Vance, D.E.; Sweeley, C.C. J. Lipid. Res., 1967, 8, 621-630.
29. Kates, M. in "Laboratory Techniques in Biochemistry and Molecular Biology" (Work, T.S.; Work, E., eds.), 1972, American Elsevier, New York, N.Y., vol.3, part II, pp.269-600.
30. Suzuki, A.; Sato, M.; Handa, S.; Muto, Y.; Yamakawa, T. J. Biochem., 1977, 82, 461-467.
31. Yamato, K.; Handa, S.; Yamakawa, T. J. Biochem., 1974, 75, 1241-1247.
32. Handa, S.; Yamato, K.; Ishizuka, I.; Suzuki, A.; Yamakawa, T. J. Biochem., 1974, 75, 77-83.
33. Young, R.W. J. Cell Biol., 1973, 57, 175-189.
34. Fleischer, B.; Zambrano, F. Biochem. Biophys. Res. Commun. 1973, 52, 951-958.

35. Keenan, T.W.; Morré, D.J.; Basu, S. J. Biol. Chem., 1974, 249, 310-315.
36. Wenger, D.A.; Petitpas, J.W.; Pieringer, R.A. Biochemistry 1968, 7, 3700-3707.
37. Subba Rao, G.; Norcia, L.N.; Pieringer, J.; Pieringer, R.A. Biochem. J., 1977, 166, 429-435.
38. Levine, M.; Narasimhan, R.; Pinteric, L.; Murray, R.K. Proced. XIth Internatl. Congress Biochem., 1979, p. 408 (abstr. no. 05-9-R49).
39. Narasimhan, R.; Levine, M.; Murray, R.K. Proced. Vth Internatl. Sympos. on Glycoconjugates, 1979, G. Thieme-Verlag, Stuttgart, in press.
40. Fluharty, A.L.; Stevens,R.L.; Miller, R.T.; Kihara, H. Biochem. Biophys. Res. Commun., 1974, 61, 348-354.
41. Yang, C.H.; Srivastava, P.N. Proc. Soc. Exp. Biol. Med., 1974, 145, 721-725.
42. Yamaguchi, S.; Aoki, K.; Handa, S.; Yamakawa, T. J. Neurochem. 1975, 24, 1087-1089.
43. Reiter, S.; Fischer, G.; Jatzkewitz, H. FEBS Lett., 1976, 68, 250-254.
44. Clermont, Y.; Perey, B. Amer. J. Anat., 1957, 100, 241-267.
45. Letts, P.J.; Hunt, R.C.; Shirley, M.A.; Pinteric, L.; Schachter, H. Biochim. Biophys. Acta, 1978, 541, 59-75.
46. Kornblatt, M.J. Canad. J. Biochem., 1979, 57, 255-258.
47. Ishizuka, I.; Yamakawa, T. J. Biochem., 1974, 76, 221-223.
48. Lingwood, C.; Murray, R.K.; Schachter, H. Proced. Soc. for Complex Carbohydrates, 1979, abstr. no. 28.
49. Lingwood, C.A.; Murray, R.K.; Schachter, H. Proced. XIth Internatl. Congress Biochem., 1979, p. 491 (abstr. no. 07-3-H99).
50. Gahmberg, C.G.; Hakomori, S. J. Biol. Chem., 1973, 248, 4311-4317.
51. Steck, T.L.; Dawson, G. J. Biol. Chem., 1974, 249, 2135-2142.
52. Lingwood, C.A. Canad. J. Biochem., in press.
53. Rapport, M.M.; Graf, L. Prog. Allergy, 1969, 13, 273-331.
54. Marcus, D.M. in "Glycolipid Methodology" (Witting, L.A., ed.), 1976, Amer. Oil Chem. Soc., Champaign, Ill., pp. 233-245.
55. Zalc, B.; Jacque, C.; Radin, N.S.; Dupouey, P. Immunochem., 1977, 14, 775-779.
56. Hakomori, S. J. Immunol., 1974, 112, 424-426.
57. Ueno, K.; Ishizuka, I.; Yamakawa, T. J. Biochem., 1975, 77, 1223-1232.
58. Nagai, Y.; Isono, Y. Japan. J. Exp. Med., 1965, 35, 315-318.
59. Steim, J.M. Biochim. Biophys. Acta, 1967, 144, 118-126.
60. Rumsby, M.G.; Rossiter, R.J. J. Neurochem., 1968, 15, 1473-1476.
61. Singh, H. J. Lipid. Res., 1973, 14, 41-49.
62. Norton, W.T. in "Basic Neurochemistry" (Albers, R.W.; Siegel, G.J.; Katzman, R.; Agranoff, B.W.,eds.), 1972, Little Brown, Boston, Mass., pp. 365-386.

63. Inoue, T.; Desmukh, D.S.; Pieringer, R.A. J. Biol. Chem., 1971, 246, 5688-5694.
64. Desmukh, D.S.; Inoue, T.; Pieringer, R.A. J. Biol. Chem., 1971, 246, 5695-5699.
65. Narasimhan, R.; Bennick, A.; Murray, R.K. Fed. Proc., 1979, 38, p. 404 (abstr. no. 925).
66. Narasimhan, R.; Bennick, A.; Palmer, B.; Murray, R.K. Proced. Soc. for Complex Carbohydrates, 1979, abstr. no. 35.
67. Karlsson, K.-A.; Samuelsson, B.E.; Steen, G.O. Eur. J. Biochem., 1974, 46, 243-258.
68. Karlsson, K.-A. in "Structure of Biological Membranes" (Abrahamsson, S.; Pascher, I., eds.), 1977, Plenum Press New York, N.Y., pp.245-274.
69. Hansson, C.G.; Karlsson, K.-A.; Samuelsson, B.E. J. Biochem., 1978, 83, 813-819.

RECEIVED December 10, 1979.

Structural Studies of Neutral Glycosphingolipids of Human Neutrophils by Electron Impact/Desorption Mass Spectrometry

BRUCE A. MACHER and JOHN C. KLOCK

Department of Medicine, University of California, San Francisco, CA 94143

Electron impact mass spectrometry has proven to be a most sensitive tool for detailed structural analysis of intact glycosphingolipids. Several kinds of information can be obained with samples of 10-300 µg, including: the number and sequence of carbohydrate residues, the major fatty acid and long chain base species, the number of branching points, and in some cases the molecular weight and   information on the position of glycosidic linkage (1-6). We have utilized a variation of electron impact mass spectrometry in the analysis of neutral glycosphingolipids of human neutrophils, which has been referred to as electron impact/desorption mass spectrometry (for a review, see ref. 7). This technique has allowed us to obtain the same type of structural information as outlined above, but with sample amounts of 1-5 µg. A lower source temperature probably leads to less thermal decomposition of the sample and thus increased sensitivity. We have been able to conclude from the spectra obtained by this method that human neutrophils containat least four neutral glycosphingolipids which have the following partial structures: Hexose-0-Cer, Hexose-0-Hexose-0-Cer, Hexosamine-0-Hexose-0-Hexose-0-Cer, Hexose-0-Hexosamine-0-Hexose-0-Hexose-0-Cer. The ceramide moiety in these four compounds is characterized as a 4-sphingenine with an N-linked palmitic, lignoceric or nervonic acid.

Materials and Methods

Isolation of human neutrophils. Leukocytes were obtained from normal donors by leukapheresis with an IBM 2997 Blood Cell Separator (8). Normal mature neutrophils were purified from this mixed leukocyte preparation by dextran sedimentation and Ficoll-Hypaque gradient centrifugation as previously described (9). The Wright-stained smears of the preparation showed that the cells were over 95% neutrophils.

0-8412-0556-6/80/47-128-127$5.00/0

Extraction and purification of neutrophil glycosphingolipids.
Purified human neutrophils were extracted with 20 volumes of
each of the following solvent mixtures: chloroform/methanol
2/1, 1/1, 1/2, (v/v). After evaporation of the organic solvents in
vacuo, the residue was dissolved in approximately 5 volumes of
chloroform/methanol/water (30/60/8, v/v) and mixed with 0.5 g
of DEAE-Sephadex A25 (Pharmacia, Piscataway, N.J.) acetate
form (10). The sample was allowed to absorb to the column
packing for 20 min and was then applied to a column of the same
material. Neutral and acidic lipid fractions were eluted as
described by Ando and Yu (11). Neutral glycosphingolipids were
further purified on a column of BioSil A, 100-200 mesh (Bio Rad,
Richmond, CA). The neutral lipid fraction was dissolved in 5-10
ml of chloroform/methanol (1/1, v/v), applied to a column of
BioSil A (2x30 cm), and eluted as 100 ml fractions with solvent
mixtures of increasing polarity (from 100% chloroform to 100%
methanol). Final purification of each neutral glycosphingolipid
was by preparative thin-layer chromatography using Silica Gel 60
High Performance Plates (EM Laboratories Inc., Cincinnati, OH)
in solvent system A (chloroform/methanol/water, 60/35/8, v/v).
Glycosphingolipids were visualized by a brief exposure to iodine,
eluted with chloroform/methanol/water (50/50/10, v/v) and
rechromatographed in solvent A or B
(chloroform/methanol/water, 100/42/6, v/v) to demonstrate
homogeneity.

Direct probe mass spectrometry. Glycosphingolipids (30-100 µg)
were permethylated as described (12). The samples (less than 5 µ
g) were subjected to electron impact/desorption analysis with a
Varian MAT CH-5 DF mass spectrometer under the following
conditions: emission current, 300µA; electron energy, 70 eV;
acceleration voltage, 3KV; ion source temperature, 160° C;
emitter wire current, programed from 0 to 35mA.

Results

Isolation of the neutral glycosphingolipids. In a typical
extraction procedure, 10^{10} purified human neutrophils yielded
100-150 mg of total glycosphingolipids. As shown in Table I,
glycosphingolipids account for approximately 10% of the total
cellular dry weight of the neutrophil. Separation of the total
neutrophil lipids by DEAE-sephadex and silicic acid column
chromatography yielded 70-100 mg of neutral glycosphingolipids
from 10^{10} cells.

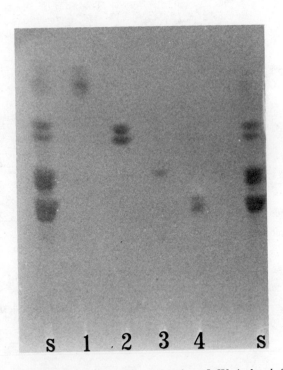

Figure 1. Thin-layer chromatography of fractions I–IV isolated from human neutrophils. The separation is on a plate of silica gel 60 (HPTLC) in solvent system A. S: erythrocyte glycosphingolipid standards; 1–4: human neutrophil glycosphingolipid fractions.

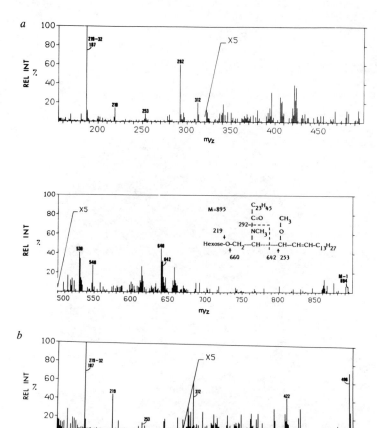

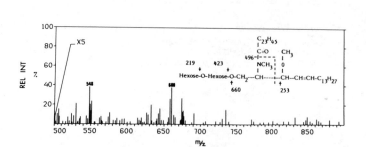

*Figure 2.   Mass spectra of the intact permethylated glycosphingolipids of fractions
I (a), II (b), III (c), and IV (d). See Materials and Methods for conditions.*

c

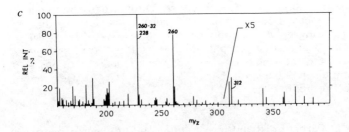

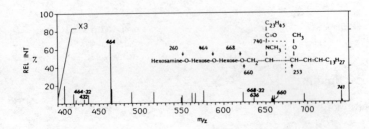

d

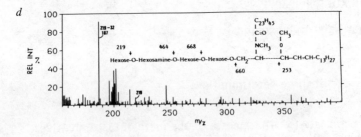

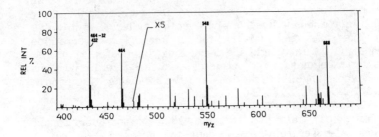

After repeated preparative thin-layer chromatography, four fractions were obtained as shown in Figure 1. When the thin-layer chromatographic mobilities of fractions I-IV were compared to standard glycosphingolipids isolated from human erythrocytes the following relationships were found: fractions I-IV had similar mobilities to GlcCer, LacCer, GbOse$_3$Cer and GbOse$_4$Cer, respectively.

Table I: Percent distribution of neutrophil lipid components

Fraction	% total cell dry weight*
Total lipid	30
Phospholipids	14.3
Neutral lipids	5.7
Total glycosphingolipids	10.0
Neutral glycosphingolipids	7.2

*average of three determinations

Direct probe analysis. The spectra of the methylated derivatives of fractions I-IV are shown in Figure 2 (a-d), together with abbreviated structural formulas and indications of some fragments (Refs. 1-6 were consulted for comparison). Only fraction I gave ions indicative of the entire molecule at m/z 894 (M-1) and m/z 863 (M-32) for a monoglycosyl-ceramide containing C$_{16:0}$ fatty acid and C$_{18:1}$ long chain base. Peaks corresponding to the permethylated carbohydrate portions of the glycosphingolipid fractions are the following: Fraction I, m/z 187, 219, 292, 278, 530, 640 and 642; Fraction II, m/z 187, 219, 422, 496, and 847; Fraction III, m/z 228, 260, 432, 464, and 636; Fraction IV, m/z 182, 187, 228, 432, 464 and 668. To our knowledge the spectrum presented for the methylated derivative of fraction III is the first to be presented for a naturally occurring compound of this structure.

Peaks corresponding to the fatty acid residue and part of the long chain were seen at m/z 294, 322, 404 and 406. Peaks for 4-sphingenine appeared at m/z 253, 294, 312 and in some scans m/z 364. The entire ceramide fragment with $C_{16:0}$ fatty acid and $C_{24:0}$ fatty acid was seen at m/z 548 and 660, respectively.

Discussion

The results presented in this report confirm and extend previous studies on the structure of neutral glycosphingolipids of human neutrophils. On the basis of carbohydrate compositional data and TLC properties structures have been prosposed for neutral glycosphingolipids prepared from whole blood including glucosylceramide (13-14) lactosylceramide (13,14,15); galabiosylceramide (15); lactoneotetraosylceramide (14); and globotetraosylceramide (13). Detailed structural analysis have not been published and therefore these structural assignments are still tentative.

The studies discussed above have all dealt with glycosphingolipids isolated from neutrophils that were derived from normal whole blood. Wherrett (16) has presented detailed structural analysis of a tetraglycosylceramide isolated from polymorphonuclear leukocytes obtained from the urine of a patient with a urinary tract infection. This glycosphingolipid was determined to be lactoneotetraosylceramide.

The data presented in this report allow the assignment of partial structures for four neutral glycosphingolipids of human neutrophils: Hexose-0-Cer, Hexose-0-Hexose-0-Cer, Hexosamine -0-Hexose-0-Hexose-0-Cer and Hexose-0-Hexosamine-0-Hexose-0-Hexose-0-Cer.

This information was obtained from samples of less than 5 µg by electron impact/desorption direct probe mass spectrometry. On the basis of complete structural analyles, to be presented elsewhere, we have been able to determine that the four fractions isolated thus far actually contain six different glycosphingolipids with the following structures:

Glcβ1→1Cer   Galβ1→1Cer
Galβ1→4Glcβ1→1Cer   Galα1→4Galβ1→1Cer
GlcNAcβ1→3Galβ1→4Glcβ1→1Cer
Galβ1→4GlcNAcβ1→3Galβ1→4Glcβ1→1Cer

In addition to these structures, human neutrophils also contain eight to twelve gangliosides and a few species of glycosphingolipids with more than four saccharide units. Purification and structural analyses are currently underway in our laboratory.

Acknowledgements

This study was supported in part by a grant from the Leukemia Research Foundation and by cancer research funds from the University of California.
Mass spectrometric analyses were provided by Mr. Lawrence R. Phillips and Ms. Betty Baltzer of the Michigan State University - NIH Mass Spectrometry Facility, and are sincerely appreciated.

Abbreviations

Glc, glucose; Gal, galactose; GlcNAc, N-acetylglucosamine; TLC, thin-layer chromatography; GlcCer, glucosylceramide; GalCer, galactosylceramide; LacCer, lactosylceramide; GbOse$_3$Cer, globotriaosylceramide; and GbOse$_4$Cer, globotetraosylceramide. Individual sugars are assumed to have the D configuration and to be in the pyranose form.

Literature Cited

1. Karlsson,K.-A. Biochemistry, 1974, 13, 3643.
2. Karlsson,K.-A. Prog. Chem. Fats Other Lipids, 1978, 16, 207.
3. Leeden,R.W.; Kundu,S.K.; Price,H.C.; Fong,J.W. Chem. Phys. Lipids. 1974, 13, 429.
4. Egge,H. Chem. Phys. Lipids, 1978, 21, 349.
5. Karlsson,K.-A. Biomed. Mass Spectrom., 1974, 1, 49.
6. Hanfland,P.; Egge,H. Chem. Phys. Lipids, 1976, 16, 201.
7. Sweeley,C.C.; Soltman,B.; Holland,J.F. in "High Performance Mass Spectrometry: Chemical Applications. Gross,M.L.Ed. American Chemical Society: New York,1978; p.209.
8. Hester,J.P.; Kellogg,R.M.; Mulzet,A.P.; Kruger,V.R.; McCredie,K.B.; Freireich,E.J. Blood, 1979, 54, 254.
9. Klock,J.C.; Bainton,D.F. Blood, 1976, 48, 149.
10. Ledeen,R.W.; Yu,R.K.; Eng,L.F. J. Neurochem. 1973, 121, 829.
11. Ando,S.; Yu,R.K. J. Biol. Chem. 1977, 252, 6247.
12. Hakomori,S.-I. J. Biochem. 1964, 55, 205.
13. Hildebrand,J.; Stryckmans,P.; Stoffyn,P. J. Lipids Res., 1971, 12, 361.
14. Narasimham,P.; Murry,R.K. Biochem. J. 1979, 326, 63.
15. Miras,C.J.; Mantzos,J.D.; Levis,G.M. Biochem. J. 1966, 98, 782.
16. Wherrett,J.R. Biochem. Biophys.Acta, 1973, 326, 63.

RECEIVED December 10, 1979.

Glycosphingolipids of Skeletal Muscle

JAW-LONG CHIEN and EDWARD L. HOGAN

Department of Neurology, Medical University of South Carolina, Charleston, SC 29403

It is currently held that glycosphingolipids are enriched in cell surface membranes and possible participants in such events as receptor interactions (1,2), permeability change (3), cellular adhesion (4) and cellular recognition (5). The likelihood of their localization in sarcolemma and possible role in myogenesis including cell fusion or in conduction of the action potential prompted us to begin their study with isolation and characterization of the gangliosides and neutral glycosphingolipids in chicken and human skeletal muscle.

Although muscle comprises approximately 40% of the body weight, there have been only a few studies of glycosphingolipids of muscle. Puro and coworkers (6) studied the qualitative and quantitative patterns of gangliosides in several extraneural tissues including skeletal and cardiac muscles of rat, rabbit and pig, but did not purify the individual gangliosides. Lassaga et al. (7) isolated four gangliosides from the hind leg and back muscle of the rabbit. One had the molar composition of hematoside (GM3) but the structures of the others – two disialo- and one trisialoganglioside – were not fully clarified. Svennerholm et al. (8) did a more complete study of human skeletal muscle. They isolated four major gangliosides and determined their composition by gas chromatography to be consistent with GM3, GM2, GDla and a sialosyltetraglycosylceramide. Recently, Levis and coworkers (9) examined the glycosphingolipids in human heart and found that human cardiac muscle contains the same gangliosides as those of human skeletal muscle. However, the distribution of gangliosides was quite different. In heart, GM3 (23%), GD3 (22%) and GM1 (16%) are the major ganlgiosides while in skeletal muscle GM3 (67%) predominates.

Neutral glycosphingolipids have also been studied in human skeletal (8) and cardiac (9) muscle. In skeletal muscle, lactosylceramide is the predominant glycolipid (38.4%) followed by globotriaosylceramide (26.3%) and globoside (12.4%); while in heart, globoside predominates (43.0%) followed by globotriaosylceramide (32.0%).

0-8412-0556-6/80/47-128-135$5.00/0

The results presented here show that the glycosphingolipids
of chicken pectoral muscle differ from those of human skeletal
and cardiac muscles. In addition, we are presenting the structures of two glucosamine-containing gangliosides which were
characterized by enzymatic hydrolysis and methylation studies.

Materials

Pectoral muscle from adult Leghorn chickens was obtained
from a local supermarket (the main source) or dissected immediately after sacrifice. Human skeletal muscle was obtained four
hours post mortem from the pectoral and iliopsosas muscles of a
70-year-old black male who died of a gunshot wound to the head.
Precoated silica gel plates (silica gel 60) were purchased
from Scientific Products. Bio-Sil A (200 - 400 Mesh) was obtained
from Bio-Rad Laboratories. Fatty acid methyl esters, sphingosine
and dihydrosphingosine were products of Supelco, Inc. as were
10% DEGS-PS, 3% SP-2340 and 3% OV-17 (all on Supelcoport support).
N-acetyl and N-glycolyl neuraminic acid, DEAE-Sephadex A50 and
neuraminidase type IX were obtained from Sigma Company. Ganglioside standards from human brain and neutral glycosphingolipid
standards from bovine erythrocytes were prepared in this laboratory.
β-galactosidase was isolated from papaya and β-hexosaminidase
was prepared from jack bean meal (10). α-N-acetylgalactosaminidase was a generous gift of Dr. Y.-T. Li of Tulane University.

Extraction of glycosphingolipids

The muscles were freed by gross dissection of extraneous
tissue which was mainly fat and peripheral nerves, and then
stored at -40°C. For an experiment, approximately 1 kg tissue
was macerated by a meat grinder and homogenized in ten volumes
of tetrahydrofuran:0.01 M KCl (4:1, v/v), stirred for 3 hours,
and filtered through a Buchner funnel. The extraction was repeated twice and the filtrates then combined and concentrated in a
rotary evaporator. One liter of chloroform-methanol (2:1, v/v)
was added to the lipid extract which has the appearance and
consistency of syrup. Gangliosides were partitioned into the
upper layer by the addition of 200 ml of water (11) and the
lower layer extracted two additional times with theoretical
upper phase containing 0.027% KCl. The combined upper layers
were then concentrated and dialyzed exhaustively at 4°C with
five changes of distilled water.

DEAE-Sephadex Column Chromatography

DEAE-Sephadex A-50 (Cl-form) was converted to the acetate
form by the following procedure: The gel was washed first with
five volumes of 0.1 N NaOH, and after rinsing with distilled

water, it was converted into the acetate form by washing with 1N
acetic acid. The sedimented gel was then rinsed repeatedly with
water until neutral, washed with methanol and packed into a column
(1.8 x 21 cm). The lipid extract obtained following dialysis
was dissolved in C:M (2:8, v/v) and applied to this column
which had been previously equilibrated in the same solvent.
The neutral lipids were eluted by the C:M (2:8, v/v) and gangliosides then eluted by methanol containing sodium acetate in the
following concentrations: 0.01 M (fraction I), 0.02 M (fraction
II), and 0.2 M (fraction III). The fractions eluted were concentrated and salt removed by dialysis.

Bio-Sil A Column Chromatography of Ganglioside Fractions

Bio-Sil A was activated at 110°C overnight, suspended in
chloroform and packed into a column (1.5 x 45 cm). Fraction I
(eluted from DEAE-Sephadex with 0.01 M sodium acetate was dissolved in C:M (2:1, v/v) and applied to the column. Gangliosides
were eluted with a C:M:H$_2$0 solvent system of increasing polarity.
We have been using the following mixtures:

Solvent I - C:M:H$_2$0 (130:70:12, v/v), 0.4 liter and solvent 2 -
C:M:H$_2$0 (120:70:14, v/v), 0.5 liter.

Fractions of 6 ml volume were collected and 50 μl aliquots
used to identify the gangliosides by TLC. Four gangliosides
have been purified from fraction I of chicken skeletal muscle
directly from the column.

Silica-gel G Column Chromatography of Neutral Glycosphingolipids

The neutral lipid fraction from the DEAE-Sephadex A-50
column was combined with the lower phase obtained after Folch
partition of the total lipid extract and the combined lipids
dried. To the same flask, 100 ml of 0.6 M NaOH in methanol was
added. The mixture was incubated at 37°C for 5 hours. Five volumes of acetone were then added and stored overnight at 4°C. The
precipitate was collected by centrifugation at 4°C and dissolved
in C:M (4:1, v/v). After application to the column (2.0 x 25 cm),
the column was washed with chloroform. Neutral glycolipids were
then eluted with tetrahydrofuran: H$_2$0 (10:1). Fractions containing neutral glycosphingolipids were pooled and their glycolipid
content examined by thin-layer chromatography.

Enzymatic Hydrolysis Employing Glycosidases

The sequence and anomeric configuration of the oligosaccharide chain was determined by step-wise hydrolysis with specific
glycosidases. The conditions of incubation for hydrolysis
are the same as those previously described (12). For the hydroly-

sis of sialic acid from gangliosides, 30 μg of the ganglioside
was dissolved in 150 μl of 0.05 M sodium acetate buffer at pH
5.0, and incubated overnight at 37°C with 4 miniunits of neuraminidase from <u>Clostridium perfringens</u>. For hydrolysis of asialo-
gangliosides and neutral glycosphingolipids, 30 μg of glycolipid
was incubated overnight with 0.3 - 1.0 unit of β-galactosidase
or β-hexosaminidase at 37°C. After the reaction was complete,
the product was partitioned to the lower layer by the addition of
5 vol. of C:M(2:1, v/v), and the upper phase washed twice with
theoretical lower phase. The combined lower layers were then
resolved by TLC.

Permethylation methods

The glycosyl linkages were determined using methylation
technique. In brief, the purified glycolipids were exposed to
dimethylsulfinyl ion and then methylated with methyl iodide (13).
The methylated derivative was applied to an LH-20 column (0.5 x
24 cm) which had been packed and eluted with acetone (14). The
combined methylated glycolipids were hydrolyzed with 0.6N H_2SO_4
in 80% aqueous acetic acid at 80°C for 18 hours, reduced and
acetylated according to Bjorndal <u>et al</u>. (15). Partially methylated galactitol and glucitol acetates were separated isothermally
at 180°C using a column packed with 3% OV-275 Supelcoport (100-120
Mesh). Amino sugar derivatives are separated by a 3% OV-17
Supelcoport (100-120 Mesh) column over a range of 180° - 200°C
with a temperature increment rate of 2°/min. (<u>16</u>, <u>17</u>, <u>18</u>).

Other Methods

Fatty acid methyl esters were extracted from the methanoly-
sate with hexane and analyzed at 190°C by GC using a 10% DEGS
column. Sphingosine bases were determined after hydrolysis (19)
as trimethylsilyl derivatives by GC using a 3% SE-30 column (20).
Sialic acid was determined by the resorcinol method (21) as
modified by Miettinen and Takki-Luukainen (22). Species of
sialic acid (NANA, NGNA, etc.) were analyzed by TLC (23) and gas
chromatography (24). Sugar composition and hexosamines were
determined as alditol acetates using GC (15).

Results

Comparison of glycosphingolipids from human and chicken
skeletal muscle. The elution of gangliosides from DEAE-Sephadex
A 50 column with these 0.01, 0.02 and 0.2 M sodium acetate concentrations separated the gangliosides into mono-, di- and poly-
sialo- fractions. The gangliosides of human muscle are shown in
Fig. 1A. The monosialogangliosides GM3, GM2 and GM1 were eluted
with 0.01 M sodium acetate in methanol (lane 2), GD3 and GD1a
with the 0.02 M solvent (lane 4) and others with 0.2 M acetate

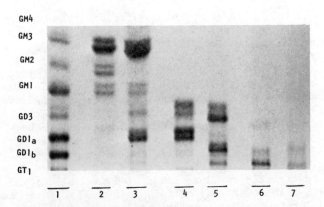

Figure 1A.   *Thin-layer chromatogram of human and chicken muscle ganglioside fractions eluted from DEAE-A 50 column*

Lane 1, standard gangliosides from human brain. Lanes 2 and 3, gangliosides eluted by 0.01M, lanes 4 and 5 by 0.02M, and lanes 6 and 7 by 0.2M sodium acetate in methanol. Lanes 2, 4, and 6 from human muscle; lanes 3, 5, and 7 from chicken muscle. Solvent system: $C:M:0.25\%$ $CaCl_2$ (60:40:9)

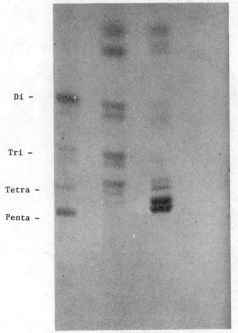

Figure 1B.   Thin-layer chromatogram of human and chicken neutral glycolipids

Lane 1 contains (from the top) standard lactosylceramide, globotriaosylceramide, neolactotetraosylceramide, and lactopentaosylceramide prepared from bovine erythrocytes. Lane 2 contains the neutral glycosphingolipids of human skeletal muscle, and lane 3 contains those of the chicken. Solvent system: $C:M:H_2O$ (60:40:9, v/v).

(lane 6). When the upper phase ganglioside fraction from chicken pectoral muscle was resolved in a similar way, four bands were eluted with the 0.01 M acetate in methanol. The major one had a mobility corresponding to GM3. The others had R_f values close to those of GM1, GD3 and GD1a respectively (Fig. 1A, lane 3). However, all contained one sialic acid per mole. The disialoganglioside fraction of chicken muscle was also considerably different from that of human. Both contained GD3, but the other major disialogangliosides in chicken migrated between GD1a and GD1b; and GD1b and GT1 respectively. Human muscle also contained appreciable trisialoganglioside of which very little was found in the chicken.

The neutral glycosphingolipids of human and chicken skeletal muscle were also remarkably different (Fig. 1B). Human muscle contained lactosylceramide as the major glycolipid followed by globotriaosylceramide and globoside (17), while in chicken muscle the major neutral glycosphingolipid (48%) migrates between nLcOse$_4$Cer and IV^{3}Gal-nLcOse$_4$Cer. It contained two moles each of galactose and N-acetylgalactosamine and one mole of glucose, and was converted to globoside by the α-N-acetylgalactosaminidase from limpet (25). Thus, it appears to be a Forssman-active glycolipid. Gas-liquid chromatographic analysis of the other neutral glycolipids was consistent with the molar composition of galactosylceramide (20%), lactosylceramide (12%), glucosylceramide (9%), globoside (8%) and globotriaosylceramide (3%).

Bio-Sil A column chromatography and glycosyl composition of the gangliosides of chicken muscle. The elution pattern of monosialogangliosides from a Bio-Sil A column is shown in Fig. 2. Under these conditions (see text), the four gangliosides separated well. The fractions containing the same ganglioside were pooled and the purity confirmed by repeat TLC. When developed with C:M:0.25% CaCl$_2$ (60:40:9, v/v), the four monosialogangliosides comigrated with GM3, GM1, GD3 and GD1a ganglioside standards isolated from human brain (Fig. 3A). But in another and alkaline solvent C:M:0.25N NaOH (60:40:9, v/v), three or all except ganglioside I (lane 2) were obviously different in mobility (Fig. 3B). Ganglioside II (lane 3) moved well ahead of brain GM1, ganglioside III (lane 4) was behind GD3, and the R_f of ganglioside IV (lane 6) was slightly less than that of the GD1a. These differences in R_f must derive from the differences in sugar composition in comparison to the standards from brain (Table I). The composition of ganglioside I was the same as brain GM3. Ganglioside II differed from GM1 in containing N-acetylglucosamine rather than N-acetylgalactosamine. Ganglioside III was a novel sialoglycolipid with a molar composition of sialic acid: N-acetylgalactosamine: galactose: glucose: sphingosine of 1:1:3:1:1. Ganglioside IV had the same sugar composition as that of sialosylhexaglycosylceramide from human spleen (26) and bovine erythrocytes (12) with a molar ratio of sialic acid: N-acetylglucosamine: galactose: glucose: sphingosine of 1:2:3:1:1.

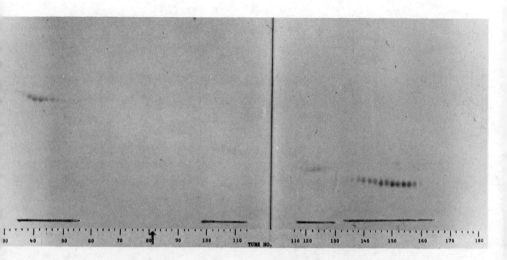

Figure 2.   Elution pattern from Bio-Sil A column of monosialogangliosides pre-pared from chicken pectoral muscle.   Column was eluted with C:M:H₂O (130:70:12, v/v) and changed to C:M:H₂O (120:70:14, v/v) at the arrow.

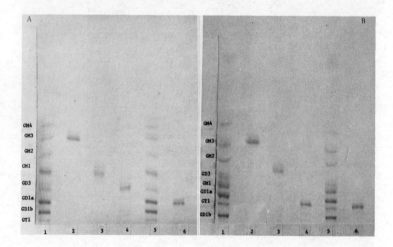

Figure 3.   Thin layer chromatograms of monosialogangliosides purified from chicken muscle

Lanes 1 and 5, human brain ganglioside standards; lanes 2, 3, 4, and 6 fractions from a Bio-Sil A colume (Figure 2).   Solvent systems: (A) CHCl₃:MeOH:0.25 CaCl₂ (60:49:9, v/v); (B) CHCl₃:MeOH:0.25M NH₄OH (60:40:9, v/v).

Table I

SUGAR COMPOSITION OF CHICKEN MUSCLE MONOSIALOGANGLIOSIDES

GANGLIOSIDE	GLc	Gal	GlcNAc	GalNAc	NANA
I	1	1.07	—	—	1.09
II	1	2.16	1.18	—	1.08
III	1	2.80	—	0.94	1.1
IV	1	2.96	1.83	—	1.04

Characterization of saccharide unit. Ganglioside I was
hydrolyzed by neuraminidase to a neutral glycolipid which was
further cleaved by β-galactosidase to become glucosylceramide.
The sequence of the two glucosamine-containing gangliosides (II
and IV) are shown in Fig. 4 and 5 respectively. Ganglioside II
was hydrolyzed by neuraminidase with no detergent to a compound
with R_f corresponding to that of Neolactotetraosylceramide which
was subsequently converted to lactriaosyl-, lactosyl-, and glucosylceramide by the consecutive actions of β-galactosidase,
β-hexosaminidase and β-galactosidase (Fig. 4). Ganglioside IV was
hydrolyzed to become a neutral hexaglycosylceramide when incubated with neuraminidase from Cl. perfringens without detergent.
The asialoglycolipid was in turn cleaved by alternate treatment
with β-galactosidase and β-hexosaminidase to yield glucosylceramide (Fig. 5). Gas-liquid chromatographic analysis of the partially methylated hexitol acetate derivatives produced 2,4,6-
tri-0-methyl-galactitol 1,3,5-tri-acetates; 2,3,6-tri-0-methyl-
glucitol-1, 4,5-triacetates and 3,6-di-0-methyl-2-deoxy-2-N-
methyl-acetamidoglucitol-1,5-diacetate. There was no 2,3,4,6-
tetra-0-methyl galacticol -1, 5-diacetate produced indicating
that the sialic acid is attached at the terminal nonreducing end
of the saccharide unit.

Fatty acids and sphingosines. The lipid composition of the
four monosialogangliosides is shown in Table II. The major
fatty acids are palmitic, stearic and oleic acids. The long
chain base is composed mainly of C-18 sphingosine with less than
20% of dihydrosphingosine.

Discussion

The structures of the glycosphingolipids of skeletal muscle
have been studied in human (8) and rabbit skeletal muscle (7)
and in human cardiac muscle (9). Qualitatively, human skeletal
and cardiac muscle contain the same neutral glycosphingolipids
and gangliosides though the gangliosides of rabbit skeletal
muscle are quite different. Rabbit does not contain GM2 or the
glucosamine-containing ganglioside reported in human skeletal
muscle.

We have used a DEAE-Sephadex column to separate gangliosides
into three groups, the mono-, di- and polysialogangliosides and
this enabled a more detailed comparison of gangliosides. Both
human and chicken skeletal muscle contain GM3 as the major ganglioside. The other monosialogangliosides of human muscle are
GM2 and GM1 but these two gangliosides were not detected in
chicken. Instead, gangliosides containing glucosamine and a sialosylpentaglycosylceramide constitute the other monosialoglyco-
lipids. Chicken muscle also differs from human in containing a
novel disialoganglioside migrating between GD1a and GD1b in alkaline conditions (Fig. 3B). The major neutral glycosphingolipid
of chicken muscle is a Forssman-hapten pentaglycosylceramide

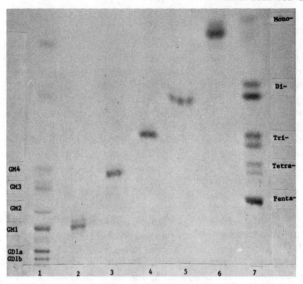

Figure 4.   Enzymatic hydrolysis of ganglioside II (lane 3 in Figure 3)

Lane 1, standard brain gangliosides. Lane 7, standard neutral glycolipids from bovine
erythrocytes. Lane 2, ganglioside II. Lane 3, 2+ neuraminidase. Lane 4, 3+ β-galactosidase. Lane 5, 4+ β-hexosaminidase. Lane 6, 5+ β-galactosidase.

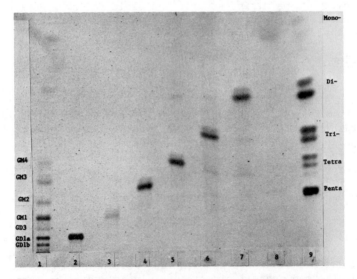

Figure 5.   Enzymatic hydrolysis of ganglioside IV (lane 6 in Figure 3)

Lanes 1 and 9, ganglioside and neutral glycosphingolipid standards as in Figure 4. Lane 2,
ganglioside IV. Lane 3, 2+ neuraminidase. Lane 4, 3+ β-galactosidase. Lane 5, 4+
β-hexosaminidase. Lane 6, 5+ β-galactosidase. Lane 7, 6+ β-hexosaminidase. Lane 8,
7+ β-galactosidase.

Table II

FATTY ACIDS AND LONG CHAIN BASES OF CHICKEN MUSCLE

MONOSIALOGANGLIOSIDES

Fatty Acids	Gangliosides			
	I	II	III	IV
C 16:0	23.2	18.1	15.4	16.6
C 16:1	1.8	2.2	3.0	2.5
C 18:0	33.5	24.7	39.2	33.4
C 18:1	21.5	38.5	31.8	24.2
C 18:2	5.2	2.1	1.2	3.7
C 20:0	2.5	1.2	2.2	3.0
C 20:1	1.5	1.8	0.5	1.4
C 21:0	1.0	--	1.0	3.5
C 22:0	2.4	1.0	--	1.1
C 22:1	5.2	2.0	1.7	2.1
C 24:1	2.2	8.4	4.0	8.5
Sphingosines				
d 18:0	18.0	16.3	14.7	10.7
d 18:1	82.0	83.7	85.3	89.3

identical to that previously described in horse kidney and spleen (27), sheep erythrocytes, (28), and canine kidney and intestine (18). In contrast, lactosylceramide predominates in human muscle and constitutes 38% of the total neutral glycolipids. Such striking differences in these compounds - both neutral and acidic - suggests that species specificity prevails over organ or tissue contraints upon the pattern or ratio of these lipids.

The use of ion-exchange chromatography enabled the subsequent greater resolution of individual gangliosides by silicic acid chromatography with Bio-Sil A and characterization of the major monosialogangliosides in chicken pectoral muscle by glycosyl composition, enzymatic sequencing and methylation analysis. The structures identified included GM3 (ganglioside I, NeuAcα2→3Galβ1→4Glc→Cer), sialosyl-lacto-N-neotetraosylceramide or IV^3 αNeuAc-nLcOse4Cer (ganglioside II, NeuAcα2→3Galβ1→4GlcNAc β1→3Galβ1→4Glc→Cer), sialosyl lacto-N-neohexaglycosylceramide or VI^3 αNeuAc-nLcOse6Cer (ganglioside IV or NeuAcα2→3Galβ1→ 4GlcNAcβ1→3Galβ1→4GlcNAcβ1→3Galβ1→4Glc→Cer), and a novel sialosylpentaglycosylceramide (ganglioside III). The sequence of this ganglioside has been determined by step-wise hydrolysis using specific glycosidases to be NeuAcα→Galβ→GalNAcβ→Galα→Gal βGlc→Cer (V NeuAc, IV Gal-GgOse4Cer) and is to our knowledge the first glycolipid of the globo-series containing sialic acid.

There are evidences for implicating glycoconjugates on the muscle cell surface in myogenesis (29, 30). Whatley et al. (29) for example found a three-fold increase in GD1a concentration just prior to fusion in a rat myoblast cell line while GM3, GM2 and GM1 did not change. McEvoy and Ellis (30) found an increased biosynthesis of several neutral glycosphingolipids and gangliosides just prior to fusion in primary cultures of chick embryo myoblasts. A role for glycolipids in such intercellular regulation is also consistent with the reported promotion of cell adhesion in Hela cells in the presence of glycolipids particularly those with tetraose chain length and a terminal galactose (5). It would seem important to consider the structures of chicken muscle glycosphingolipids in relation to the study of myogenesis in view of the wide usage of embryonic chick muscle as a model system.

Acknowledgment

This work was supported in part by the Muscular Dystrophy Association. We thank Mrs. Eve Thrasher for secretarial support and typing this manuscript.

Literature Cited

1. Mullin, B. R., Fishman, P. H., Lee, G., Aloj, S. M.,
    Ledley, F. D., Winand, R. J., Kohn, L. D. and Brady, R. O.
    Proc. Natl. Acad. Sci. U.S.A., 1976, 73, 842–846.
2. Lee, G., Aloj, S. M., Brady, R. O. and Kohn, L. D.
    Biochem. Biophys. Res. Comm., 1976, 73, 370–377.
3. Glick, J. L. and Githens, S. Nature, 1965, 208, 88.
4. Roseman, S. Chem. Phys. Lipids, 1970, 5, 270–297.
5. Huang, R. T. C. Nature, 1978, 276, 624–626.
6. Puro, K., Maury, P. and Huttunen, J. K. Biochim. Biophys.
    Acta, 1969, 187, 230–235.
7. Lassaga, F. E., Lassaga, A. G. and Caputto, R. J. Lipid
    Research, 1972, 13, 810–815.
8. Svennerholm, L., Bruce, A., Mansson, J.-E., Rynmark, B.-M.
    and Vanier, M.-T. Biochim. Biophys. Acta, 1972, 280,
    626–636.
9. Levis, G. M., Karli, J. N. and Moulopoulos, S. C. Lipids,
    1979, 14, 9–14.
10. Li, S.-C. and Li, Y.-T. J. Biol. Chem., 1970, 245, 5153–
    5160.
11. Folch, J., Lee, M. and Sloane Stanley, G. H. J. Biol.
    Chem., 1957, 226, 497–509.
12. Chien, J.-L., Li, S.-C., Laine, R. A. and Li, Y.-T. J.
    Biol. Chem., 1978, 253, 4031–4035.
13. Hakomori, S. J. Biochem. (Tokyo), 1964, 55, 205–208.
14. Yang, H. and Hakomori, S. J. Biol. Chem., 1971, 246,
    1192–1200.
15. Bjorndal, H., Lindberg, B. and Svensson, S. Acta
    Chem. Scand., 1967, 21, 1802–1804.
16. Stellner, K., Saito, H. and Hakomori, S. Arch.
    Biochem. Biophys., 1973, 155, 464–472.
17. Stoffel, W. and Hanfland, P. Hoppe-Seyler's Z. Physiol.
    Chem., 1973, 354, 21–31.
18. Sung, S.-S., Esselman, W. J. and Sweeley, C. C. J. Biol.
    Chem., 1973, 248, 6528–6533.
19. Gaver, R. C. and Sweeley, C. C. J. Am. Oil. Chem. Soc.,
    1965, 42, 294–298.
20. Carter, H. E. and Gaver, R. C. J. Lipid Res., 1967, 8,
    391–395.
21. Svennerholm, L. Biochem. Biophys. Acta, 1957, 24, 604–
    611.
22. Miettinen, T. and Takki-Luukainen, I. T. Acta Chem.
    Scand., 1954, 13, 856–859.
23. Granzer, E. Z. Physiol. Chem., 1962, 328, 277–279.
24. Yu, R. K. and Ledeen, R. W. J. Lipid Res., 1970, 11,
    506–515.
25. Uda, Y., Li, S.-C. and Li, Y.-T. J. Biol. Chem., 1977,
    252, 5194–5200.
26. Wiegandt, H. Eur. J. Biochem., 1974, 45, 367–369.

27.   Siddiqui, B. and Hakomori, S.   J. Biol. Chem., 1971,
       246, 5766-5769.
28.   Fraser, B. A. and Mallette, M. F.   Immunochemistry, 1974,
       11, 581-585.
29.   Whatley, R., Ng, K. D., Rogers, J., McMurray, W. C. and
       Sanwal, B. D.   Biochem. Biophys. Res. Comm., 1976, 70,
       180-185.
30.   McEvoy, F. A. and Ellis, D. E.   Biochem. Soc. Trans., 1977,
       5, 1719-1721.

RECEIVED December 10, 1979.

Glycosphingolipids and Glyceroglucolipids of Glandular Epithelial Tissue

BRONISLAW L. SLOMIANY and AMALIA SLOMIANY

Department of Medicine, New York Medical College, New York, NY 10029

A major problem encountered in the analysis of glycolipids is the assurance that glycolipids are removed from the tissue in high yield. Utilization of the classical methods for the isolation of lipids have introduced some limitations with regard to the extractibility of highly polar glycosphingolipids and hence have led to many false statements and to misconceptions that the glycosphingolipid compositions are well explored.

Our systematic investigations into the nature of ABH antigens of gastric mucosa have resulted in the isolation and identification of a number of fucolipids, Forssman glycosphingolipid variants and sulfated glycosphingolipids, which have differences in their internal composition, anomeric configuration, length of oligosaccharide chains and degree of complexity. The successful isolation of these glycolipids was the result of a new methodological approach that considered the effect of carbohydrate moiety on the solubility of glycosphingolipids and their tight association with other membrane components. Our extensive studies on glycosphingolipids of gastric mucosa indicate that in order to obtain complete solubilization of this class of compounds, entirely different methodological approaches must be considered.

In spite of the assumption that the mucous glycolipids and glycoproteins are similar to, or possibly derived from those found on cell surfaces, glycosphingolipids have not been found to be constituents of mucus secretions. However, the presence of a new type of glycolipid (glyceroglucolipid) has been demonstrated. This implies that glycosphingolipids are confined to membranous structures of the cell in which they may vary in composition, content and expression, and that this may be essential for certain specialized functions of the cell. A protective role of glyceroglucolipids in the cell may be speculated from their localization in the gastric mucous barrier and their resistance to chemical and biological degradation in the most obnoxious of environments.

0-8412-0556-6/80/47-128-149$7.00/0

© 1980 American Chemical Society

In this article we review developments in methodological
approaches for isolation of glycosphingolipids in high yields; demonstration of glycosphingolipid complexity as well as species,
individual and organ specificity; demonstration of distinctive
features of the epithelial tissue versus its secretion; and description of a new group of glycolipids which are confined to mucous secretions.

The Glycosphingolipids of Gastric Mucosa and Salivary Glands

In early attempts to isolate blood group ABH antigens the
idea of "lipid-hapten" has been criticized since the antigens were
not extractable by ether or alcohol-ether mixtures (1,2), but instead the ABH activity was found in more polar solvents (3). The
presence of complex glycosphingolipids in animal tissues and their
extractibility were not known, hence the solubility properties
were sufficient to support the idea that ABH blood group antigens
are glycoproteins. In early studies, the immunologically active
lipids were obtained by solubility differences in organic solvents
and by precipitation as metal complexes (4,5). More recently, the
isolation and characterization of glycosphingolipids has been
greatly advanced, but the problem with complex components has not
been completely solved. In contrast to the well-known blood group
active glycoproteins of gastric mucosa and gastric secretion,
little was known about these glycosphingolipids except for the
early work of Masamune and Siojima (6).

Extraction with Chloroform/Methanol. Extraction of hog
gastric mucosa with the conventional mixture of chloroform/metha-
nol (2/1, v/v) resulted in the isolation of several glycosphingolipids. The blood-group active glycosphingolipids purified from
this extract contained up to seven carbohydrate residues in the
molecule (Table 1, structures 1,2,3,4). The heptahexosylceramides
1 and 2 exhibited strong blood group A activity and differed from
each other in linkages of the subterminal galactose to N-acetylglucosamine residues (type 1 and type 2 chains) and in fatty acid
composition (7,8). Glycosphingolipid 1 with the type 1 chain had
11.4% hydroxylated acids and 15.7% of C_{22}-C_{24}, whereas glycosphingolipid 2 with type 2 chain had 35.4% hydroxylated fatty
acids and 44.4% of C_{22}-C_{24} fatty acids. The differences in fatty
acids apparently were sufficient to affect chromatographic mobility
of these compounds, and permitted their isolation as two distinct
bands. This enabled us to show for the first time the existence of
two types of chains in A-active glycosphingolipids, since only
type 2 chains were found in A and H glycosphingolipids of the
erythrocytes (9,10). Also, the isolated glycosphingolipids differed
from those of human erythrocytes by having an additional galactose
residue adjacent to the lactosyl portion of the carbohydrate chain.
Further studies of blood group activity of various isolated
glycosphingolipids led to isolation of components 3 and 4

(Table I). The hexahexosylceramide 3 was A-active but lacked N-acetylglucosamine; its activity in A- anti-A system was somewhat diminished as compared to that of the heptahexosylceramides 1 and 2 (11). The absence of N-acetylglucosamine was also detected in tetrahexosylceramide 4, which manifested H activity(12) but again exhibited weaker antigenic potency than that shown for H-active penta-, octa- and decahexosylceramides of the erythrocytes (10).This decrease in antigenic activity apparently results from the proximity of the antigenic determinant and hydrophobic portion of these glycosphingolipids, ot it may be due to the absence of N-acetylglucosamine, which in some subtle way influences the activity of the antigenic determinants.

The fucose-containing glycosphingolipids, so abundant in hog gastric mucosa, were not detected in rat sublingual and submaxillary glands (13), although both tissues are functionally similar. Only traces of fucose were found in crude glycosphingolipid fractions prior to thin-layer chromatography. Curiously enough, only traces of fucose were also found in the glycoproteins of rat sublingual (14) and submaxillary glands (15). The absence of fucose-containing glycosphingolipids supports the findings of Kent and Sanders (16), who have studied the distribution of blood group A antigen throughout the digestive tract of rat and found its highest content in large intestine. Their data, together with our results, suggest gradient distribution of fucose-containing glycosphingolipids and glycoproteins throughout the digestive tract of the rat and possibly of other mammalian species. The neutral glycosphingolipids (Table II) contained glucose,galactose and N-acetylgalactosamine. N-acetylglucosamine was found only in very small amounts in the ganglioside fractions of the glands. The submaxillary and especially sublingual glands exhibited a high content of the sulfated glycosphingolipids. These were composed of mono- and di-hexose sulfatide, with the former being predominant in both types of glands. The high content of sulfated glycosphingolipids is in agreement with histological studies of Pritchard and Rusen (17) and Pritchard (18) on the distribution of radiosulfate in rat salivary glands. The abundance of sulfated glycosphingolipids in salivary glands may indicate that they participate in the secretory processes of these glands.

Buffered Tetrahydrofuran. In 1973, Tettamanti et al. (19) described an improved procedure for the extraction, separation and purification of brain gangliosides. In this method, the brain tissue was subjected to homogenization and extraction with buffered (potassium phosphate buffer, pH 6.8) tetrahydrofuran. Following centrifugation, diethyl ether was added and the mixture separated into organic and aqueous phase. The gangliosides, recovered exclusively in the aqueous phase, were then freed of residual phospholipids and other minor contaminants (i.e. peptides)by column chromatography on silica gel.This procedure, as shown by the authors,was superior to the commonly used chloroform/methanol

Table I.

Structures of the glycosphingolipids characterized from hog and
dog gastric mucosa.

Glycolipid	Structure
1.	GalNAcα1→3Galβ1→3GlcNAcβ1→3Galβ1→4Galβ1→4Glcβ1→1Cer 　　　　2 　　　　↑ 　　　1αFuc
2.	GalNAcα1→3Galβ1→4GlcNAcβ1→3Galβ1→4Galβ1→4Glcβ1→1Cer 　　　　2 　　　　↑ 　　　1αFuc
3.	GalNAcα1→3Galβ1→3Galβ1→4Galβ1→4Glc1→1Cer 　　　　2 　　　　↑ 　　　1αFuc
4.	Fucα1→2Galα1→3Galβ1→4Glc1→1Cer
5.	GalNAcα1→3Galβ1→4GlcNAcβ1→3Galβ1→4Glcβ1→1Cer 　　　　2　　　3 　　　　↑　　　↑ 　　　1αFuc　1αFuc
6.	GalNAcα1→3Galβ1→3/4GlcNAc1→3Gal1→4Glc1→1Cer 　　　　2 　　　　↑ 　　　1αFuc
7.	Gal1→4GlcNAc1→3Gal1→4Glc1→1Cer 　　　3 　　　↑ 　　　1Fuc
8.	GalNAcα1→3Gal1→3Gal1→4Glc1→1Cer 　　　　2 　　　　↑ 　　　1αFuc
9.	GalNAcα1→3GalNAcβ1→3Galα1→4Galβ1→4Glc1→1Cer

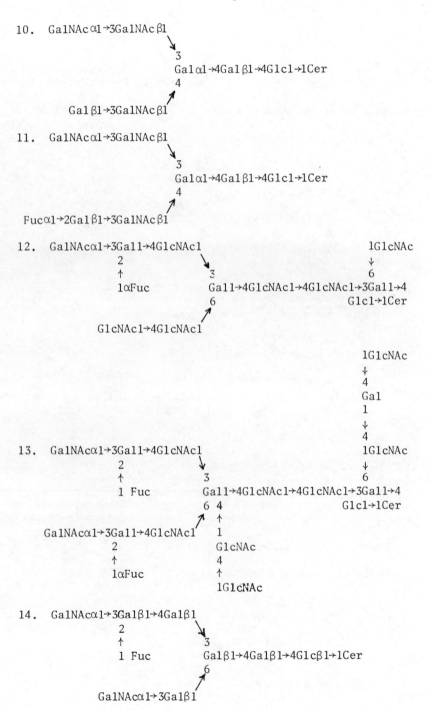

10. GalNAc α1→3GalNAc β1
                          ↘
                           3
            Gal α1→4Gal β1→4Glc1→1Cer
                           4
                          ↗
      Gal β1→3GalNAc β1

11. GalNAcα1→3GalNAc β1
                          ↘
                           3
            Gal α1→4Gal β1→4Glc1→1Cer
                           4
                          ↗
  Fucα1→2Gal β1→3GalNAc β1

12. GalNAcα1→3Gal1→4GlcNAc1   1GlcNAc
                2   ↓
                ↑   ↘   6
              1αFuc   3
                           Gal1→4GlcNAc1→4GlcNAc1→3Gal1→4
                                6   Glc1→1Cer
                               ↗
      GlcNAc1→4GlcNAc1

                                                   1GlcNAc
                                                      ↓
                                                      4
                                                     Gal
                                                      1
                                                      ↓
                                                      4
13. GalNAcα1→3Gal1→4GlcNAc1   1GlcNAc
                2   ↘   ↓
                ↑   3   6
              1 Fuc   Gal1→4GlcNAc1→4GlcNAc1→3Gal1→4
                                6 4   Glc1→1Cer
                               ↗ ↑
  GalNAcα1→3Gal1→4GlcNAc1   1
                2   GlcNAc
                ↑   4
              1αFuc   ↑
                             1GlcNAc

14. GalNAcα1→3Galβ1→4Galβ1
                2   ↘
                ↑   3
              1 Fuc   Galβ1→4Galβ1→4Glcβ1→1Cer
                                6
                               ↗
      GalNAcα1→3Galβ1

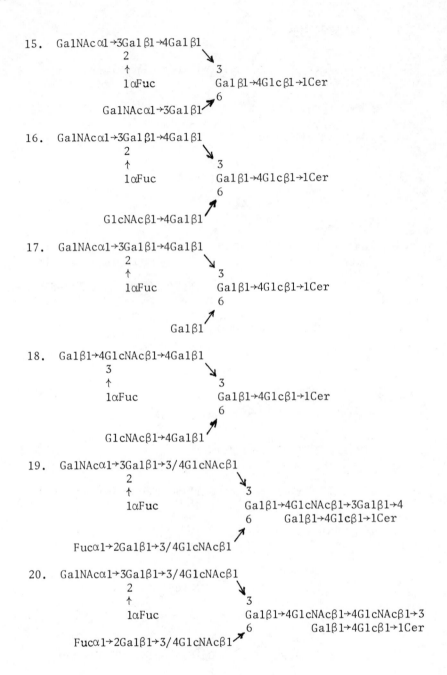

15. GalNAcα1→3Galβ1→4Galβ1
         2   ↘
         ↑   3
       1αFuc   Galβ1→4Glcβ1→1Cer
                                      6
       GalNAcα1→3Galβ1↗

16. GalNAcα1→3Galβ1→4Galβ1
         2   ↘
         ↑   3
       1αFuc   Galβ1→4Glcβ1→1Cer
                                      6
                                      ↗
       GlcNAcβ1→4Galβ1↗

17. GalNAcα1→3Galβ1→4Galβ1
         2   ↘
         ↑   3
       1αFuc   Galβ1→4Glcβ1→1Cer
                                      6
                                     ↗
               Galβ1↗

18. Galβ1→4GlcNAcβ1→4Galβ1
         3   ↘
         ↑   3
       1αFuc   Galβ1→4Glcβ1→1Cer
                                      6
                                     ↗
       GlcNAcβ1→4Galβ1↗

19. GalNAcα1→3Galβ1→3/4GlcNAcβ1
         2   ↘
         ↑   3
       1αFuc   Galβ1→4GlcNAcβ1→3Galβ1→4
                                      6   Galβ1→4Glcβ1→1Cer
                                      ↗
       Fucα1→2Galβ1→3/4GlcNAcβ1↗

20. GalNAcα1→3Galβ1→3/4GlcNAcβ1
         2   ↘
         ↑   3
       1αFuc   Galβ1→4GlcNAcβ1→4GlcNAcβ1→3
                                      6   Galβ1→4Glcβ1→1Cer
       Fucα1→2Galβ1→3/4GlcNAcβ1↗

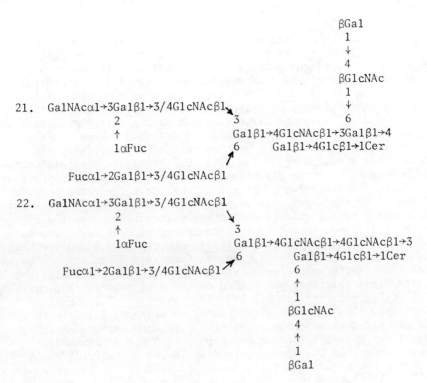

Table II

The Composition and Molar Ratios of Carbohydrates of Rat Sublingual and Submaxillary Glycosphingolipids.

Glycosphingolipid	Glc		Gal		GlcNAc		GalNAc	
	RSL	RSM	RSL	RSM	RSL	RSM	RSL	RSM
Glucosylceramide	1.0	1.0						
Galactosylceramide			1.0	1.0				
Dihexosylceramide	1.0	1.0	0.97	0.99				
Trihexosylceramide	1.0	1.0	1.95	1.91				
Tetrahexosylceramide	1.0	1.0	1.92	1.90			1.01	0.94
Pentahexosylceramide	1.0	1.0	3.02	2.95			0.98	0.97
Monosialoganglioside	1.0	1.0	1.02	0.97	0.28	0.06	0.31	0.05
Disialoganglioside	1.0	1.0	0.98	0.95	tr.	0.10	tr.	0.09
Monohexose sulfatide			1.0	1.0				
Dihexose sulfatide	1.0	1.0	0.96	0.98				

RSL, rat sublingual: RSM. rat submaxillary; tr., traces
(From Ref. 13)

/water partition systems (20,21).
Application of buffered tetrahydrofuran extraction to gastric mucosa, followed by careful examination of the aqueous phase for various glycosphingolipids, indicated that this phase contained sialoglycosphingolipids and considerable quantities of neutral glycosphingolipids (22). These were separated from the acidic glycosphingolipids by DEAE-Sephadex column chromatography (23). The neutral glycosphingolipid fraction of hog gastric mucosa was shown to consist mostly of fucolipids (24,25), whereas that of dog gastric mucosa exhibited a high content of N-acetylgalactosamine-containing glycosphingolipids (26). Of the eight fucolipids purified from the neutral glycolipid fraction of the aqueous phase of buffered tetrahydrofuran extract of hog gastric mucosa, four were found to be identical with those (Table I, compounds 1-4) isolated previously by the chloroform/methanol extraction procedure (7,8,11,12) and the elucidated structures of four new compounds (5-8) are listed in Table I (24,27). Fucolipids 5,6 and 8 exhibited blood group A-activity, whereas fucolipid 7 was not active in the A anti-A, B anti-B or H anti-H systems. In fucolipid 6, the subterminal galactose was linked to the next sugar in the chain, N-acetylglucosamine by both 1→3 (40%) and 1→4 (60%) linkages. Fucolipid 8 was structurally related to compound 3 and, although it exhibited blood group A-activity, its carbohydrate chain was devoid of N-acetylglucosamine. Fucolipid 7 had a carbohydrate chain identical in structure to that in the glycolipid from normal and human adenocarcinoma tissue (28). Thus, it became apparent that this glycosphingolipid is not only present in human glandular tissue but also in glandular tissue of other species and may not necessarily be a specific antigen of malignant cells.
The carbohydrate chain of fucolipid 5 contained seven sugar residue and differed from the known A-active glycosphingolipids by the presence of a second fucose residue, linked to C-3 of the internal N-acetylglucosamine. Identification of this glycosphingolipid provided the first evidence for the existence of difucosyl blood group A-active glycosphingolipids with carbohydrate structures identical to difucosyl oligosaccharides of glycoprotein origin (29), suggesting that the same carbohydrate chains may be linked to a lipid or protein core. A similar glycolipid was isolated later from dog intestine (30).
Examination of the neutral glycolipids in the aqueous fraction of a buffered tetrahydrofuran extract of dog gastric mucosa indicated presence of glycosphingolipids containing significant amounts of N-acetylgalactosamine, but only traces of fucose (26). A similar conclusion as to the content of fucolipids in dog gastric mucosa was reached earlier by McKibbin and Lyerly (31).
Extensive purification of the glycosphingolipids present in the neutral fraction resulted in the isolation of three distinct glycolipids exhibiting Forssman antigenic activity (Table I, compounds 9,10,11). Thin-layer chromatographs of these glycolipids is illustrated in Fig. 1. Chemical analyses of the purified

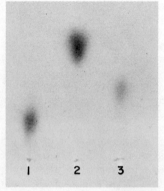

FEBS Letters

Figure 1. Thin-layer chromatography of the glycolipids with Forssman activity

(1) Glycolipid 11, Table I; (2) glycolipid 9, Table II; (3) glycolipid 10, Table I. Solvent system: chloroform/methanol/water (60/35/8, v/v/v). Visualization: orcinol reagent. (26)

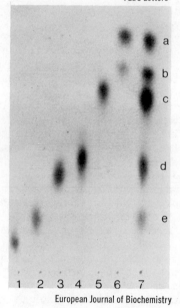

European Journal of Biochemistry

Figure 2. Thin-layer chromatogram of purified glycolipid 11 (Table I) and its enzymatic hydrolysis products

(1) Native glycolipid; (2) glycolipid obtained by treatment of the native compound with α-fucosidase; (3) glycolipid obtained by the sequential treatment of the native compound with α-fucosidase, β galactosidase and β-N-acetylhexosaminidase; (4) glycolipid obtained by treatment of the defucosylated compound (from line 2) with α-N-acetylgalactosaminidase and β-N-acetylhexosaminidase; (5) glycolipid obtained by treatment of the defucosylated compound (from line 2) with β-galactosidase, α-N-acetylgalactosaminidase and β-N-acetylhexosaminidase; (6) glycolipid from line 5 after incubation with α- and β-galactosidase. Standards: (7a) glucosylceramide; (7b) lactosylceramide; (7c) triglycosylceramide; (7d) glycolipid 9, Table I; (7e) glycolipid 10, Table I. Solvent system: chloroform/methanol/water (65/35/8, v/v/v). Visualization: orcinol reagent. (33)

compounds indicated that the carbohydrate moiety of glycolipid 9 consisted of glucose, galactose and N-acetylgalactosamine in a molar ratio of 1:2:2. The same carbohydrates, but in a molar ratio of 1:3:3, were present in glycolipid 10, whereas glycolipid 11 contained glucose, fucose, galactose and N-acetylgalactosamine in a molar ratio of 1:1:3:3 (26). Further structural studies (32,33) revealed that the carbohydrate moiety of glycolipid 9 is chemically identical with that of Forssman hapten, characterized previously from kidney and intestine of dog (34,35,36) and from spleen and kidney of horse (37,38). The results of chemical and enzymatic analyses of glycolipid 10 suggested that this compound contains two terminal sugar residues, galactose and N-acetylgalactosamine, and thus has a branched structure. Susceptibility of glycolipid 10 to glycosidase degradation in the sequence: β-galactosidase, β-N-acetylhexosaminidase, and the sequence: α-N-acetylgalactosaminidase, β-N-acetylhexosaminidase indicated that the side chains are composed of βGal→βGalNAc and αGalNAc→β GalNAc disaccharides. Parallel studies on permethylated fragments of such enzymic degradation products established that the above disaccharide chains are linked by 1→4 and 1→3 bonds, respectively, to the galactose residue adjacent to lactosylceramide of the glycolipid core. The presence of two side chains, αGalNAc(1→3)βGalNAc and αFuc(1→2)βGal(1→3)βGalNAc, in glycolipid 11 was clearly demonstrated with the aid of glycosylhydrolases (Fig. 2) and permethylation analysis. Furthermore, the native glycolipid 11 exhibited both Forssman and H antigenic activities. Defucosylation of this glycolipid (0.1 M trichloroacetic acid, 100°C for 2 h) resulted in the loss of its H-activity but had no effect on its reactivity with Forssman anti-serum or on its ability to inhibit hemagglutination in the A/anti-A system. The latter activity, shared by all three compounds (9,10,11), is thought to be due to the presence of a terminal α-N-acetylgalactosamine residue in the Forssman antigen and blood group A determinant (39).

It has been suggested earlier (39) that Forssman antigen may not be a single compound. This view was supported by Gahmberg and Hakomori (40) who isolated two polymorphic variants of Forssman glycolipid from hamster fibroblasts. Both variants, however, shared the common terminal structure composed of three sugar residues, GalNAc(α1→3)GalNAc(β1→3)Gal. Our data indicate that this terminal structure is not only common for the Forssman glycolipid variants containing straight carbohydrate chains, but also can be located on the termini of glycolipids with branched structures which carry more than one antigenic determinant. Isolation of the Forssman hapten variants from the aqueous phase of buffered tetrahydrofuran lipid extracts indicates that glycolipids bearing Forssman antigen may exhibit considerable water solubility. This behavior may be directly related to the relatively strong antigenic properties of Forssman hapten under the physiological conditions. In accord with these results the term "Forssman antigen" should refer to glycosphingolipids bearing a terminal

GalNAc(α1→3)GalNAc structure and should not be used with reference
to one particular chemical entity, i.e. globopentaglycosylceramide.
Sulfated glycosphingolipids (41,42) from the buffered tetrahydrofuran lipid extract were also investigated in our laboratory.
In lipid extracts of hog gastric mucosa these glycolipids were
found mainly in the organic phase. After rigorous purification,
three sulfated glycosphingolipids were obtained in a homogeneous
form. These were identified as galactosylceramide sulfate, lactosylceramide sulfate and triglycosylceramide sulfate. The structures of these compounds are presented in Table III. The presence
of galactosyl and lactosylceramide sulfates in gastric mucosa and
the small intestine has been reported earlier (31), whereas the
isolated triglycosylceramide sulfate (compound 3, Table III), not
reported heretofore, provided the first indication that sulfated
carbohydrates also occur in more complex glycosphingolipids. Successful isolation of this sulfated glycolipid represents another
example of the superiority of buffered tetrahydrofuran extraction
over the conventional chloroform/methanol procedure.

  Butanol Extraction. Development of a butanol extraction procedure for the isolation of complex glycosphingolipids
from erythrocyte membrane (43) prompted us to apply this method,
with modification, to hog gastric mucosa. The aqueous phase, after
n-butanol extraction, was subjected to alkaline treatment to degrade the glycoproteins susceptible to the β-elimination reaction.
The products of alkaline degradation were dialyzed and the proteins separated from the glycolipids by chromatography on Cellex-
P column (44). The glycolipid fraction was then acetylated, chromatographed on a Florisil column and purified to homogeneity
(in the acetylated form) by thin-layer chromatography. Although
several glycolipid bands were detected, only two individual compounds were successfully purified to homogeneity (Table I, compounds 12,13). Both glycolipids exhibited blood group A-activity
and their carbohydrate portions were highly enriched in N-acetylglucosamine. Results of chemical analyses (44) indicated that
glycolipid 12 contained twelve sugar units, and glycolipid 13 contained eighteen sugar units. In glycolipid 12 one residue of fucose, one residue of N-acetylgalactosamine and two out of six residues of N-acetylglucosamine were located at non-reducing termini.
Glycolipid 13 contained two terminal residues of fucose, two residues of N-acetylgalactosamine and two terminal residues of N-
acetylglucosamine. Analysis of the glycolipid fragments recovered
after three complete steps of Smith degradation of glycosphingolipids 12 and 13 showed, in both glycolipids, the presence of glucose, galactose and N-acetylglucosamine in the molar ratios of
1:2:2. Partial acid hydrolysis of these fragments resulted mainly
in the formation of Gal→Glc→ceramide, GlcNAc→Gal→Glc→ceramide and
GlcNAc→GlcNAc→Gal→Glc→ceramide (44). This suggested that the sequential arrangement of the sugar units in the saccharide chains
adjacent to the ceramide core in both glycolipids was

160

Table III

Sulfated Glycosphingolipids of Gastric Mucosa

Glycolipid Structure
1. $SO_3H \rightarrow 3Gal \rightarrow$ceramide
2. $SO_3H \rightarrow 3Gal1 \rightarrow 4Glc \rightarrow$ceramide
3. $SO_3H \rightarrow 3Gal1 \rightarrow 4Gal1 \rightarrow 4Glc \rightarrow$ceramide
4. $SO_3H \rightarrow 6GlcNAc\beta1 \rightarrow 3Gal\beta1 \rightarrow 4Glc \rightarrow$ceramide
5. $Gal\beta1 \rightarrow 4GlcNAc(6 \leftarrow SO_3H)\beta1 \rightarrow 3Gal\beta1 \rightarrow 4Glc \rightarrow$ceramide

Gal→GlcNAc→GlcNAc→Gal→Glc→ceramide. In each glycolipid, one residue of galactose present in the backbone pentasaccharide was involved in branching. Among other features, noted for the first
time in glycosphingolipids was the occurrence of di-(N-acetyl)chi-
tobiose. This sequence, originally reported in the carbohydrate
chains of porcine (45,46) and horse (47) blood group active glycoproteins, was also recently found in the complex glycosphingolipids of erythrocyte membrane (48). The presence of glycolipids
with carbohydrate structures identical to those found in oligosaccharides of glycoprotein origin lent further support for the
existence of a common pathway for the biosynthesis of blood group-
active glycoproteins and glycosphingolipids.

    Sodium Acetate Extraction. In our further studies of
fucolipids of hog gastric mucosa, we have found that the residue
left after very thorough extraction of lipids (chloroform/methanol,
2/1, twice for 24 h at room temperature) still contained considerable quantities of more complex glycosphingolipids, which
were extractable with a mixture of methanol/chloroform/water containing sodium acetate. Accordingly, we have developed a procedure
which involves initial pre-extraction of mucosa scrapings with
chloroform/methanol (2/1, v/v) to remove simple glycolipids, followed by extraction of the residue with sodium acetate in methanol/
chloroform/water (60/30/8, v/v/v). The highest yield of glycolipids was obtained with 0.4 M sodium acetate in the above metha-
nol/chloroform/water system (49). Glycosphingolipids recovered in
such extracts included neutral glycolipids containing fucose as
well as acidic glycolipids containing both sialic acid and sulfate.
Separation of these glycolipids into neutral and acidic components
was accomplished by DEAE-Sephadex column chromatography (23). The
neutral glycolipid fraction was then peracetylated and chromatographed on a Florisil column. The fucolipids were contained mainly
in the 1,2-dichloroethane/acetone (1/1, v/v) eluate. This fraction,
after extensive purification on thin-layer plates, yielded five
individual fucolipids (49,50), four of which exhibited blood group
A-activity (Table I, compounds 14-17) and one (compound 18) inactive in the ABH system. Common features of all five fucolipids
were a carbohydrate chain with two branches and high enrichment
of galactose. In fucolipids 14-17, one of the branches was terminated by the blood group A-determinant, while the others terminated either with α-N-acetylgalactosamine (compounds 14 and 15),
β-N-acetylglucosamine (compound 16) or β-galactose (compound 17).
Fucolipid 18, which lacked ABH blood group determinants, also contained two branches, one terminating with β-galactose and the
other with β-N-acetylglucosamine.
    The fact that only one type of complex glycosphingolipid (en-
riched in galactose) was obtained may have reflected the procedure
of purification, especially the choice of a Florisil column and
the solvents used for elution of acetylated compounds. This possibility became obvious when the neutral glycolipid fraction of the

sodium acetate extracts of hog gastric mucosa was subjected (in
the acetylated form) to chromatography on a silicic acid column
(51,52). The 1,2-dichloroethane/acetone (1/1, v/v) eluate from
this column contained two additional fucolipids (each 12 sugar
residues long), whereas the acetone fraction contained fucolipids
with 14 sugar units. The subsequent fraction, eluted with acetone/
methanol (1/1, v/v), contained fucolipids with 18-24 sugar units;
and the last fraction, eluted with methanol/chloroform/water
(90/10/2), consisted of fucolipids with 28-36 sugar residues (53).
The isolated fucolipids in their native form, did not migrate on
thin-layer plates in the solvent systems used for purification of
the previously described blood group ABH fucolipids (22,54). However, in the acetylated form all of the compounds studied exhibited good mobilities in several solvent systems (Fig. 3 and 4). The
proposed structures for glycolipids purified from 1,2-dichloro-
ethane/acetone (compounds 19,20) and acetone (compounds 21,22)
fractions are given in Table I. The most interesting features of
these four fucolipids were the presence of two antigenic determinants (A and H) on the same glycolipid molecule and the similarity of the oligosaccharide chains to those present in the
blood group (A+H) active glycoproteins (45,46).

The carbohydrate and sphingosine composition of the major
fucolipids purified from the acetone/methanol and methanol/chloro-
form/water fractions are given in Table IV. Fucolipids 23-25 were
present in the acetone/methanol fraction, whereas the methanol/
chloroform/water eluate contained fucolipids 26-28 (53). In hema-
gglutination-inhibition assays all six compounds were potent inhibitors of agglutination of human group A-cells by anti-A serum
(1.5-3.1 µg/0.1 ml) and human O-cells by anti-H lectin (2.1-4.5
µg/0.1 ml), indicating that the carbohydrate chain of each fucolipid bears two types of blood group determinant, A and H. Although
the structures of these fucolipids are not yet elucidated, certain
features of the saccharide chains can be suggested on the basis of
carbohydrate analysis, immunological assays and the susceptibility
of the native and defucosylated glycosphingolipids to the action of
specific exoglycosidases. These data indicate that the carbohydrate
chain of fucolipid 23 contains four branches, two terminated by
β-galactose, one by the blood group A (GalNAcα1→3[Fucα1→2]Gal-)
antigenic determinant and one by the blood group H (Fucα1→2Gal-)
determinant; fucolipid 24 contains two branches terminated by the
blood group A determinant, one by the H and one by β-galactose;
fucolipid 25 contains two branches terminated by the A determinant,
one by H and two by β-galactose; fucolipid 26 contains three
branches terminated by the A determinant, one by H and two by
β-galactose; fucolipid 27 contains three branches terminated by
the A determinant, one by H and three by β-galactose; and fucolipid 28 contains three branches terminated by the A determinant,
two by H, two by β-galactose and one by β-N-acetylglucosamine.

Lipids extracted from hog gastric mucosa with 0.4 M sodium
acetate in methanol/chloroform/water were also investigated for

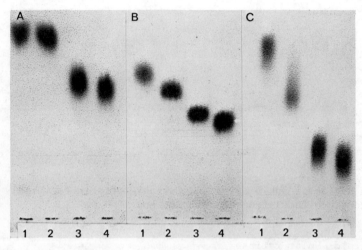

European Journal of Biochemistry

Figure 3.   Thin-layer chromatography of the acetylated blood group (A+H) complex fucolipids

(1) Fucolipid 19, Table I; (2) fucolipid 20, Table I; (3), fucolipid 21, Table I; (4) fucolipid 22, Table I. Solvent system: chloroform/acetone/methanol/water (52/40/00/4, by volume), plate A; 1,2-dichloroethane/methanol/water (80/25/2, v/v/v), plate B; 1,2-dichloroethane/acetone/methanol/water (50/40/10/4, by volume), plate C. Visualization: orcinol reagent. (52)

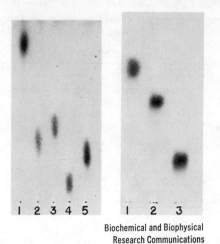

Figure 4.   Thin-layer chromatography of the acetylated highly complex fucolipids from hog gastric mucosa

Left plate, developed in chloroform/methanol/2M NH₄OH (40/15/1.5, v/v/v). (1) Fucolipid 19, Table I; (2) fucolipid 24, Table IV; (3) fucolipid 23, Table IV; (4) fucolipid 26, Table IV; (5) fucolipid 25, Table IV. Right plate, developed in chloroform/methanol/water (60/40/10, v/v/v). (1) Fucolipid 26, Table IV; (2) fucolipid 27, Table IV; (3) fucolipid 28, Table IV. Visualization: orcinol reagent (53)

Biochemical and Biophysical
Research Communications

Table IV.

Molar Ratios of Sphingosine and Carbohydrates in the Highly
Complex Fucolipids from Gastric Mucosa.

Fucolipid	Molar ratios[a]						No. of sugar residues
	Fuc	Gal	Glc	GlcNAc	GalNAc	Sphingosine	
23	2.01	7.86	1.0	5.92	1.05	1.0	18
24	2.84	7.50	1.0	6.77	1.78	0.9	20-21
25	2.90	9.81	1.0	7.85	2.02	0.8	24
26	3.81	9.77	1.0	9.68	3.12	0.8	28
27	3.85	11.78	1.0	11.89	2.79	0.9	32
28	4.63	12.40	1.0	13.60	3.04	0.7	35-36

[a]Relative to Glc=1

(From Ref. 53)

the presence of sulfated glycosphingolipids. For this, the acidic glycolipids, eluted from DEAE-Sephadex with sodium acetate in methanol/chloroform/water, were acetylated and separated on a silicic acid column into several fractions (55,56). The 1,2-dichloroethane and 1,2-dichloroethane/acetone eluates contained mainly sialoglycosphingolipids, together with traces of the di- and trihexose sulfatides described previously (41). Fractions eluted with more polar solvents contained several new sulfated glycosphingolipids. Some of these glycolipids contained sulfate and sialic acid. Whether these compounds represent homogeneous glycosphingolipids containing both sialic acid and sulfate on the same molecule or are a mixture of sulfated and sialylated glycosphingolipids remains to be established. However, two of the characterized sulfated glycosphingolipids (55,56) were devoid of sialic acid and contained glucose, galactose, N-acetylglucosamine and sulfate in molar ratios of 1:1:1:1 and 1:2:1:1, respectively. The proposed structures of these glycolipids are shown in Table III (compounds 4 and 5).

These newly identified compounds differ from previously characterized sulfated glycosphingolipids (41,42) with respect to sugar composition, length of the carbohydrate chain and the site of sulfation. Results of periodate oxidation and permethylation analyses showed that both compounds contain N-acetylglucosamine 6-sulfate. To our knowledge, sulfated glycosphingolipids containing sulfated N-acetylglucosamine have not been previously described in mammalian gastric mucosa or other tissues. However, N-acetylglucosamine 6-sulfate was found in blood group (A+H) sulfated glycoproteins of hog gastric mucosa (45,46). This again indicates that in glandular epithelial tissue the same oligosaccharides may be linked to a lipid or protein core.

New Approach to Isolation of Glycosphingolipids. Progressive discoveries of more complex glycosphingolipids, revealingly similar in structure to glycoproteins, indicate that current techniques for the isolation of glycosphingolipids are inadequate and do not permit complete recovery of all constituents by any one procedure. Size and complexity of the carbohydrate portion governs extractibility and lends to some of these glycosphingolipids the properties of glycoproteins. Therefore, they are either left behind during the extraction or are classified as glycoproteins. To overcome the problem of glycoprotein-like properties of complex glycosphingolipids and at the same time to isolate short-chain glycosphingolipids which may be in strong association with other components of the cell membrane, we have recently introduced a new approach for the isolation of glycosphingolipids (unpublished). In this procedure, gastric mucosa is homogenized in solubilizing buffer (sodium sulfite) and treated sequentially with RNA-ase and DNA-ase to decrease the viscosity of the homogenate, and then subjected to alkaline degradation (β-elimination) and pronase digestion. The resultant tissue digest is extracted with chloroform/

methanol (2/1, v/v) to remove short-chain glycosphingolipids and the aqueous phase is adjusted to 1% with a zwitterionic detergent. After centrifugation, the clear supernatant fraction is subjected to gel filtration (Bio-Gel P-60) and chromatography on DEAE-Sephadex. Following molecular sizing the Bio-Gel P-4 and/or P-6 columns, the glycolipids are acetylated and purified to individual components by chromatography on thin-layer plates or on Bio-Beads SX-1 columns. Since the entire process of isolation is conducted in a solute phase and in the presence of a detergent, the artifactual entrapment of glycosphingolipids is eliminated.

Glycolipids of Mucous Secretion

The oral, gastrointestinal, bronchial, pulmonary and reproductive tracts of higher animals secrete copious quantities of viscous mucus which functions mainly as a lubricant and protective agent. The viscous properties of the mucous secretions are the result of the presence of high molecular weight glycoproteins called mucins (57). These glycoproteins have been studied extensively (see for review ref. 57,58); however, until recently no information was available on glycolipids of mucous secretions. Furthermore, the general assumption was that both mucous glycoproteins and glycolipids are similar to, or possibly derived from, those found on the cell surfaces (59). To provide data on the nature of glycolipids of mucous secretions, we have analyzed glycolipid constituents of the lipid extracts derived from gastric secretion, gastric mucous barrier, saliva and alveolar lavage.

Analyses of the lipid extracts from human gastric secretion revealed that glycolipids constitute about 30% of the lipid fraction (60), whereas in gastric secretions from dog Heidenhain pouch and from ligated rat stomach, the glycolipid fraction comprises up to 50% of the total lipids (61). On thin-layer chromatography, the glycolipid fraction from human secretion could be separated into nine individual components, five glycolipid components were present in the gastric secretion of dog, and four in the gastric secretion of rat (61,62). Each of the purified glycolipids contained fatty acids, glucose and glyceryl- monoethers. In addition, two glycolipids from human gastric secretion were sulfated. None of these glycolipids contained sphingosine, phosphorus or alkenyl ethers (61,63). All of the glycolipids were susceptible to deacylation under mild alkaline conditions, indicating the presence of ester-linked fatty acids, and the sulfated compounds were also susceptible to acid solvolysis (Fig. 5). Results of structural analyses performed on the major glycolipid components of human gastric secretion indicated that the glycolipids of gastric secretion are composed of one or more glucose residues linked to a monoalkylmonoacylglycerol lipid core (64,65). The proposed structures for glycolipids of human gastric secretion are presented in Table V.

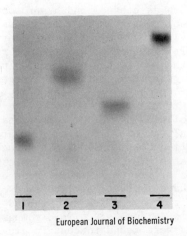

Figure 5. Thin-layer chromatogram of the major sulfated glycolipid from human gastric secretion

(1) Native glycolipid, compound 4, Table V; (2) desulfated glycolipid; (3) desulfated and deacylated glycolipid; (4) digalactosyl diglyceride standard. Solvent system: chloroform/ methanol/water (65/25/4, v/v/v). Visualization: orcinol reagent. (64)

European Journal of Biochemistry

Table V.

Glyceroglucolipids of Human Gastric Secretion.

Glycolipid	Structure
1.	Glcα1→3-1,(3)-0-alkyl-2-0-acylglycerol
2.	Glcα1→6Glcα1→6Glcα1→6Glcα1→6Glcα1→6Glcα1→3-1,(3)-0-alkyl-2-0-acylglycerol
3.	Glcα1→6Glcα1→6Glcα1→6Glcα1→6Glcα1→6Glcα1→6Glcα1→3-1,(3)-0-alkyl-2-0-acylglycerol
4.	SO₃H-6Glcα1→6Glcα1→6Glcα1→3-1,(3)-0-alkyl-2-0-acylglycerol
5.	SO₃H-6Glcα1→6Glcα1→6Glcα1→6Glcα1→3-1,(3)-0-alkyl-2-0-acylglycerol

Our studies on the origin of glyceroglucolipids in gastric
secretion established that these compounds are present not only in
the soluble portion of gastric secretion (dissolved mucin), but
also in the gastric mucous barrier and in the preformed intracellular mucus contained within the secretory granules of the epithelial cells (66). Furthermore, we have demonstrated that insti-
llation of various noxious agents such as ethanol and hyperosmotic
NaCl causes depletion of glyceroglucolipids from gastric mucous
barrier (67). Similar depletion of glyceroglucolipids was observed
in various gastrointestinal disorders (gastritis, gastric ulcers)
(68). These data clearly establish the importance of glyceroglucolipids as an essential component of gastric secretion and suggest
the possibility of their involvement in the defense mechanism
against the injury of the mucosal surfaces.

In further studies on the glycolipids of mucous secretions,
we have directed our attention to saliva (69,70). Since glycoproteins of salivary and gastric secretion bear considerable structural and immunological similarities (71,72), it was of interest
to determine whether the glycolipids of saliva resemble those of
gastric secretion. Accordingly, we have isolated a glycolipid
fraction from lipid extracts of whole human saliva and studied the
composition and structure of seven individual glycolipid components (Fig. 6). All seven compounds were found to contain glucose, fatty acids and glyceryl-monoethers. One of the glycolipids
also contained sulfate (70). Results of chemical analyses (Table
VI), indicated that these glycolipids are structurally related to
those of gastric secretion, i.e. they contain polyglucosyl carbohydrate chains linked to monoalkylmonoacylglycerol. Again, glycosphingolipids were not detected. Our data are consistent with the
results of earlier studies on the biosynthesis of carbohydrate-
containing substances in the salivary glands of mice (73), in
which stimulation with isoproterenol increased the synthesis of
glycolipid of glyceroglycolipid type. Also, Pritchard's studies
(74) on sulfolipid formation in rat submandibular glands have
demonstrated the presence of a sulfotransferase catalyzing the
transfer of labelled sulfate from 3'-phosphoadenosine-5'-phospho-
sulfate to an endogenous lipid acceptor. This radio-labelled sulfolipid produced by submandibular gland was shown to be of the
glyceroglycolipid type. Our recent studies (75,76) on the origin of
glyceroglucolipids in the saliva indicate that these compounds are
elaborated by the parotid and submandibular glands and that their
levels are elevated in the salivary secretions derived from individuals with a high rate of salivary calculus formation. Whether
there is a direct association between the glyceroglucolipid content of the saliva and the development of plaque, calculus and
periodontal disease remains to be established.

For the analysis of extracellular glycolipids of respiratory
tract, we have chosen the acellular material lining the alveoli
of mammalian lungs. This unique lipid-protein mixture, responsible
for the reduction of alveolar surface forces during respiration,

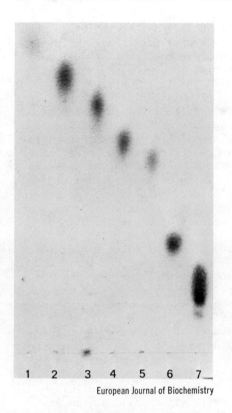

1   2   3   4   5   6   7

European Journal of Biochemistry

Figure 6.   Thin-layer chromatogram of the glycolipids purified from human saliva (see Table VI for structures)

(1) Glycolipid 1; (2) glycolipid 2; (3) glycolipid 3; (4) glycolipid 4; (5) desulfated glycolipid 5; (6) glycolipid 6; glycolipid 6; (7) glycolipid 7. Solvent system: chloroform/methanol/water (65/35/8, v/v/v). Visualization: orcinol reagent. (70)

Table VI.

Glyceroglucolipids of Human Saliva.

Glycolipid	Structure
1.	Glcα1→3-1,(3)-0-alkyl-2-0-acylglycerol
2,3.	Glcα1→6Glcα1→3-1,(3)-0-alkyl-2-0-acylglycerol
4.	Glcα1→6Glcα1→6Glcα1→3-1,(3)-0-alkyl-2-0-acylglycerol
5.	SO_3H-6Glcα1→6Glcα1→6Glcα1→3-1,(3)-0-alkyl-2-0-acylglycerol
6.	Glcα1→6Glcα1→6Glcα1→6Glcα1→6Glcα1→6Glcα1→3-1,(3)-0-alkyl-2-0-acylglycerol
7.	Glcα1→6Glcα1→6Glcα1→6Glcα1→6Glcα1→6Glcα1→6Glcα1→6Glcα1→3-1, (3)-0-alkyl-2-0-acylglycerol

includes the surface-active phospholipids and other moieties such as neutral lipids, proteins and carbohydrates (77,78,79). Investigations on the nature of the carbohydrate component of pulmonary surfactant indicated that this material is not only associated with a protein but also is present in the lipid extract (80). Analyses of the lipid extracts from alveolar lavage of rabbit, per formed in our laboratory (81,82), showed that the carbohydrate component associated with lipids consists exclusively of glucose. About 60% of this carbohydrate was associated with neutral glyco-lipids and 40% with acidic glycolipids. Extensive purification of the glycolipids present in these fractions resulted in the isola-tion of four individual components. Three of these glycolipids contained glucose, fatty acids and glycerl-monoethers, whereas the major acidic glycolipid, in addition to the above components, con-tained sulfate ester (82). The structures of these glycolipids are shown in Table VII.

Our data (81,82) on glycolipids of the alveolar lining layer of rabbit lungs clearly show that these compounds, as those of gastric secretion and saliva, belong to the glyceroglucolipid class. Thus, it appears that an acellular glycolipids in the se-cretions of the alimentary tract and in the alveolar lining layer of mammalian lungs are entirely different from those found in cell membranes. The physiological importance of secretory glycolipids is still unknown. Glyceroglucolipids present in mucous secretions of the alimentary tract are part of the protective lining of the surface epithelial cells and in saliva they may be involved in the process of tooth pellicle formation, whereas in the acellular material lining the surfaces of alveoli glyceroglucolipids may participate in spreading of the surfactant layer within the al-veolus.

The Nature of ABH Blood Group Antigens in Mucous Secretion

The occurrence and nature of blood specific antigens in tis-sue and in mucous secretions has been studied by a number of in-vestigators (83,84,85,86,87); the early data suggested that mucous secretions contain water-soluble antigens whereas red cells and most of the other tissues contain only the alcohol-soluble anti-gens. In spite of that, the discovery of blood group-active glyco-sphingolipids and glycoproteins from the same source (see for re-view ref. 22,54,58,88,89,90) led to the proposal of their co-existence in the tissues and to the assumption (59) that secre-tions represent also a mixture of blood group-active glycosphingo-lipids and glycoproteins. Furthermore, evidence was presented on the glycoprotein nature of ABH antigens of erythrocytes (91,92, 93), which were known to contain antigens of the glycosphingo-lipid character only.

Our studies on glycolipids of gastric secretion (62,63,64,65) and saliva (69,70) showed that these secretions do not contain glycosphingolipids; instead glyceroglucolipids were found. To

Table VII.

Glyceroglucolipids of Alveolar Lavage from Rabbit.

Glycolipid	Structure
1.	Glcα1→3-1,(3)-0-alkyl-2-0-acylglycerol
2.	Glcα1→6Glcα1→6Glcα1→6Glcα1→6Glcα1→3-1,(3)-0-alkyl-2-0-acyl-glycerol
3.	Glcα1→6Glcα1→6Glcα1→6Glcα1→6Glcα1→6Glcα1→3-1,(3)-0-alkyl-2-0-acylglycerol
4.	SO$_3$H-6Glcα1→6Glcα1→6Glcα1→6Glcα1→3-1,(3)-0-alkyl-2-0-acyl-glycerol

Table VIII.

ABH blood group activity in human saliva and gastric secretion.

Assay	Material	Activity*
1.	Native gastric secretion	+
2.	Native saliva	+
3.	Delipidated gastric secretion	+
4.	Delipidated saliva	+
5.	Native and delipidated gastric secretion treated with 0.5 M NaOH, (60 h, room temperature)	-
6.	Native and delipidated saliva treated with 0.5 M NaOH (60 h, room temperature)	-
7.	Alkaline degradation of gastric secretion in the presence of A-active glycosphingolipid	+
8.	Alkaline degradation of saliva in the presence of A-active glycosphingolipid	+
9.	Lipid extract of gastric secretion	-
10.	Lipid extract of saliva	-
11.	Glycolipid fraction of saliva	-
12.	Glycolipid fraction of gastric secretion	-
13.	Blood group A-active glycosphingolipid in the presence of lipid extract from saliva or gastric secretion	+

* (+) signifies inhibition of hemagglutination
  (-) indicates hemagglutination

determine the nature of blood group ABH antigens in saliva and
gastric secretion, the native and delipidated samples, total lipids, and purified glycolipids were tested for antigenic activity.
The lack of inhibition of agglutination with total lipids and with
purified glycolipids clearly indicated that the antigenic properties of saliva and gastric secretion were not related to the lipid
portion of these secretions (94,95). The native activities of saliva and gastric secretion were abolished by treatment with alkali
which is known to destroy blood group-active glycoproteins, but
is completely ineffective in degradation of glycosphingolipids.
However, neither alkali nor the presence of native glycolipids
from saliva or gastric secretion were capable of diminishing the
antigenic potency of added blood group A glycosphingolipids
(Table VIII). Also, the removal of lipids prior to the hemagglu-
tination-inhibition assay did not decrease the native activity of
the samples; to the contrary, a slight increase in potency per mg
of residue was noted.

These data clearly indicate that glycoproteins (water-soluble
antigens) are responsible for the blood group activity of the secretions and their presence in secretory tissue is only temporary,
whereas glycosphingolipids thus far isolated from a number of tissues represent antigens which are an integral part of the cell
membranes (94,95). This distinctive feature of epithelial-secre-
tory tissue versus its secretion does not explain the origin of
blood group antigens of the erythrocytes. The coexistence of glycosphingolipid and glycoprotein ABH antigens is still disputed.
According to Koscielak et al. (96) the erythrocyte stroma is only
equipped with antigens of glycosphingolipid nature. This is
strongly opposed by others (91,92,93) who have provided evidence
that erythrocyte membrane antigens are of dual origin. It is possible that rigorous separation of blood group-active glycoproteins
and glycosphingolipids between secretion and secretory tissue is
not applicable to erythrocytes, which represent an unusual type of
tissue entirely.

Acknowledgement. This study has been supported by Grants
AM No. 21684-02 and 25372-01 from National Institute of Arthritis,
Metabolism and Digestive Diseases, National Institutes of Health,
United States Public Health Service.

Literature Cited

1. Hallauer, C., Z. Immunitaetsforsch. Exp.Ther., 1934, 83, 114.
2. Kossjakow, P.N. and Tribulew, G.P., Z. Immunitaetsforsch. Exp. Ther., 1940, 98, 261.
3. Stepanov, A.V., Jusin, A., Makajeva, Z. and Kossjakow, P.N., Biokhimiya, 1940, 5, 547.
4. Hamasata, Y. Tohoku J. Exp. Med., 1950, 52, 17.
5. Masamune, H., Maekara, T. and Hakomori, S., Tohoku J. Exp.Med. 1954, 59, 225.
6. Masamune, H. and Siojima, S., Tohoku J. Exp. Med., 1951, 54, 319.
7. Slomiany, A. and Horowitz, M.I., J. Biol. Chem. 1973, 248, 6232.
8. Slomiany, A., Slomiany, B.L. and Horowitz, M.I., J. Biol.Chem. 1974, 249, 1225.
9. Hakomori, S., Stellner, K. and Watanabe, K., Biochem.Biophys. Res. Commun., 1972, 49, 1061.
10. Stellner, K., Watanabe, K. and Hakomori, S., Biochemistry, 1973, 12, 656.
11. Slomiany, B.L., Slomiany, A. and Horowitz, M.I., Biochim.Bio-phys. Acta, 1973, 326, 224.
12. Slomiany, B.L., Slomiany, A. and Horowitz, M.I., Eur. J. Bio-chem., 1974, 43, 161.
13. Slomiany, A., Annese, C. and Slomiany, B.L.,Biochim.Biophys. Acta, 1976, 441, 316.
14. Moschera, J. and Pigman, W., Carbohydr. Res., 1975, 40, 53.
15. Keryer, G., Herman, G. and Rossignol, B., Biochim. Biophys. Acta, 1973, 306, 446.
16. Kent, S.P. and Sanders, E.M., Proc. Soc. Exp. Biol. Med.,1969, 132, 645.
17. Pritchard, E.T. and Rusen, D.R., Arch. Oral. Biol., 1972, 17, 1619.
18. Pritchard, E.T., Arch. Oral Biol., 1973, 18, 1.
19. Tettamanti, G., Bonali, F., Marchesini, S. and Zambotti, V., Biochim. Biophys. Acta, 1973, 296, 160.
20. Folch-Pi, J., Lees, M. and Sloane-Stanley, G.H., J. Biol. Chem., 1957, 226, 497.
21. Suzuki, K., J. Neurochem., 1965, 12, 629.
22. Slomiany, A., Slomiany, B.L. and Horowitz, M.I., in Glycolipid Methodology (Witting, L.A., ed.) Am. Oil Chem. Soc., Cham-paign, IL., 1976, pp. 49-74.
23. Yu, R.K. and Ledeen, R.W., J. Lipid Res., 1972, 13, 680.
24. Slomiany, A. and Slomiany, B.L., Biochim. Biophys. Acta., 1975, 388, 135.
25. Slomiany, B.L., Slomiany, A. and Horowitz, M.I., Eur. J. Bio-chem., 1975, 56, 353.
26. Slomiany, A., Slomiany, B.L. and Annese, C., FEBS Lett.,1977, 81, 157.
27. Slomiany, B.L., Slomiany, A. and Horowitz, M.I., Eur. J. Bio-chem., 1975, 56, 353.

28. Yang, H.J. and Hakomori, S.I., J. Biol. Chem., 1971,246, 1192.
29. Lloyd, K.O., Kabat, E.A. and Rosenfield, R.E., Biochemistry, 1966, 5, 1502.
30. Smith, E.L., McKibbin, J.M., Karlsson, K.A., Pascher, I. and Samuelsson, B.E., Biochim. Biophys. Acta, 1975, 398, 84.
31. McKibbin, J.M. and Lyerly, D.F., Ala. J. Med. Sci., 1973, 10, 299.
32. Slomiany, A. and Slomiany, B.L., Eur. J. Biochem., 1977, 76, 491.
33. Slomiany, B.L. and Slomiany, A., Eur. J. Biochem., 1978,83, 105.
34. Esselman, W.J., Ackerman, J.R. and Sweeley, C.C., J. Biol. Chem., 1973, 248, 7310.
35. Sung, S.J., Esselman, W.J. and Sweeley, C.C., J. Biol. Chem., 1973, 248, 6528.
36. Smith, E.L., McKibbin, J.M., Karlsson, K.A., Pascher, I. and Samuelsson, B.E., Biochim. Biophys. Acta, 1975, 388, 171.
37. Siddiqui, B. and Hakomori, S.I., J. Biol. Chem., 1971, 246, 5766.
38. Karlsson, K.A., Leffler, H. and Samuelsson, B.E., J. Biol. Chem., 1974, 249, 4819.
39. Rapport, M.M. and Graf, L., Progr.Allergy, 1969, 13, 273.
40. Gahmberg, C.G. and Hakomori, S.I., J. Biol. Chem., 1975, 250, 2438.
41. Slomiany, B.L., Slomiany, A. and Horowitz, M.I., Biochim.Bio- phys. Acta, 1974, 348, 386.
42. Slomiany, B.L., Slomiany, A. and Badurski, J., Post. Biochem., 1975, 21, 319.
43. Gardas, S. and Koscielak, J., FEBS Lett., 1974, 42, 101.
44. Slomiany, B.L. and Slomiany, A., FEBS Lett., 1977, 73, 175.
45. Slomiany, B.L. and Meyer, K., J. Biol. Chem., 1972, 247, 5062.
46. Slomiany, B.L. and Meyer, K., J. Biol. Che., 1973, 248, 2290.
47. Newman, W. and Kabat, E.A., Arch. Biochem. Biophys., 1976, 172, 535.
48. Zdebska, E. and Koscielak, J., Eur. J. Biochem., 1978, 91,517.
49. Slomiany, B.L. and Slomiany, A., Biochim. Biophys. Acta, 1977, 486, 531.
50. Slomiany, B.L. and Slomiany, A., Chem. Phys. Lipids, 1977,20, 57.
51. Slomiany, A. and Slomiany, B.L., FEBS Lett., 1978, 90, 293.
52. Slomiany, B.L. and Slomiany, A., Eur. J. Biochem., 1978, 90, 39.
53. Slomiany, B.L., Slomiany, A. and Murty, V.L.N., Biochem. Bio- phys. Res. Commun., 1979, 88, 1092.
54. McKibbin, J.M., J. Lipid Res., 1978, 19, 131.
55. Slomiany, B.L., Slomiany, A. and Annese, C., J. Am. Oil Chem., 1978, 55, 239A.
56. Slomiany, B.L. and Slomiany, A., J. Biol. Chem., 1978, 253, 3517.
57. Herp. A., Wu, A.M. and Moschera, J., Mol. Cel. Biochem., 1979,

23, 27.
58. Glass, G.B.J. and Slomiany, B.L., in Mucus in Health and Disease, (Elstein, M. and Parke, D.V., eds.), Plenum Publishing Corp., New York, 1977, pp. 311-347.
59. Pigman, W. and Moschera, J., in Biology of the Cervix (Blandau, R.J. and Moghissi, K., eds.), University of Chicago Press, Chicago, 1973, pp. 143-173.
60. Slomiany, B.L., Slomiany, A. and Glass, G.B.J., Fed. Proc., 1977, 36, 978.
61. Slomiany, A. and Slomiany, B.L., J. Am. Oil Chem. Soc., 1978, 55, 239A.
62. Slomiany, A. and Slomiany, B.L., Biochem. Biophys. Res. Commun. 1977, 76, 115.
63. Slomiany, B.L., Slomiany, A.and Glass, G.B.J., FEBS Lett., 1977, 77, 47.
64. Slomiany, B.L., Slomiany, A. and Glass, G.B.J., Eur. J. Biochem., 1977, 78, 33.
65. Slomiany, B.L., Slomiany, A. and Glass, G.B.J., Biochemistry, 1977, 16, 3954.
66. Slomiany, A., Yano, S., Slomiany, B.L. and Glass, G.B.J., J. Biol. Chem., 1978, 253, 3785.
67. Slomiany, A., Patkowska, M.J., Slomiany, B.L. and Glass,G.B.J. Internatl. J. Biol. Macromol., 1979, in press.
68. Slomiany, B.L. and Slomiany, A., IRCS Med. Sci., 1979, 7,373.
69. Slomiany, B.L. and Slomiany, A., Biochem. Biophys. Res. Commun 1977, 79, 61.
70. Slomiany, B.L., Slomiany, A. and Glass, G.B.J., Eur. J. Biochem., 1978, 84, 53.
71. Kent, S.P. and Sanders, E.M., Proc. Soc.Exp. Biol. Med., 1969, 132, 645.
72. Lambert, R., Andre, C. and Berard, A., Digestion, 1971, 4, 234.
73. Galanti, N. and Baseraga, R., J. Biol. Chem., 1971, 246,6814.
74. Pritchard, E.T., Biochem. J., 1977, 166, 141.
75. Slomiany, A., Slomiany, B.L. and Mandel, I.D., Submitted for publication.
76. Slomiany, B.L., Slomiany, A. and Mandel, I.D., Submitted for publication.
77. Scarpelli, E.M., Clutario, B.C. and Taylor, F.A., J. Appl. Physiol., 1967, 23, 880.
78. Sanderson, R.J., Paul, G.W., Vatter, A.E. and Filley, G.F., Res. Physiol., 1976, 27, 379.
79. Godinez, R.J., Sanders, R.L. and Longmore, W.J., Biochemistry, 1975, 14, 830.
80. Colacicco, G., Buckelew, A.R. and Scarpelli, E.M., J. Appl. Physiol., 1973, 34, 743.
81. Slomiany, B.L., Smith,F.B. and Slomiany, A., Biochim. Biophys. Acta, 1979, in press.
82. Slomiany, A., Smith, F.B. and Slomiany, B.L., Eur. J. Biochem. 1979, 98, 47.

83. Schiff, F. and Adelsberger, L., Z. Immunitaetsforsch. Exp.
    Ther., 1924, 40, 335.
84. Eilser, M. and Mortisch, P., Z. Immunitaetsforsch. Exp.Ther.,
    1928, 57, 421.
85. Oppenheimer, C., "Handbuch der Biochemie", Gustav von Fischer
    Jena, 1930.
86. Friedenreich, V. and Hartmann, G., Z. Immunitaetsforsch.Exp.
    Ther., 1938, 92, 141.
87. Hartmann, G., "Group Antigens in Human Organs", Munksgaard,
    Copenhagen, 1941.
88. Slomiany, B.L. and Slomiany, A., in Progress in Gastroentero-
    logy, (Glass, G.B.J., ed.), Grune & Stratton, Inc., New York,
    1977, 3, 349.
89. Hakomori, S.I. and Kobata, A., in The Antigens, (Sela, M.,
    ed.), Academic Press, New York, 1974, 2, 79.
90. Rauvala, H. and Finne, J., FEBS Lett., 1979, 97, 1.
91. Fucuda, M. and Osawa, T., J. Biol. Chem., 1973, 248, 5100.
92. Takasaki, S. and Kobata, A., J. Biol. Chem., 1976, 251, 3610.
93. Takasaki, S., Yamashita, K. and Kobata, A., J. Biol. Chem.,
    1978, 253, 6086.
94. Slomiany, A., Slomiany, B.L. and Glass, G.B.J., Biochim.Bio-
    phys. Acta, 1978, 540, 278.
95. Slomiany, B.L. and Slomiany, A., Eur. J. Biochem., 1978, 85,
    249.
96. Koscielak, J., Miller-Podraza, H., Krauze, R. and Piasek, A.,
    Eur. J. Biochem., 1976, 71, 9.

RECEIVED December 10, 1979.

11

Fucolipids and Gangliosides of Human Colonic Cell Lines

BADER SIDDIQUI and Y. S. KIM

Veterans Administration Medical Center and Department of Medicine,
University of California, San Francisco, CA 94121

Carcinoma of the large bowel is a major hazard in most
affluent countries. In the United States alone, 100,000 persons
get colonic cancer each year and half of them die from it. Can-
cerous growth of tissues appears to be the result of cells not
following the normal differentiation pathway towards formation
and maintenance of normal functional organs. One approach to
treatment of cancer is to redirect the cellular differentiation
pathway toward normal growth with chemical agents. Several chemical agents have been used to modify the differentiation process
in cultured tumor cell lines. A variety of chemical compounds,
including cyclic AMP, sodium butyrate, dimethylformamide,
dimethylsulfoxide, 5-bromodeoxyuridine, and tri-fluoro-methyl-2-
deoxyuridine can affect morphological and biochemical properties
of cells. Some reports demonstrate that the tumorigenicity of
cancer cells is markedly reduced or completely abolished by these
agents. (See review by Prasad and Sinha, 1.)

Butyrate treated Hela cells (2) and KB cells showed marked
increases in the amounts of G_{M3} gangliosides and elevated levels
of the enzyme, CMP:sialic acid: lactosylceramide sialosyltrans-
ferase, required for its synthesis.

Human colonic mucosa and colonic tumors are rich in glycolipids including gangliosides and several fucolipids. These
lipids are important because they often determine blood group and
other surface properties of cells. To understand better the
effects of differentiating agents on tumor cells, we have been
concentrating our efforts on the effect of agents like sodium
butyrate or dimethylsulfoxide on colonic tumor cell lines.
Previous studies in our laboratory have dealt with some of the
effects of sodium butyrate on two colonic tumor cell lines,
SW-480 and SW-620.

This study describes the effect of sodium butyrate on glycolipids from four human colonic tumor cell lines, SKCO-1, HT-29,
SW-480 and SW-620 and a human fetal intestinal line, FHS.

0-8412-0556-6/80/47-128-177$5.00/0
© 1980 American Chemical Society

Materials and Methods

Cell Lines. The human fetal intestinal cell line (FHS) was
kindly supplied to us by Dr. Walter A. Nelson-Rees at the Naval
Bioscience Laboratory, Oakland, California. The human colonic
cell line, SKCO-1 developed by Drs. G. Trempe and L.F. Olds, was
obtained from Dr. Jorgen Fogh, Sloan Kettering Institute, Rye, New
York. The HT-29 cell line was developed and obtained from Dr. J.
Fogh. SW-480 and SW-620 were developed at Scott and White Clinic
in Temple, Texas and were obtained from Col. A. Liebovitz.
All of the cell lines are routinely maintained as monolayer
in Dulbecco's modified Eagle's medium supplemented to 10% with
fetal bovine serum, 100 units/ml of penicillin and 100 mg/ml of
streptomycin.

Labelling of Cells. For labelling experiments, Dulbecco's
modified Eagle's medium was used, except it contained only 1%
glucose. Cells were seeded at 1 to 2 x 10^6 cells/75 cm^2 flask in
medium at 37° and allowed to attach for 20-24 hours. The medium
was then replaced with fresh medium containing sodium butyrate,
1.0 mM in case of SKCO-1 cells or 2.5 mM for the other cell lines.
Medium was changed every 3-4 days. After 8 days, medium containing
50µCi of [^{3}H]-galactose (specific activity 9.1 Ci/m mole, (New
England Nuclear Corporation, Boston, Massachusetts)) or [3H]-
fucose (specific activity 13.2 mCi/m mole, (NEN)) was added with
or without butyrate. The cells were further incubated for 20-24
hours. The cells were harvested with 10 mM phosphate-buffered,
0.15 M saline, pH 7.4, containing 2 mM EDTA and washed three times
with cold phosphate-buffered saline. Cells were collected by
centrifugation.

Isolation of Labelled Glycolipids. Cells were sonicated in a
small volume of saline and the total protein was determined on an
aliquot by the method of Lowry et. al. (5). Total lipids were
extracted with 20 volumes of chloroform; methanol (2:1) filtered,
and the residue re-extracted with 10 volumes of chloroform:
methanol: water (1:2:0.15). Extracts were combined and concentrated at 40° under vacuum and dialyzed against distilled water
for 2 days at 4°. The dialyzate was dried and applied on a
1 x 10 cms DEAE-Sephadex column (6). Labelled neutral glycolipids,
along with other lipids, were eluted with 50 ml chloroform:
methanol: water (30:60:8) and the ganglioside fraction, also
containing sulfoglycolipids, was eluted with chloroform: methanol:
0.8 M sodium acetate (30:60:8).
In some experiments, total lipids were separated into upper
phase and lower phase. Each phase was applied separately to
columns containing DEAE-Sephadex to isolate three classes of
glycolipids: neutral glycolipids, sulfoglycolipids and gangliosides (7). Gangliosides and sulfoglycolipid fractions were
dialyzed and lyophilized. Glycolipids were resolved by thin layer
chromotography.

Thin Layer Chromotography. Unless otherwise stated, all
thin layer chromotography was on plates coated with silica Gel G
(E. Merck, Dramstadt). All the solvents were mixed on a volume
basis. Neutral glycolipid fractions were developed in chloroform:
methanol: water (60:35:6.5). Labelled fucolipid fractions were
developed in chloroform: methanol: water (40:40:10). For separation of gangliosides, chloroform: methanol: 2.5N aqueous NH_4OH
(60:40:9) was used.

Fluorography of TLC Plates. TLC plates were developed in the
appropriate solvent system and dried at 50° for 10-15 minutes.
The plates were impregnated with the scintillating medium by
dipping them into 20% 2,5,Diphenyloxazole (PPO) in toluene, dried
and exposed to X-ray film (Kodak, X-Omat R XR_2) for several days
at -70°. Fluorgraphs were then developed as described (8).

Results

Effect of Sodium Butyrate on Morphology and Cell Growth. FHS,
SKCO-1, HT-29 did not show any significant morphological changes
with sodium butyrate. SW-480 and SW-620 cells produce angular
cells rich in cellular membranes. These processes were pronounced
with SW-620 cell lines.
   Cells were seeded at 1 to 2 X 10^6 cells/75cm^2 flask with
sodium butyrate concentrations from 0.5 to 5.0 mM and without
butyrate in growth medium. After 8 days, the cells were harvested and protein was determined. Figure 1 shows total milligram
protein/T-75cm^2 flasks as plotted against sodium butyrate concentrations. The cell protein per flask of SKCO-1 decreased
sharply with increased concentrations of butyrate when compared
with control culture cells. With SW-480 and SW-620 culture cells,
protein was decreased against butyrate concentrations, but the
decrease was more pronounced with SW-620 cells. Cell protein of
FHS and HT-29 cultures were unaffected (Fig. 1).

TLC of Ganglioside. Figure 2 TLC patterns of gangliosides
obtained from fetal cell lines and three colonic cancer cell lines.
The fetal cell lines (track 1) contained uncharacterized gangliosides, a through f; SW-480 (track 4) contained uncharacterized
gangliosides, a through h; HT-29 gangliosides (track 3) have a
simpler pattern; G_{M3} is the major ganglioside in the SKCO-1 line
(track 2).

Labelled Fucolipids. Figure 3 shows fluorograms which were
obtained from cells labelled with [^{3}H]-fucose with and without
butyrate treatment. Fucolipids were not found in fetal cells
and, therefore, are not shown here. Figure 3A, track 2, shows
fucolipid patterns of SW-480 cells without butyrate. Although

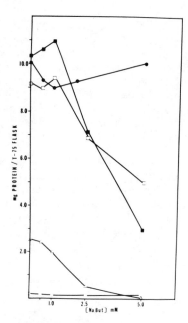

Figure 1. Total milligrams of protein/ T-75 cm flasks vs. millimolar concentrations of sodium butyrate in growth medium. Total cell protein was determined after cells were incubated in growth medium with sodium butyrate for eight days at 37°C. ○, FHS; △, SKCO-1; ●, HT-29; □, SW-480; ■, SW-620.

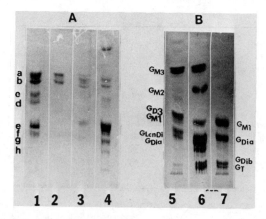

Figure 2.   TLC chromatogram of gangliosides in chloroform:methanol:2.5N NH₄OH (60:40:9)

1, FHS; 2, SKCO-1; 3, HT-29; 4, SW-480 cell lines; 5, small intestine gangliosides used as standards; 6, 7 contain human brain standard gangliosides. Apparent discrepancy in mobilities among A and B is because they were obtained from different runs and conditions vary slightly. Gangliosides visualized by spraying with resorcinol, followed by heating at 130°C for 15–25 min.

G_{M3}, $NeuAc\alpha2 \rightarrow 3Gal\beta1 \rightarrow 4GlB1 \rightarrow 1'Cer$
G_{D3}, $NeuAc\alpha2 \rightarrow 8NeuAc\alpha2 \rightarrow 3Gal\beta1 \rightarrow 4Glc\beta1 \rightarrow 1'Cer$
G_{M2}, $GalNAc\beta1 \rightarrow 4Gal(3 \leftarrow 2\alpha NeuAc)\beta1 \rightarrow 4Glc\beta1 \rightarrow 1'Cer$
G_{M1}, $Gal\beta1 \rightarrow 3GalNAc\beta1 \rightarrow 4Gal(3 \leftarrow 2NeuAc)\beta1 \rightarrow 4Glc\beta1 \rightarrow 1'Cer$
G_{D1a}, $NeuAc\alpha2 \rightarrow 3Gal\beta1 \rightarrow 3GalNAc\beta1 \rightarrow 4Gal(3 \leftarrow 2NeuAc)\beta1 \rightarrow 4Glc\beta1 \rightarrow 1'Cer$
G_{D1b}, $Gal\beta1 \rightarrow 3GalNAc\beta1 \rightarrow 4Gal(3 \leftarrow 2\alpha NeuAc8 \leftarrow 2\alpha NeuAc)\beta1 \rightarrow 4Glc1\beta \rightarrow 1'Cer$
G_{LcnDi}, $NeuAc\alpha 2 \rightarrow 8NeuAc\alpha2 \rightarrow 3Gal\beta1 \rightarrow 4GlcNac\beta1 \rightarrow 3Gal\beta1 \rightarrow 4Glc\beta1 \rightarrow 1'Cer$

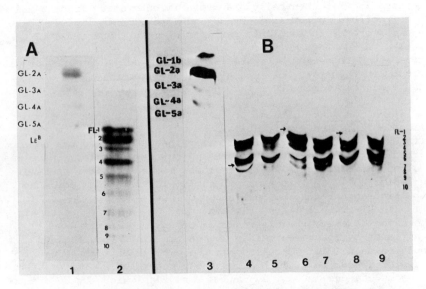

Figure 3.   Thin-layer chromatographic autoradiograms of labeled fucolipids

A: 1, labeled standard glycolipids GL-2a through Le[b]; 2, labeled fucolipids from SW-480 control cells. B: 3, labeled standard glycolipids GL-1b through Gl-5a. Labeled fucolipids from cells grown in butyrate-free medium are: HT-29, track 4; SW-480, track 6; SW-620, track 8. Fucolipids from cells grown in butyrate are: HT-29, track 5; SW-480, track 7; SW-620, track 9. A was developed in chloroform:methanol:water (60:35:8); B was developed in chloroform:methanol:water (40:40:10). In B equal amounts of fucolipid activity from both control cultures and butyrate-treated cells were applied. A is the result of direct radioautography of glycolipids. TLC plate B was dipped in 20% PPO in toluene and dried prior to exposure to x-ray film for several days at −70°C. Arrows show reactions of faint bands.

GL-1b, Galβ1 → 1'Cer
GL-2a, Galβ1 → 4Glcβ1 → 1'Cer
GL-3a, Galα1 → 4Galβ1 → 4Glcβ1 → 1'Cer
GL-4a, GalNAcβ1 → 3Galα1 → 4Galβ1 → 4Glcβ1 → 1'Cer
GL-5a, GalNAcα1 → 3GalNAcβ1 → 3Galα1 → 4Galβ1 → 4Glcβ1 → 1'Cer

the pattern of SKCO-1 cells are now shown here, there was no
change in the fucolipid patterns of these cells with and without
butyrate. Figure 3B shows fucolipid patterns of cells with and
without butyrate treatment. Tracks 4 and 5 show fucolipid
patterns of HT-29 cells with and without butyrate treatment. In
the HT-29 cell line fucolipid FL-1 is not present but there is a
decrease in FL-5 when cells are grown in butyrate. Track 6 and 8
show fucolipid patterns of SW-480 and SW-620 cell lines. There
is a difference in fucolipid patterns between these two cell
lines although they were both derived from the same patient. On
treatment with sodium butyrate, FL-1 is markedly decreased or disappears in these two cell lines (Fig. 3B, Track 7 and 9) and
reappearance of slow-migrating fucolipids (FL-7 through FL-9).

Labelled Gangliosides. Figure 4 shows fluorograms of gangliosides labelled with $[^3H]$-galactose from cells grown with or
without butyrate. In the fetal cell line (FHS) there was no
marked difference between treated and untreated cells. There was
a slight difference in the intensities between the two spots of
G_{M3} (Fig. 4A). G_{M3} is a major ganglioside in SKCO-1 cells and
labelling appeared to be unaffected by butyrate treatment (Fig.
4B). In HT-29 cell lines, the amount of G_{M3} appeared to remain
the same; however, the distribution of G_{M3} components was
   affected by butyrate. Minor changes in other gangliosides
could be seen (Fig. 4C). Although the overall pattern of ganlio-
sides of SW-480 cells with and without butyrate is similar, there
are some changes in G_{M3}, G_{M2} and G_{M1} regions which may be due to
alterations in the lipid moieties (Fig. 4D). Similar results are
also observed with SW-620 cells, as shown in Figure 4E.

Labelled Neutral Glycolipids. Neutral glycolipids were
labelled with $[^3H]$-galactose. As was seen with the ganglioside,
the butyrate affected the neutral glycolipid patterns but the
most marked alterations appeared to be due to changes in the
lipid moeities.

Discussion

   In the present study, sodium butyrate had a differentiated
effect on cell morphology. Sodium butyrate caused the SW-620
lines to become markedly angular with extension of many membran-
eous processes. These effects were also seen with the SW-480
cell lines but were less pronounced. No morphological changes
were observed when SKCO-1, HT-29 and FHS cell lines were cultured
in sodium butyrate.
   The concentration of sodium butyrate was observed to have a
differential effect on cell growth in colonic cell lines. After
culturing for 8 days with 5 mM sodium butyrate, the cell protein
per flask of the SCKO-1 line was decreased to less than 10% of
the control cultures. In the SW-620 culture, cell protein per

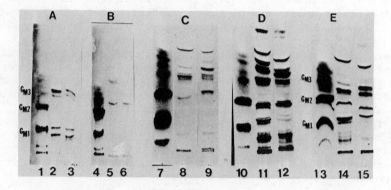

Figure 4.   Thin-layer chromatographic fluorograms of labeled gangliosides

The plates were developed in chloroform:methanol:2.5N NH_4OH (60:40:9). There are apparent discrepancies in the mobilities among the fluorograms because each plate was obtained from different runs and the conditions varied slightly. Standard [^{3}H]-gangliosides G_{M3}, G_{M2}, and G_{M1} were used in tracks 1, 4, 7, 10, and 12. A: labeled gangliosides obtained from FHS cell lines with (3) and without (2) butyrate treatment. B: gangliosides of SKCO-1 with (6) and without (5) butyrates. C: gangliosides of HT-29 with (9) and without (8) butyrate. D: gangliosides of SW-480 with (12) and without (11) butyrate. E: gangliosides of SW-620 cell lines with (15) and without (14) butyrate. Bands above G_{M3} in C, D, and E are sulfoglycolipids.

was reduced to 20-25% of the control while the related line SW-480
showed a 50% reduction in cell protein. Cell protein of FHS and
HT-29 cultures was unaffected.

When cells were cultured in labelled fucose or galactose in
the presence or absence of butyrate, alterations in the labelled
glycolipids were observed. Treatment of all of the cell lines
with butyrate did not markedly affect the incorporation of $[^3H]$
galactose in ganglioside per milligram of cell protein.

In all of the lines except SW-480, butyrate caused a decrease
in monoglycosylceramide compared to diglycosylceradmie; however,
the changes were not as distinct as the changes in gangliosides.
When SW-480 and SW-620 cell lines were grown in the presence of
butyrate, the fastest migrating fucolipid disappeared concomitant with the appearance of slow-migrating fucolipids.

Although there were few qualitative changes in the gangio-
side patterns of the SKCO-1 and FHS lines, there were marked
alterations of gangliosides in the HT-29, SW-480 and SW-620 cell
lines. The major changes were seen within the components compil-
ing the G_{M3} fraction. In HT-29, SW-480, and SW-620, there was a
shift in G_{M3} to less polar components suggesting that the carbohydrates may be unchanged but the lipid moieties are altered.
Alternatively, there may an acetylation of a hydroxyl group in
the carbohydrate moiety since it has been shown that the butyrate
increases the amount of acetylated histones in Friend erythroleukemic cells (9). The butyrate-induced shift to less polar
components is also seen in the G_{M2} fraction.

The shift in G_{M2} and G_{M3} components may be important in dis-
turbing cell surface properties. SW-480 and SW-620 showed
dramatic morphological alterations when cultured in butyrate, and
these cells had obvious shifts to less polar components within
the G_{M2} and G_{M3} fractions. In the SKCO-1 and FHS lines, these
shifts were not observed and thus there were no morphological
changes in these two cell lines in butyrate. However, since
HT-29 cells did not change morphology in butyrate but also demonstrated the polarity shift in G_{M2} and G_{M3} components, the correlation between the two may be more complex, such as being dependent upon concentration or distribution of these components on
the cell surface. We are currently exploring the effects of
butyrate on ganglioside components of other cell lines to determine if this glycolipid shift is related to morphological alterations and to the malignant properties of cells.

Summary

     In the present study, we examined the pattern of fucolipids
and gangliosides in cultured cell lines and alterations produced
by a differentiating agent. A human fetal intestinal line (FHS),
and four human colonic tumor lines (SKCO-1, HT-29, SW-480 and
SW-620) were used. Cells were grown with or without sodium butyrate, (1.0 mM in SKCO-1 and 2.5 mM in all other cell lines) in

growth medium. After 8 days medium containing 50μCi of [³H]-galactose or [³H]-fucose was added with or without butyrate, followed by incubation for another 20-24 hours. Glycolipids were purified by column chromatography, characterized by thin-layer chromatography and were detected by radioautography or by conventional staining. Each tumor line revealed a distinct pattern of labelled fucolipids consisting of at least 10 components. No labelled fucolipids were detected in the FHS cell lines. The butyrate treated SKCO-1 cells did not show any change in fucolipid patterns. In HT-29 cell lines, there was a decrease of fucolipid FL-5 when the cells were grown in butyrate. There is a difference in fucolipid patterns between SW-480 and SW-620 cell lines. On treatment with sodium butyrate FL-1 (fast moving fucolipid) is markedly decreased or disappears, and there is appearance of slow migrating fucolipids (FL-7 through FL-9).

Gangliosides were labelled with galactose. In the fetal cell lines (FHS) and SKCO-1 there was no marked difference between treated and untreated cells. In HT-29, SW-480, and Sw-620 cell lines, the amounts of G_{M3} appeared to remain the same, but the distribution of G_{M3} components was affected by butyrate. These changes, might be due to alterations in the lipid moieties or, alternatively, there might be an acetylation of a hydroxyl group in the carbohydrate moiety, since it has been shown that the butyrate increases the amount of acetylated histones.

Acknowledgements

This work was supported in part by the United States Public Health Service Grant CA-14905 from the National Cancer Institute through the National Large Bowel Cancer Project, and by the Veterans Administration Medical Research Service.

We are indebted to Dr. J. S. Whitehead for his critical review and valuable discussions in the preparation of this manuscript. We also appreciate the technical assistance of Mr. James Bennett.

Literature Cited

1. Prasad, K.N., and Sinha, P.K. (1978) *In* "Cell Differentiation and Neoplasia" (G.F. Saunders, ed.), pp 111-141. Raven Press, New York.
2. Fishman, P.H., Bradley, R.M., and Henneberry, R.C. (1976) Arch. Biochem. Biophys. 172, 618-626.
3. Macher, B.A., Lockney, M., Moskal, J.R., Fung, Y.K., and Sweeley, C.C. (1978) Exptl. Cell Res. 117, 95-102.
4. Kim, Y.S., Tsao, D., Siddiqui, B., Whitehead, J.S., Arnstein, P., Bennett, J., and Hicks, J. Cancer, In press.
5. Lowry, O.H., Rosebrough, N.J., Farr, A.L., and Randall, R.J. (1951) J. Biol. Chem. 193, 265-275.

6. Ledeen, R.W., Yu, R.K., and Eng, L.F. (1973) J. Neurochem.
    21, 829-939.
7. Siddiqui, B., Whitehead, J.S., and Kim, Y.S. (1978) J. Biol.
    Chem. 253, 2168-2175.
8. Mills, A.D., and Laskey, R.A. (1975) Eur. J. Biochem. 56,
    335-341.
9. Reeves, R., and Cserjesi, P. (1979) J. Biol. Chem. 254,
    4283-4290.

RECEIVED December 10, 1979.

Biosynthesis of Blood-Group Related Glycosphingolipids in T- and B-Lymphomas and Neuroblastoma Cells

MANJU BASU and SUBHASH BASU

Department of Chemistry, University of Notre Dame, Notre Dame, IN 46556

MICHAEL POTTER

National Cancer Institute, Bethesda, MD 20014

Glycosphingolipids (GSLs) are principal constituents of all eukaryotic cell membranes. Four different classes of glycosphingolipids ($\underline{1}$) are commonly found in animal cells: a) GSLs containing mono- or disaccharides; b) GSLs containing a core structure, GalNAcβ-Galβ1-4Glc-Cer (Gg series); c) a core structure of Galα-Galβ1-4Glc-Cer (Gb series); d) a core structure of GlcNAcβ-Galβ1-4Glc-Cer (Lc series).

Short-chain GSLs of the first three families appear to be ubiquitous among eukaryotic cells. However, long-chain GSLs of the latter two families are important constituents of the plasma membranes of numerous animal cells ($\underline{2},\underline{3}$). Cell surface GSLs of the globoside family (Gb series) and blood group family (Lc series) have been implicated in the processes of cell-cell recognition and growth regulation ($\underline{4},\underline{5}$), receptor function ($\underline{6},\underline{7}$), malignant transformation ($\underline{8}$-$\underline{10}$), and blood group specificity ($\underline{11}$-$\underline{17}$). In recent years specific blood group-active glycosphingolipids (A, B, H, Le^a, Le^b, P$_1$, and I) have been identified in human erythrocytes ($\underline{11}$-$\underline{17}$). The neolactotetraosylceramide, nLcOse$_4$Cer (Gal$\beta\overline{1}$-4GlcNAcβ1-3Galβ1-4Glc-Cer) exists as a common structure in these GSLs. The possibility that nLcOse$_4$Cer is a tumor-associated surface antigen in NIL polyoma-transformed tumor cells was suggested by Hakomori and his co-workers ($\underline{18},\underline{19}$). Irrespective of blood type, nLcOse$_4$Cer has also been identified in normal human erythrocytes ($\underline{20}$) and in elevated quantities in the erythrocyte stroma of patients with congenital dyserythropoietic anemia type II ($\underline{21}$). Are these changes in the content of nLcOse$_4$Cer due to blocked synthesis of higher chain length blood group glycosphingolipids ($\underline{22}$) or to elevated activity of UDP-Gal:LcOse$_3$Cer (β1-$\overline{4}$) galactosyltransferase (EC 2.4.1.86) ($\underline{24}$)? In search of answers

to this question we have studied biosynthesis in vitro
of nLcOse$_4$Cer and its conversion to GMlb(GlcNAc) ganglioside or blood group H$_1$- and B-active GSLs (see Fig.
1) in different tumor cells of primate and non-primate
origin.

It is now well established that the T-lymphocytes
that develop in the thymus and released in the circulation have different physiological immune responses
from the antigen-stimulated, immunoglobulin-secreting
B-lymphocytes (Fig. 2). In recent years tumors of the
mouse lymphoreticular system have become the best
models for the study of myeloma proteins (23). In the
present report we have compared the biosynthetic
routes of the blood group-related glycosphingolipids
mentioned above in mouse lymphoreticular tumors and
neuroblastomas. The binding of lectins and toxins to
some of these tumor cells has also been studied to
obtain information about the nature of glycoconjugates
present on the cell surfaces.

Materials and Methods

Mouse Lymphoreticular Tumors. Since 1971
National Cancer Institute has been freezing and stor-
ing transplantable mouse lymphoreticular tumors. More
than 1000 different tumors have been deposited. The
most common tumor types available are lymphocytic
tumors of bone marrow and thymic origin, Abelson
virus-induced lymphosarcomas and plasmacytomas, and
chemically induced plasmacytomas. The tumors under
investigation in our laboratory are listed in Table I.
Some data on the cell surface markers have been
described recently by Mathieson et al. (24). The
plasmacytomas are also frequently checked for immunoglobulin production and antigen-binding activity.
Since all of the tumors at Cancer Institute are enter-
ed in a computer bank (as mentioned in Table I), a
descriptive identification system has been adopted
(Table II). The computer name of a tumor contains
four pieces of information: i) strain of origin; ii)
mode of induction; iii) cell type; and iv) accession
number.

Abelson Virus-Induced Lymphocytic Tumors. Abelson
virus (A-MuLV or MuLV-A) is a type C RNA virus (25)
and exists in the murine leukemia virus complex. This
is a defective virus and contains the Moloney leukemia
virus helper component and a replication-defective
Abelson component (23) that transforms lymphocytes and
3T3 fibroblasts. While Abelson virus is not

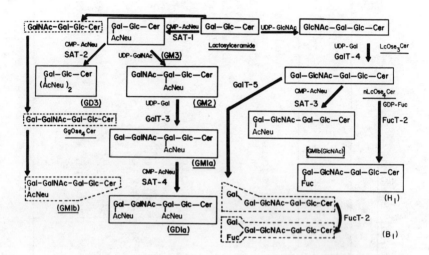

Figure 1.   *Proposed pathways for glycosphingolipid biosynthesis*

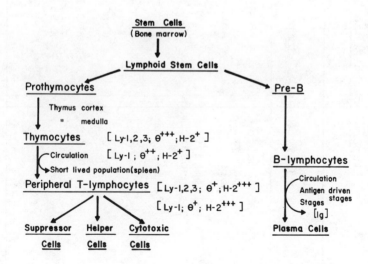

Figure 2.   *Development of B- and T-lymphocytes*

Table I

Tumors of the Mouse Lymphoreticular System

Lymphocyte	Tumor Type	Computer Name	Characteristics	Induction Condition
B	Bone marrow lymphocytic tumors	ABLS-1	θ^- (leukogenic)	Abelson virus (C-type RNA)
		ABLS-140		
	Plasmacytic lymphosarcomas	ABPL-2	θ^-	TE/109 days AB/89 days
	Plasmacytomas	TEPC-15	IgA polymer (α)	Mineral oil or alkane
		TEPC-824		
		CBPC-101	IgG ($\gamma_2 a$)	
		BPC-1	IgH ($\gamma_2 b$)	
		X-5563	Unknown	Spontaneous ileocecal

Table I. (contd.)

Lymphocyte	Tumor Type	Computer Name	Characteristics	Induction Condition
T	Thymic lympho-cytic neoplasms	SAKRTLS-13	$Ly-1^+$	Spontaneous
		L-4946	$Ly-1^+(2^+)$	
	Lymphocytic tumors of thymic origin	BALENTL-3	θ^+, $Ly-2,3^+$	Chemically induced
		BALENTL-5	θ^+, $Ly-2,3^+$	Ethylnitro-sourea/173 days
		P-1798	θ^+, $Ly-1,2,3^+$	Estrogen pellet/521 days

Table II

Abbreviations used to Identify Tumor Lines of Mouse Lymphoreticular System

Strain	Induction	Cell Type
BAL = BALB/C	AB = Abelson Virus	PC = plasma cell
CB = (C57B1/6XBALB/C)F$_1$	TE = Pristane[a]	LS = lymphosarcoma
AKR		TL, TS = thymic lympho-cytic neoplasm
X = C$_3$H		MS = mastocytomas

[a]Pristane = 2,6,10,14-tetramethylpentadecane.

infectious in a mouse colony, leukemias can be induced in adult BALB/C mice within 21 to 30 days after injection.

Lymphosarcomas and Plasmacytic Lymphosarcomas. These tumors arise in bone marrow cavities or lymph nodes under the influence of Abelson virus. For the production of plasmacytic lymphosarcomas (PL), a pre-incubation period of 2 to 3 months with pristane is necessary (26). PL cells are distinguished from other plasmacytomas by their size and lymphoid character.

Lymphocytic Tumors of Thymic Origin. The appearance of spontaneous tumors (e.g. SAKRTLS-13) in AKR and C58 mice (27) is quite common. Other thymic tumors can also be induced chemically (28) (BALENTL-3, -5 or P-1798; Table I) or virally (Moloney leukemia virus (29)). In the early stage of tumor development they are confined to the thymus, but later the tumor is metastasized to the spleen, liver, kidney, and lymph nodes.

Plasmacytomas. The transplantable plasmacytomas are derived from tumors induced in BALB/C mice. These tumors were induced by intraperitoneal (IP) implantation of plastic materials (Lucite discs or Millipore diffusion chambers) or by the IP injection of mineral oils (light and heavy medicinal mineral oils, Bayol, F, Drakeol GVR) and alkanes such as pristane. Plasmacytomas arise in peritoneal tissues and require a mineral oil environment during their early development.

Cell Culture. Human neuroblastoma IMR-32 (passaged through nude mice; the cells were donated by Dr. Steven E. Brooks, Kingsbrook Jewish Medical Center, Brooklyn) and mouse neuroblastoma clones NIE-115, NS-20, and N-18) (donated by Dr. Shraga Makover, Hoffmann LaRoche, Inc., Nutley, New Jersey) were maintained in our laboratory as described previously (30,31). Confluent monolayers (6 to 8 x 10^6 cells per 250-ml Falcon plastic flask) were harvested for enzymatic studies with phosphate-buffered saline [7.0 mM potassium phosphate/0.14 M NaCl - buffer, pH 7.2 (Pi/NaCl)] containing 0.1% EDTA.

A clone of guinea pig tumor cells, 104Cl (the cells were donated by Dr. Charles H. Evans, National Cancer Institute, Bethesda, MD), was maintained in our laboratory (7) on RPMI-1640 medium (Gibco) supplemented with 10% fetal bovine serum (Gibco). Cultures were grown in Falcon T-flasks (75 cm^2) containing 15

ml of medium and incubated under a water-saturated 95%
air/5% CO_2 atmosphere at 37°C. The medium was changed
once before harvesting, and cells were harvested when
they reached a population density of 5 to 8 x 10^6 cells
per T-flask.

   Glycosphingolipids. Acceptor GSLs were isolated
from various animal tissues (32). Lactosylceramide
and GM3 ganglioside were isolated from bovine spleen
(33), GM1 and GM2 gangliosides from human brains (34),
and B-active neolactopentaosylceramide (nLcOse5Cer;
Galα1-3Galβ1-4GlcNAcβ1-3Galβ1-4Glc-Cer)from rabbit
erythrocytes (35,36) and bovine erythrocytes (37).
Neolactotetraosylceramide (nLcOse4Cer;Galβ1-4GlcNAcβ1-
3Galβ1-4Glc-Cer) and lactotriaosylceramide (LcOse3Cer;
GlcNAcβ1-3Galβ1-4Glc-Cer) were prepared from
nLcOse5Cer by sequential degradation with fig α-galac-
tosidase (38,39) and papaya β-galactosidase (33,40).
GgOse4Cer was prepared from bovine brain gangliosides
by mild acid hydrolysis according to a previously
published method (41). The purified glycosphingolipids were analyzed before use as substrates by gas
chromatography-mass spectrometry (42).

   Glycosphingolipid: Glycosyltransferase Assays. A
25-33% (vol/vol) homogenate of mouse tumors or harvested cells in 0.32 M sucrose containing 0.1% 2-mer-
captoethanol and 0.001 M EDTA (pH 7.0) was used as
enzyme source. Membrane fractions for glycolipid:gly-
cosyltransferase assays were isolated at the junction
of 0.32 M and 1.2 M on a discontinuous sucrose density
gradient (32,43).

   i) Galactosyltransferase Assays. The complete
incubation mixture contained the following components
(in micromoles) in final volumes of 0.045 ml: glycosphingolipid acceptors, 0.025; Triton X-100, 100 µg;
cacodylate-HCl buffer, pH 7.3, 10; $MnCl_2$, 0.125; UDP-
[^{14}C]Gal, 25,000 cpm (1.3 x 10^6 cpm per µmole) and
homogenate of tumor or cells, 0.3 to 0.5 mg of protein
(estimated by the method of Lowry et al. (44) using
bovine serum albumin as standard). After 2 hours at
37°C, the reaction was stopped by adding 2.5 µmoles of
EDTA (pH 7.0).

   ii) Sialyltransferase Assays. The complete incubation mixture contained the following components (in
micromoles) in final volumes of 0.065 ml: glycosphingolipid acceptors, 0.05; Triton CF-54 and Tween-80
(2:1), 200 µg; cacodylate-HCl buffer, pH 6.4, 9; $MgCl_2$,
0.25; CMP-[^{14}C]AcNeu, 61,000 cpm (2.6 x 10^6 cpm per

μmole); and homogenate of tumor or cells, 1.0 to 1.5 mg of protein. After 2 hours at 37°C, the reaction was stopped by adding 10 μl of chloroform-methanol (2:1).

iii) Fucosyltransferase Assays. The complete incubation mixture contained the following components (in micromoles) in final volumes of 0.037 ml: glycosphingolipid acceptors, 0.025; detergent, G-3634A (Atlas Chemical), 100 μg; cacodylate-HCl buffer, pH 6.4, 10; MgCl$_2$, 0.125; EDTA (pH 7.0), 0.5; GDP-[^{14}C]Fuc, 24,000 cpm (222 μCi per μmole and 3.5 x 10^6 cpm per μmole); and homogenate of tumor or cells, 0.12 to 0.28 mg of protein. After 1 hour at 37°, the reaction was stopped by adding 10 μl of chloroform-methanol (2:1).

The incubation mixtures were assayed by the double chromatographic method ([33](http://),[40](http://)) or by a combination of high voltage borate electrophoresis and reverse flow chromatography ([32](http://),[40](http://)) on Whatman 3MM paper using chloroform-methanol-H$_2$O (50:40:10) as solvent system. The appropriate areas of each chromatogram were determined quantitatively in a toluene scintillation system with a Beckman scintillation counter (model LS-3133T).

Binding of [^{125}I]Lectin and [^{125}I]Toxin to Cell Surfaces. Falcon T-flasks (75 cm^2) containing confluent populations of cultured cells (IMR-32, NIE-115, NS-20, N-18, or 104Cl) were washed with PBS (2 x 10 ml) at 15-20°C and incubated with [^{125}I]lectin or [^{125}I] toxin (specific activities are mentioned in Tables VII and VIII) in 3 ml of serum free medium (Eagle's MEM for human neuroblastoma IMR-32 cells; Dulbecco's MEM for mouse neuroblastoma NIE-115, N-18; and NS-20; and RPMI-1640 for guinea pig 104Cl cells) for 15 minutes at 37° C. The medium was removed; the cell layer was washed gently with PBS (2 x 10 ml) at 15-20°C and then kept in an incubator for 10-15 minutes at 37°C in the presence of 5 ml of 0.1% EDTA in PBS (pH 7.2). The loose cells were finally dispersed and transferred to a 15-ml graduated centrifuge tube with a disposable Pasteur pipette. An aliquot (0.5 to 1 ml) was taken and filtered through borosilicate fiber discs (Whatman GF/A, porosity, 1.0 μm; diameter, 2.4 cm) in a Millipore apparatus. The discs were washed with cold 5% trichloroacetic acid (TCA) or 5% TCA followed by chloroform-methanol (2:1) and dried at 100°C for 15 minutes. [^{125}I] content was quantitatively determined in a toluene scintillation system in the presence and absence of PCS (Amersham/Searle) with a Beckman LS-3133T counter. Purified cholera toxin was purchased from Schwarz/Mann and labeled with Na^{125}I in the presence of Chloramine-T according to the method of

Table III

Glycolipid: Galactosyltransferase Activities in Mouse and Human Tumor Cells

Tumor Type	[^{14}C]Galactose Incorporated		
	LcOse$_3$Cer (β1-4)	nLcOse$_4$Cer (α1-3)	GM2 (β1-3)
	pmol/mg protein/2 hr		
Mouse			
B-lymphocytic: ABLS-140	1,251	115	194
TEPC-15	5,282	1,642	3,132
T-lymphocytic: L-4946	15,570	886	2,075
BALENTL-3	3,465	533	244
Human			
Neuroblastoma IMR-32	3,270	595	82

Cuatrecasas (45). The Dolichos biflorus lectin was a
gift from Dr. Marilynn Etzler (46). Ulex europeus and
Bandeirea simplicifolia lectins were purified in our
laboratory and iodinated with Na^{125}I (carrier-free) by
Sepharose 4B-bound lactoperoxidase (47)and diazotized
iodoaniline coupling (48) procedures, respectively.

Results and Discussion

A. Blood Group-Related Glycosphingolipid Synthesis
in Tumor Cells. Mouse lymphoreticular tumors contain
at least three different glycolipid:galactosyltrans-
ferase activities, which can be distinguished by their
acceptor specificities (Table III). Recently we have
shown (22) that cultured cells (TSD) from the cerebrum
of a Tay-Sachs-diseased fetus contain high activity of
UDP-Gal:LcOse$_3$Cer (β1-4) galactosyltransferase (Fig. 1,
GalT-4 or EC 2.4.1.86) (35),[1] but very little activity
of UDP-Gal:GM2(β1-3) galactosyltransferase (Fig. 1,
GalT-3 or EC 2.4.1.62) (49). The present study indicates that the levels of these two galactosyltransferase activities depend on the differentiated cell
types. In plasmacytomas (TEPC-15) the levels of these
two enzymatic activities are high and are almost comparable, whereas in B-lymphocytic tumor obtained by
induction with Abelson virus (lymphosarcoma, ABLS-140)
the level of GalT-3 was only 15% that of GalT-4 activity. It appears that mouse lymphoreticular tumors
(B- or T-lymphocytic origin) induced virally (ABLS-
140, Table I) or chemically (BALENTL-3, Table I) show
a lower content of GalT-3 activity, whereas highest
activity is present in tumors obtained by mineral oil
induction (TEPC-15). Among all the tumors tested, the
activities of GalT-4 and GalT-5 (Fig. 1, UDP-Gal:
nLcOse$_4$Cer(α1-3) galactosyltransferase or EC 2.4.1.87)
(38) are highest in spontaneous thymic lymphocytic
neoplasms (L-4946). The activities of these three
glycolipid:galactosyltransferases were also compared
(under present conditions in vitro) in human neuroblastoma IMR-32 cells. Synthesis of GM1 ganglioside
from GM2 is unusually low compared with the synthesis
of neolactotetraosylceramide (nLcOse$_4$Cer) or B-active
neolactopentaosylceramide (nLcOse$_5$Cer), as in ABLS-140
and BALENTL-3 mouse lymphoreticular tumors. From our
previous studies it appears that synthesis of neolactotetraosylceramide is ubiquitous among various normal
(40,53) and tumor cells (7,22,37,54,55) grown in
culture. It is important to find out whether these
neutral blood group core structures are then transformed to any type-specific antigenic sialic acid- or
fucose-containing glycosphingolipid on the surface of
tumor cells.

Sialyltransferase activities using three specific
glycolipid substrates were also tested in these mouse
lymphoreticular tumors. The activity of CMP-AcNeu:GM3
(α2-8)sialyltransferase (50) was highest in ethylnitro-
sourea-induced thymic lymphocytic tumor BALENTL-3 but
almost negligible in B-lymphocytic tumors (Table IV).
The transfer of sialic acid to the terminal galactose
of gangliotetraosylceramide is catalyzed efficiently
by a sialyltransferase present in embryonic chicken
(41,51) and rat (52) brains. A Golgi-rich membrane
preparation isolated from bovine spleen (32,33) also
catalyzes the reaction efficiently. Recently we have
shown that both embryonic chicken brain (43) and bovine
spleen (33,43) also catalyze the transfer of sialic
acid (AcNeu) from CMP-[^{14}C]AcNeu to neolactotetraosylceramide to form AcNeu-nLcOse$_4$Cer, or GM1b(GlcNAc). We
have also characterized and established the sialic
linkage present in the ^{14}C-labeled product obtained
from this enzymatic reaction.[2]

Studies with glycoprotein substrate specificities
of porcine submaxillary sialyltransferase suggest (56)
that the enzyme catalyzing the transfer of sialic acid
to Galβ1-3GalNAc-R$_1$ glycoprotein acceptor may not
catalyze sialic acid transfer to Galβ1-3GalNAc-R$_2$ glycolipid acceptor (i.e. GM1) or Galβ1-4GlcNAc-R$_3$
acceptor. Using glycosphingolipids as acceptors
(nLcOse$_4$Cer and GgOse$_4$Cer) for the substrate competition studies with embryonic chicken brain and bovine
spleen Golgi-rich membrane systems (33,43),[2,3] it
appears that both reactions might be catalyzed by the
same enzyme. In the present studies, except in TEPC-
15 and IMR-32, the activities with both substrates
(nLcOse$_4$Cer and GgOse$_4$Cer) are almost comparable in all
other tumor cell lines. Further kinetic studies of
these two activities are under way. The natural occur-
ence of AcNeu-GgOse$_4$Cer (GM1b) in rat ascites tumor
cells has been reported recently (57).

Different glycolipid:fucosyltransferase activities
have been reported (32,55,58) to catalyze the addition
of fucose to position C-2 of the terminal D-galactose
and position C-3 of the internal N-acetylglucosamine
of nLcOse$_4$Cer to form blood group H or human tumor-
specific lipid (10). Incorporation of fucose into
GgOse$_4$Cer was also reported recently by Taki et al.
(59) in rat ascites hepatoma cells (AH 7974F). From
our present studies it appears that GDP-Fuc:nLcOse$_4$Cer
(α1-2)fucosyltransferase activity (Fig. 1, FucT-2 or
EC 2.4.1.89) is 10 to 20 times more active in mouse
B-lymphocytic tumors (ABLS-140 and TEPC-15) than in
T-lymphocytic tumors (L-4946 and BALENTL-3). From

Table IV

Glycolipid: Sialyltransferase Activities in Mouse and Human Tumor Cells

Tumor Type	[^{14}C]AcNeu Incorporated		
	nLcOse$_4$Cer	GgOse$_4$Cer	GM3
	(α2-3)/(α2-6)	(α2-3)/(α2-6)	(α2-8)
	pmol/mg protein/2 hr		
Mouse			
B-lymphocytic: ABLS-140	316	401	< 10
TEPC-15	307	623	< 10
T-lymphocytic: L-4946	1,081	992	90
BALENTL-3	534	617	1,554
Human			
Neuroblastoma IMR-32	92	213	< 10

Table V

Glycolipid: Fucosyltransferase Activities in Mouse and Human Tumor Cells

Tumor Type	[14C]Fucose Incorporated		
	nLcOSe$_4$Cer (α1-2)/(α1-3)	nLcOSe$_5$Cer (α1-2)/(α1-3)	GM1 (α1-2)
	cpm x 10^{-2}/mg protein/hr		
Mouse			
B-lymphocytic: ABLS-140	73	222	< 0.1
TEPC-15	103	140	8
T-lymphocytic: L-4946	9	40	
BALENTL-3	5	3	< 0.1
Human			
Neuroblastoma IMR-32	(124)[a] (310)[b] 31	(121)[a] (363)[b] 29	(15)[a] < 0.1

Table V (contd.)

a,b The complete incubation mixtures contained the same components as described in the text except that higher concentration (0.01 μmole per 50 μl incubation volume) and lower specific activity of GDP-[14C]fucose (3.6 x 10⁶ cpm per μmole) was used for a) IMR-32 homogenate and b) membrane fraction isolated at the junction of 0.32 and 1.2 M sucrose gradient. The mixtures were incubated for 2 hr at 37°C.

Values not given in parentheses represent incorporation of [14C]fucose into respective substrates using high specific activity GDP-[14C]fucose (189 mCi/mmole; New England Nuclear) and cell homogenates as described in the text.

competition studies with a bovine Golgi-rich membrane fraction (60) and an IMR-32 membrane fraction it appears that the transfer of [^{14}C]fucose to nLcOse$_4$Cer and nLcOse$_5$Cer is probably catalyzed by the same enzyme.[4] Activities with these two substrates in mouse lymphoreticular tumors and human neuroblastoma cells are almost comparable. However, little or no fucose transfer to GM1 ganglioside or GgOse$_4$Cer (unpublished) has been observed with these tumor cells under our present assay conditions. Higher activities of glyco-protein: (α-2) or (α-3) fucosyltransferase activities in human cancer tissues and in sera of cancer patients have been reported from different laboratories (61-64) but the function of these fucosylated glycoconjugates on malignant cell surfaces is still unknown.

  B. Binding of [^{125}I]Lectins and [^{125}I]Toxin to Neuroblastoma Cells of Primate and Nonprimate Origin. In order to obtain some idea about the nature of gly-coconjugates and their gross topographical orientation on the cell surfaces, we measured the binding of ^{125}I-labeled lectins and toxin to human neuroblastoma IMR-32 and mouse neuroblastoma N1E-115, NS-20, and N-18 clones (Tables VII and VIII). The "5% TCA Wash" column (Table VII) represents ^{125}I-labeled lectin or toxin bound to both glycoprotein and glycolipid. The "5% TCA plus chloroform-methanol 2:1 Wash" column represents tightly bound lectin or toxin to the cell surfaces. These results suggest that, in addition to GM1 ganglio-side, IMR-32 cells may contain some globoprotein and ganglioprotein (65) with terminal N-acetylgalactosamine (Table VII) and α-fucose residues (Table VIII) (65). Although we found very little activity of FucT-2 (Fig. 1) in mouse neuroblastoma clones (55,60) it appears that N-18 has the highest Ulex europeus [^{125}I]lectin binding ability of all clonal lines tested. The bio-synthesis in vitro of non-fucose B-active neolactopen-taosylceramide (37,53-55) in cultured mouse (37,54), guinea pig (53),and human tumor cells (55,60) has been established in our laboratory. It is important to see changes in its appearance on the tumor cell surfaces during chemically induced differentiation. The bind-ing studies with B. simplicifolia [^{125}I]lectin to N1E-115 cells (after chemical differentiation) showed no marked difference when compared with control cells (Fig. 3). In the presence of cytochalasin-B the cell volume increases, a change that may represent increased binding sites or reduced internalization of the [^{125}I] lectin due to the alterations of microtubules.

  C. Role of Glycosphingolipids as Cell Surface Receptors. Both B- and T-lymphocytes emerge from bone marrow into the lymphatic tissues and enter the blood

Table VI

Sugar Specific Binding of Lectins and Toxins

Name	Molecular Wt. (Subunits)	Sugar Specificity	Blood Group Type
Bandeiraea simplicifolia	114,000 (4)	$Gal(\alpha 1-3)Gal-R_1$	B
Cholera toxin	84,000 (2)	$Gal(\beta 1-3)GalNAc-Gal-R_2$ $\mid$ $AcNeu$	GM1
Ulex europeus	170,000 (n.d.)	$Fuc(\alpha 1-2)Gal-R_3$	H
Dolichos biflorus	113,000 (4)	$GalNAc(\alpha 1-3)Gal-R_4$	A

[a] R_1, R_2, R_3, or R_4 = oligosaccharides, glycolipid, or glycoprotein.

Table VII

Binding of [125I]Lectin and [125I]Toxin to IMR-32 Cell Surfaces

[125I]Lectin/Toxin	Concentration	[125I]Lectin or Toxin Bound	
		cpm per 10^6 cells	
		5% TCA wash	5% TCA + Chloroform-methanol (2:1) wash
	µg/ml		
Dolichos biflorus	2.5	248	222
	5.0	436	419
	10.0	1,387	1,407
Cholera toxin	1.5	2,175	1,372
	3.0	3,662	2,816
	12.0	11,572	8,731

The specific activities of [125I]-labeled Dolichos biflorus lectin and cholera toxin were 1.12×10^8 cpm and 3.1×10^7 cpm per mg of protein, respectively.

Table VIII

Ulex Europeus [125I]Lectin Binding to Neuroblastoma Cell Surfaces

$[^{125}I]$Lectin Added	Human Neuroblastoma	Mouse Neuroblastoma		
	IMR-32	NS-20	NIE-115	N-18
μg/ml	$ng/10^6$ cells			
20	14	42	14	36
40	33	61	44	46
80	60	110	116	216

The specific activity of Ulex europeus $[^{125}I]$lectin was 0.5×10^6 cpm per mg of protein.

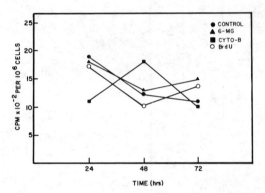

Figure 3. Effect of chemical differentiating agents on radioactive Bandeiraea simplicifolia *[^{125}I]lectin binding to mouse neuroblastoma (N1E-115) cell surfaces*

Falcon T-flasks containing 104Cl cells (1.5–2.5 × 10^6 cells per flask) were incubated with nothing (●), 4 μM BrdUrd (○), 4 μM 6-mercaptoguanosine(▲), or 1 μg Cytochalasin-B (40) per ml of DMEM containing 10% fetal bovine serum (GIBCO) for the indicated periods. Binding of B. simplicifolia *[^{125}I]lectin (9.0 × 10^6 cpm per mg protein) to cell surfaces of each flask was studied according to the method described in text except that 20 μg of lectin/3mL of serum-free DMEM was used and [^{125}I]lectin-bound cells were harvested from each flask in 5.0 mL of PBS.*

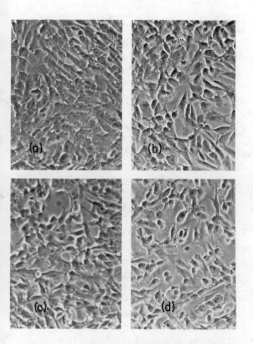

Figure 4.   Phase contrast micrographs of 104Cl guinea pig tumor cells grown in culture (×100)

(a): Control cells grown in 250-mL Falcon T-flask as described in text. (b): Cells incubated with GM1 ganglioside (10 μg/mL) as described in text. (c): Cells incubated 30 min at 37°C with cholera toxin (5.0 μg/mL) in serum-free RPMI-1640 medium. (d): Cells incubated 30 min at 37°C with cholera toxin after treatment with GM1 ganglioside.

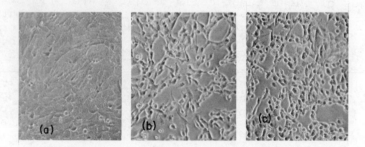

Figure 5.   Phase contrast micrographs of 104Cl guinea pig tumor cells grown in culture

(a): Control cells grown in Falcon T-flasks (75 cm²) (see text). (b): Cells incubated 25 min at 37°C with Dolichos biflorus [¹²⁵I]lectin (12 μg/mL) is RPMI-1640. (c): Cells incubated with lectin as in (b), then washed with PBS (2 × 10mL).

stream as free, nonadhesive cells carrying specific
antigen-binding receptors on their surfaces. The mouse
embryonic thymus has θ-positive lymphocytes by day 12
of gestation; by day 15 the embryonic spleen contains
both Ig-positive and θ-positive cells. However these
fetal spleen cells do not yield antibody-producing
cells when grown in culture. It is believed that antigen selects programmed cells in the adult animal, but
the mechanism of expression of these programmed cells
is not known. It is also not known whether a specific
class of glycoconjugate is expressed during development of B- or T-lymphocytes. Our preliminary investigation into the biosynthesis of blood group-related
glycosphingolipids in mouse lymphoreticular tumors at
different stages (Fig. 2) may answer some of these
questions. Makita and his co-workers have reported
enhanced activities of four glycolipid: glycosyltransferases (involved in the synthesis of Forssman hapten)
in hamster thymuses bearing advanced stages of lymphoma growth (66).
   On the basis of [^{125}I]lectin and [^{125}I]toxin binding to tumor cell surfaces we proposed (7) the existence of α-GalNAc-linked (Forssman like) and GM1-like
receptors (glycolipid or glycoprotein) on the surfaces
of human neuroblastoma IMR-32 and guinea pig tumor
cells (104Cl). Exogenously added Forssman glycolipid
or GM1 ganglioside increased binding of Dolichos
biflorus [^{125}I]lectin and [^{125}I]cholera toxin (7),
respectively, in addition to the marked morphological
changes of these tumor cells (Figs. 4 and 5). At this
stage it is merely speculative to propose that antigenic glycolipids (Forssman hapten, nLcOse$_4$Cer, or
nLcOse$_5$Cer) or acidic glycolipids (GM3, GM1, GM1b, or
GM1b(GlcNAc)) detected on tumor cell surfaces play some
role in cell growth and behavior. It is possible that
the specific proteins involved in growth regulation
(e.g., cholera toxin-mediated adenylate cyclase activation (67) or DNA replication (31)) are attached to
tumor cell surfaces through specific terminal sugar
residues of glycosphingolipids or glycoproteins.

Acknowledgments

   This work was supported by U.S. Public Health
Grants NS-09541 and CA-14764 and a grant-in-aid from
Miles Laboratories, Inc., Elkhart, Indiana to S.B.
We would like to thank Dr. Prabir Bhattacharya, Dr.
Kathleen A. Presper, Dr. Joseph R. Moskal, Mr. Alex
Vuckovic, and Mr. William G. Shanabruch for help in
the lectin binding studies and cell culture work.

Footnotes

[1]IUPAC-IUB Commission on Biochemical Nomenclature
(1976) "Enzyme Nomenclature" (and amendments thereto)
Biochim. Biophys. Acta, 429, 1-45.
[2]Chien, J. L., Basu, M., Basu, S. and Stoffyn, P.,
manuscript in preparation.
[3]Basu, M., Basu, S. and Stoffyn, P., manuscript in
preparation.
[4]Presper, K. A., Basu, M. and Basu, S., manuscript in
preparation.

Literature Cited

1. IUPAC-IUB Commission on Biochemical Nomenclature
   "The Nomenclature of Lipids", Lipids, 1977, 12,
   455-468.
2. Hakomori, S. I. Biochem. Biophys. Acta, 1975, 417.
   55-89.
3. Yamakawa, T.; Nagai, Y. Trend Biochem. Sci., 1978,
   3, 128-131.
4. Gahmberg, C. G.; Hakomori, S. I. Proc. Nat. Acad.
   Sci. U.S.A., 1973, 70, 3329-3333.
5. Lingwood, C. A.; Hakomori, S. I. Exptl. Cell Res.,
   1977, 108, 385-391.
6. Gahmberg, C. G.; Kiehn, D.; Hakomori, S. I.
   Nature, 1974., 248, 413-415.
7. Basu, M.; Basu, S.; Shanabruch, W. G.; Moskal,
   J. R.; Evans, C. H. Biochem. Biophys. Res. Commun.,
   1976, 71, 385-392.
8. Baumann, H.; Nudelman, E.; Watanabe, K.; Hakomori,
   S. Cancer Res., 1979, 39, 2637-2643.
9. Steiner, S.; Melnick, J. L.; Kit, S.; Somers, K. D.
   Nature, 1974, 248, 682-684.
10. Hakomori, S. I. Prog. Biochem. Pharmacol., 1975,
    10, 167-196.
11. Handa, S. Japan J. Exp. Med., 1963, 33, 347-360.
12. Hakomori, S. I.; Watanabe, K. "Glycolipid
    Analysis", Witting, L. A., Ed.; American Oil
    Chemists' Society, Champaign, Ill., 1976, p. 13-
    47.
13. Gardas, A.; Kosceilak, J. Eur. J. Biochem., 1973,
    32, 178-187.
14. Ando, S.; Yamakawa, T. J. Biochem., 1973, 73,
    387-396.
15. Hakomori, S. I.; Strycharz, G. D. Biochemistry,
    1968, 7, 1279-1285.

16. Naiki, M.; Fong, J.; Ledeen, R.; Marcus, D. M. Biochemistry, 1975, 14, 4831-4837.
17. Watanabe, K.; Hakomori, S. I.; Childs, R. A.; Feizi, T. J. Biol. Chem., 1979, 254, 3221-3228.
18. Gahmberg, C. G.; Hakomori, S. I. J. Biol. Chem., 1975, 250, 2438-2446.
19. Sundsmo, J. S.; Hakomori, S. I. Biochem. Biophys. Res. Commun., 1976, 68, 799-806.
20. Ando, S.; Kon, K.; Isobe, M.; Yamakawa, T. J. Biochem., 1973, 73, 893-895.
21. Joseph, K. C.; Gockerman, J. P. Biochem. Biophys. Res. Commun., 1975, 65, 146-152.
22. Basu, M.; Presper, K. A.; Basu, S.; Hoffman, L. M.; Brooks, S. E. Proc. Nat. Acad. Sci., U.S.A., 1979, 76, 4270-4274.
23. Potter, M. in "Mosbacher Colloquium", Melchers, F.; Rajewsky, K., Eds.; Springer-Verlog, New York, 1976; p. 141-172.
24. Mathieson, B. J.; Campbell, P. S.; Potter, M.; Asosky, R. J. Exptl. Med., 1978, 147, 1267-1279.
25. Abelson, H. T.; Rabstein, L. S. Cancer Res., 1970, 30, 2208-2212; ibid., 2213-2222.
26. Potter, M.; Sklar, M. D.; Rowe, W. P. Science, 1973, 182, 592-594.
27. Rowe, W. P. Cancer Res., 1973, 33, 3061-3068.
28. Kirschbaum, A.; Liebelt, A. G. Cancer Res., 1955, 10, 689-692.
29. Moloney, J. R., Federation Proc., 1962, 21, 19-31.
30. Basu, S.; Moskal, J. R.; Gardner, D. A. in "Ganglioside Function: Biochemical and Pharmacological Implications", Porcellati, G.; Ceccareli, G.; Tettamanti, G., Eds.; Plenum Press, New York, 1976; Vol. 71, p. 45-63.
31. Bhattacharya, P.; Simet, I.; Basu, S. Proc. Nat. Acad. Sci., U.S.A., 1979, 76, 2218-2221.
32. Basu, S.; Basu, M.; Chien, J. L. J. Biol. Chem., 1975, 250, 2956-2962.
33. Chien, J. L. Ph.D. Thesis, University of Notre Dame, Notre Dame, Indiana, 1975.
34. Svennerholm, L. in "Comprehensive Biochemistry", Florkin, M.; Stolz, E. H., Eds.; Elsevier, Amsterdam, 1969; Vol. 18, p. 201-227.
35. Basu, M.; Basu, S. J. Biol. Chem., 1972, 247, 1489-1495.
36. Basu, M. D.Sc. Thesis, University of Calcutta, Calcutta, India, 1974.
37. Moskal, J. R. Ph.D. Thesis, University of Notre Dame, Notre Dame, Indiana, 1977.
38. Basu, M.; Basu, S. J. Biol. Chem., 1973, 248, 1700-1706.

39. Li, Y. T.; Li, S. C. Methods Enzymol., 1972, 28, 714-720.
40. Basu, S.; Basu, M.; Moskal, J. R.; Chien, J. L.; Gardner, D. A. in "Glycolipid Methodology", Witting, L. A., Ed.; American Oil Chemists Society Press, Champaign, Ill., 1976; p. 123-139.
41. Basu, S. Ph.D. Thesis, University of Michigan, Ann Arbor, Michigan, 1966.
42. Bjorndal, H.; Hellerqvist, C. G.; Linderberg, B.; Svensson, S. Angew. Chem. Int. Ed. Engl., 1970, 9, 610-619.
43. Basu, S.; Basu, M.; Chien, J. L.; Presper, K. A. in "Ganglioside Structure and Function", Svennerholm, L.; Mandel, P., Eds.; Plenum Press, New York and London; 1979, in press.
44. Lowry, O. H.; Rosenbrough, N. J.; Farr, A. L.; Randall, R. J. J. Biol. Chem., 1951, 193, 264-275.
45. Cuatrecasas, P. Biochemistry, 1973, 12, 3567-3577.
46. Etzler, M. E.; Kabat, E. A. Biochemistry, 1970, 9, 869-877.
47. David, G. S.; Reisfeld, R. A. Biochemistry, 1974, 13, 1014-1021.
48. Hays, C. E.; Goldstein, I. J. Anal. Biochem., 1975, 67, 580-584.
49. Basu, S.; Kaufman, B.; Roseman, S. J. Biol. Chem., 1965, 240, 4115-4117.
50. Kaufman, B.; Basu, S.; Roseman, S. J. Biol. Chem., 1968, 243, 5804-5806.
51. Kaufman, B.; Basu, S.; Roseman, S. Methods Enzymol., 1966, 8, 365-368.
52. Stoffyn, A.; Stoffyn, P.; Yip, M. C. M. Biochim. Biophys. Acta, 1975, 409, 97-103.
53. Basu, M.; Moskal, J. R.; Gardner, D. A.; Basu, S. Biochem. Biophys. Res. Commun., 1975, 66, 1380-1388.
54. Moskal, J. R.; Gardner, D. A.; Basu, S. Biochem. Biophys. Res. Commun., 1974, 61, 751-758.
55. Presper, K. A.; Basu, M.; Basu, S. Proc. Nat. Acad. Sci., U.S.A., 1978, 75, 289-293.
56. Rearick, J. I.; Saddler, J. E.; Paulson, J. C.; Hill, R. L. J. Biol. Chem., 1979, 254, 4444-4451.
57. Hirabayashi, Y.; Taki, T.; Matsumoto, M. FEBS Lett., 1979, 100, 253-257.
58. Pacuszka, T.; Kosceilak, J. Eur. J. Biochem., 1976, 64, 499-506.
59. Taki, T.; Hirabayashi, Y.; Matsumoto, M.; Kojima, K. Biochim. Biophys. Acta, 1979, 572, 105-112.
60. Presper, K. A., Ph.D. Thesis, University of Notre Dame, Notre Dame, Indiana, 1979.
61. Bauer, C. H.; Kottgen, E.; Reutter, W. Biochem. Biophys. Res. Commun., 1977, 76, 488-494.

62. Ip, C.; Das, T. Cancer Res., 1978, 38, 723-728.
63. Khilanani, P.; Chou, T. H.; Lomen, P. L.; Kessel,
    D. Cancer Res., 1977, 37, 2557-2559.
64. Chatterjee, S. K.; Bhattacharya, M.; Barlow, J. J.
    Cancer Res., 1979, 39, 1943-1951.
65. Tonegawa, Y.; Hakomori, S. I. Biochem. Biophys.
    Res. Commun., 1977, 76, 9-17.
66. Ishibashi, T.; Atsuta, T.; Makita, A. J. Nat.
    Cancer Inst., 1975, 55, 1433-1436.
67. O'Keefe, R.; Cuatrecasas, P. J. Membrane Biol.,
    1978, 42, 61-79.

RECEIVED December 10, 1979.

Altered Glycolipids of CHO Cells Resistant to Wheat Germ Agglutinin

PAMELA STANLEY

Department of Cell Biology, Albert Einstein College of Medicine, Bronx, NY 10461

In recent years this laboratory has isolated and partially characterized a variety of CHO cell mutants selected for resistance to different plant lectins (1,2). Mutants which fall into three different genetic complementation groups have been selected by virtue of their resistance to the cytotoxicity of wheat germ agglutinin (WGA). One of the mutants which may be selected with WGA falls into complementation group I and, if isolated from a WGA selection, it is termed Wga^{RI}. Mutants in complementation group I have been shown to lack a specific N-acetylglucosaminyltransferase activity which appears to provide the biochemical basis of lectin resistance in this genotype (see 3). The other previously described Wga^R mutants (termed Wga^{RII} and Wga^{RIII}) fall into complementation groups II and III respectively, and have been shown to possess decreased sialylation of glycoproteins and the ganglioside GM_3 at the cell surface (4,5). However, an enzymic basis for these genotypes has not been uncovered.

A fourth type of Wga^R CHO cell mutant has now been isolated (Stanley, manuscript in preparation). This mutant is more highly resistant to WGA than the previously described mutants and it has been shown to belong to a new complementation group (group VIII). In this paper, the glycolipids of Wga^{RVIII} cells are compared with those of parental CHO cells and the other Wga^R CHO cell mutants.

Materials and Methods

Alpha medium (containing ribonucleosides and deoxyribonucleosides) and fetal calf serum (FCS) were obtained from Grand Island Biological Co., U.S.A.. Reagent grade chloroform, methanol and hydrochloric acid were obtained from Fisher Scientific Co., U.S.A. and redistilled before use. Pre-coated silica gel 60 plates (0.25mm) were obtained from E.M. Laboratories, Germany. Alphanaphthol and resorcinol were obtained from Fisher Scientific Co., U.S.A. Resorcinol was twice recrystallized before use. Dowex 1 x 4 (100-200 mesh) chloride form was obtained from Bio Rad Laboratories, U.S.A. and converted to the acetate form according to

0-8412-0556-6/80/47-128-213$5.00/0

the manufacturers instructions. Neuraminidase of V.cholerae was
obtained from Calbiochem, U.S.A.
     N-acetyl (4,5,6,7,8,9-^{14}C) neuraminic acid was obtained from
Amersham, U.S.A. All other chemicals were reagent grade. Purified glycolipid standards and reagents for GLC analysis were kind
gifts from Dr. Samar Kundu, Albert Einstein College.

Cell Culture.

     Cells were cultured in alpha medium 10% FCS at 37° in suspen-
sion. The experiments reported were performed with the following
cell lines: Gat$^-$2 (a glycine-adenosine-thymidine requiring auxo-
troph – the parent from which each of the mutants was derived);
Gat$^-$2Wga^{RI}1N; Gat$^-$2Wga^{RII}4C; Gat$^-$2Wga^{RIII}6F; and Gat$^-$2Wga^{RVIII}1-3.
The nomenclature of these lines is simplified in the text and
figures to P for parental (Gat$^-$2), and to Wga^{RI}, Wga^{RII}, Wga^{RIII}
and Wga^{RVIII} for the respective mutant lines.

Extraction of Glycolipids.

     Exponentially-growing cells ($\sim$2–3x10^9) were washed twice with
50–100 volumes of phosphate buffered saline (PBS). After resuspension in 50ml PBS, aliquots were taken for cell counting and for
protein determination. The cells were centrifuged, resuspended in
40ml 10mM Tris-HCl pH 7·4 and centrifuged at 2000 rpm at 4° in an
International PR2 centrifuge. The pellet was extracted with 20
volumes redistilled chloroform-methanol (C:M) 2:1 by mixing 2 min
at low speed in a Waring blender. The mixture was filtered through
a sintered glass funnel and the residue subsequently extracted with
10 volumes C:M (1:2) – based on original cell pellet volume. The
combined filtrates were rotoevaporated, redissolved in C:M (1:1)
at 10^8 cell equivalents per ml and stored at -20°.
     Cells which were treated with neuraminidase prior to lipid
extraction were washed twice in PBS and resuspended at 10^8 cells
per ml in PBS containing 50 units neuraminidase per ml or in PBS
alone (control). The cell suspensions were incubated 5 min at 37°,
centrifuged at 1200 rpm for 10 min at 4° in an International PR2
and washed once with cold PBS. The cells were resuspended in
hypotonic Tris-HCl, centrifuged and extracted with chloroform-
methanol in the manner described above.

Thin Layer Chromatography.

     Lipid extracts (0.3ml) were dried under nitrogen, resuspended
in $\sim$15µl C:M (1:1) and spotted on activated silica gel 60 plates
with purified glycolipid standards GM$_3$, GM$_2$, lactosylceramide (LC)
and glucosylceramide (GC) containing $\sim$10µg sialic acid each.
Lipid extracts treated with neuraminidase were dried under N$_2$ and
incubated at 37° with 50 µl (25 units) V.cholerae neuraminidase.
After 16 hr, 1 ml C:M (2:1) was added, the samples were incubated

15 min at 23° and centrifuged at 3000 rpm for 15 min in a PR2
International centrifuge. The supernatant was dried under N_2, redissolved in ∿15 µl C:M (1:1) and spotted on activated TLC plates.
Plates were developed by ascending chromatography in chloroform:
methanol:0.02% $CaCl_2$ (60:40:9). Dried plates were subsequently
stained by α-naphthol/sulphuric acid to detect carbohydrate or
resorcinol to detect sialic acid-containing glycolipids (6,7).

Determination of Sialic Acid in Lipid Extracts.

   Free and lipid-bound sialic acid were determined in each lipid
extract by the thiobarbituric acid (TBA) method (8) following partial
purification of free sialic acid on Dowex 1 x 4 (100-200 mesh) acetate
form (9). Lipid extracts containing ∿20-30µg sialic acid were dried
under nitrogen and resuspended in 2ml H_2O. About 10,000 cpm [14]C-
sialic acid was added to each sample. The samples were redissolved
in 1.0ml redistilled methanol and heated ∿10 sec at 80°. For the determination of total sialic acid, samples were treated with 0.05N
HCl (by adding 1.0ml 0.1NHCl to samples in methanol) for 1 hr at
80°C. The pH of the hydrolyzed samples was adjusted to pH∿8.0 with
NaOH, and they were heated at 56° for 5 min to destroy lactones
which might have formed during the hydrolysis. These preparations
were filtered through ∿2cm glass wool and subsequently loaded onto
a 3 cm column of Dowex (acetate form). The sample eluate and a
4 ml wash of distilled deionized water (DDW) were collected together and a 1 ml aliquot taken for scintillation counting. The
column was then eluted with 7.5ml 1N formic acid. These eluates
were dried on an evapomix (or lyophilized), reconstituted in DDW
and assayed for ^{14}C-sialic acid and unlabelled sialic acid by the
TBA assay (performed on duplicate or triplicate samples). The determination of free sialic acid in the lipid extracts was made on
samples which had been dried and reconstituted in methanol (1.0ml),
0.1NHCl and 0.1NNaOH (1ml each, added together) and adjusted to
pH 8.0. These samples were heated at 56° for 5 min and then passed
over Dowex (acetate) exactly as described above.
   Detection of sialic acid in the form of CMP-sialic acid was
preliminarily examined using the assay of Kean and Roseman (10).
Aqueous samples (0.2ml) were incubated with 30 µl cold sodium
borohydride (100 mg/ml) with agitation for 15 min before the addition
of 30µl acetone. After a further 15 min at room temperature, the
samples were assayed for sialic acid by the TBA method.

Results

   The glycolipids of parental CHO cells and the four different
Wga[R] CHO cell lines were compared by thin-layer chromatography of
lipid extracts. As described previously by this laboratory (5)
and by others (11,12), the major glycolipid in CHO cells is the
ganglioside GM3 which has the structure sialic acid $\underline{\alpha 2,3}$ galactose
$\underline{\alpha 1,3}$ glucose-ceramide. This is indicated in Figs. 1,3 and 4 by
the co-migration of the major carbohydrate-containing band of

216

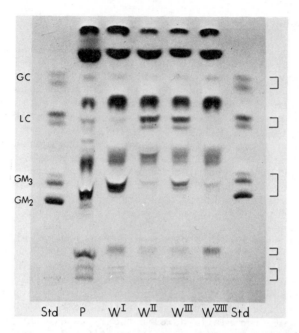

Figure 1. Glycolipids of parental and WgaᴿCHO cells stained with α-naphthol

Lipid extracts from ~3 × 10⁷ cells were compared by TLC (see Methods). Areas that stained blue after the α-naphthol/sulfuric acid are bracketed. The individual glycolipids in the mixture of purified glycolipid standards are also identified (Std). Cell extracts are identified as P (parental cell extract) and Wᴵ, Wᴵᴵ, Wᴵᴵᴵ and Wᵛᴵᴵᴵ for Wgaᴿᴵ, WgaᴿᴵᴵWgaᴿᴵᴵᴵ and Wgaᴿᵛᴵᴵᴵ. The major glycolipid band from parental cells ran between authentic GM₂ and GM₃. However, in other experiments this band was shown to co-migrate with GM₃ (see Figures 2–4). Bands that occur near the origin and precede GM₃ have been shown to co-migrate with free sugars (sialic acid and neutral sugars) and do not correspond to known glycolipids.

parental CHO cells with authentic GM3 marker glycolipid. Further evidence that this band is GM3 is provided in Figs. 2 and 3. In Fig. 2 it is shown that neuraminidase treatment of a parental cell lipid extract converts most of the band co-migrating with authentic GM3 to two bands which co-migrate with authentic lactosylceramide (LC;gal-gluc-ceramide). The enzyme also converts the GM3 standard to LC but does not change the position of authentic GM2. In Fig. 3 it is shown that the band in CHO cell extracts which co-migrates with GM3 also stains with the resorcinol reagent which is specific for sialic acid. Taken together, the data in Figs. 1-3 show that GM3 is the major glycolipid in CHO cells and that other gangliosides, if present, are in small amounts not detected by these methods.

The glycolipid pattern of each Wga^R CHO mutant is also given in Figs. 1-3. Three of the four mutants exhibit altered glycolipids. As described previously (5), Wga^{RII} and Wga^{RIII} cells possess low amounts of GM3 and increased amounts of LC compared with parental CHO cells. By contrast, Wga^{RI} cells exhibit a glycolipid pattern identical with parental cells. Even the acessibility of GM3 on the surface of Wga^{RI} cells to neuraminidase appears to be similar to that of parental CHO cells (Fig. 4). This is particularly interesting in view of the fact that the majority of the "acidic" or "complex" asparagine-linked carbohydrate moieties of CHO membrane glycoproteins are altered in Wga^{RI} cells to a partially-processed intermediate of the structure Manα1,6 [Manα1,3] - Manα1,6 [Manα1,3]-Manβ1,4GlcNAcβ1,4 GlcNAc Asn peptide (13; Etchison and Summers, manuscript in preparation). Thus it might have been expected that steric protection of membrane GM3 molecules would be reduced at the Wga^{RI} cell surface compared with parental CHO cells.

The data in Figs. 1-4 were obtained from mutants selected independently from those described in our previous experiments (5) and demonstrate that the altered glycolipid patterns expressed by these mutants are a stable phenotypic property distinctive of each genotype. The new Wga^R mutant (Wga^{RVIII}) also exhibits a unique glycolipid pattern (Figs. 1-3). Like Wga^{RII} cells, Wga^{RVIII} cells possess very low amounts of GM3. However in contrast to Wga^{RII} (and Wga^{RIII}) mutants, Wga^{RVIII} cells exhibit no concomitant increase in the amounts of LC or GC visible on the chromatograms. This suggests that these mutants may be making less GM3 due to a defect prior to the addition of the first glucose moiety to ceramide.

Since the major glycolipid of CHO cells is GM3 and since this ganglioside contains one mole of sialic acid per mole, the differences in GM3 contents between the Wga^R CHO mutants may be quantitated by determining the amount of glycosidically-bound sialic acid in chloroform:methanol cell extracts. The results of such an analysis are given in Table I. Parental and Wga^{RI} CHO cells contain about 1.0 μg glycosidically-bound sialic acid per mg cell protein. Lipid extracts of these cell lines also contain a small amount of free sialic acid (∼0.1μg per mg cell protein). Each of the remaining Wga^R mutants exhibits decreased levels of

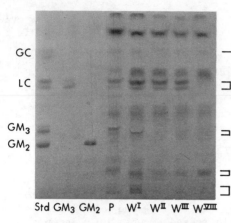

GC

LC

GM₃

GM₂

Std GM₃ GM₂ P   Wᴵ Wᴵᴵ Wᴵᴵᴵ Wᵛᴵᴵᴵ

Figure 2. *Glycolipids of parental and Wgaᴿ CHO cells after neuraminidase treatment*

Extracts from ~3 × 10⁷ cells and glycolipids GM₃ and GM₂ were treated with neuraminidase and analyzed by TLC in parallel with mixed glycolipid standards (see Methods). Plate was stained with α-naphthol/sulfuric acid; areas that turned blue are bracketed.

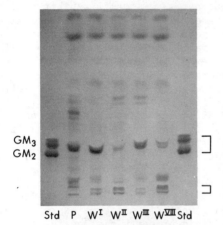

GM₃
GM₂

Std P Wᴵ Wᴵᴵ Wᴵᴵᴵ Wᵛᴵᴵᴵ Std

Figure 3. *Gangliosides of parental and Wgaᴿ CHO cells*

Extracts from ~3 × 10⁷ cells and a mixture of GM₃ and GM₂ were compared by TLC after staining with resorcinol. To improve sensitivity of technique, the spray was applied heavily, giving rise to some nonspecific staining. Only those bands that reproducibly stained the characteristic blue of gangliosides are bracketed.

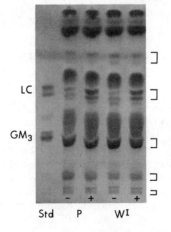

LC

GM₃

        –   +   –   +
Std   P   Wᴵ

Figure 4. *Accessibility of GM₃ in parental and Wgaᴿᴵ cell membranes to action of neuraminidase*

Cells were washed, and half were treated with neuraminidase (see Methods). Other half were treated identically in the absence of enzyme. Lipid extracts later made from neuraminidase-treated and control samples in the usual way, and extracts from ~6 × 10⁷ cells were compared by TLC. Plates stained with α-naphthol/sulfuric acid; blue areas are bracketed. (—): Controls; (+): treated with neuraminidase.

Table I
Sialic Acid Contents of Lipid Extracts of Parental
and Wga^R CHO Cells

Cell Line	Bound Sialic Acid (ng per mg cell protein)		Free Sialic Acid (ng per mg cell protein)		Ave % Bound Sialic Acid Wga^R/Parental Cells
	Expt. 1	Expt. 2	Expt. 1	Expt. 2	
Parental	1030	910	125	90	
Wga^{RI}	1290	990	169	75	117%
Wga^{RII}	94	55	439	435	7%
Wga^{RIII}	420	560	73	80	50%
Wga^{RVIII}	110	310	741	520	20%

Sialic acid was partially-purified from lipid extracts after hydrolysis in 0.05NHCl (total sialic acid) or without hydrolysis (free sialic acid) as described in Methods. The amounts of glycosidically-bound sialic acid (total minus free) and free sialic acid were determined by the TBA assay. The hydrolysed samples were also analyzed using gas liquid chromatography by the method of Yu and Ledeen (15) and shown to possess essentially identical amounts of total sialic acid. However, the crude lipid extracts were too impure to make GLC the method of choice in the absence of extensive purification of the glycolipids.

glycosidically-bound sialic acid as would be predicted. Wga^{RIII} cells possess $\sim$0.5µg bound sialic acid per mg cell protein (a 50% decrease compared with parental cells) but exhibit similar levels of free sialic acid to that found in parental cell extracts. Wga^{RII} and Wga^{RVIII} cells exhibit the most marked decrease in bound sialic acid having only approximately 10% of that found in parental CHO cells. Surprisingly, however, both these cell lines possess high levels of free sialic acid ($\sim$4-8-fold the amounts found in parental CHO cells).

The observation of high levels of free sialic acid in lipid extracts of Wga^{RII} and Wga^{RVIII} cells prompted us to examine whether it might be sialic acid complexed with the nucleotide cytidine monophosphate (i.e. CMP-sialic acid).

CMP-sialic acid is detected as free sialic acid in the TBA assay (10). This question was investigated in preliminary experiments by comparing the sensitivity of the free sialic acid in Wga^{RII} and Wga^{RVIII} lipid extracts to borohydride. Kean and Roseman (10) have shown that borohydride treatment will destroy free sialic acid while leaving CMP-sialic acid intact and therefore capable of reacting normally with the TBA reagent. In fact, when the aqueous preparations of free sialic acid from Wga^{RII} and Wga^{RVIII} cell extracts were treated with borohydride, all reactivity with the TBA reagent was abolished (data not shown). Thus it would appear that very little (if any) of the TBA positive material in these preparations is in the form of CMP-sialic acid.

Discussion

Wheat germ agglutinin (WGA) may be used to select at least four distinct mutations in CHO cells, three of which exhibit different alterations in glycolipid metabolism. In this paper we have shown that independent isolates of the previously described mutants Wga^{RI}, Wga^{RII} and Wga^{RIII} exhibit identical glycolipid patterns to other members of their respective complementation groups. In addition we have described a new Wga^{R} mutant (Wga^{RVIII}) which exhibits yet another glycolipid pattern. Wga^{RVIII} cells synthesize reduced amounts of GM$_3$ (the major glycolipid in CHO cells) and do not synthesize increased amounts of precursor molecules such as LC or GC. This mutant also exhibits marked alterations in resistance to a variety of lectins (Stanley, manuscript in preparation) suggestive of extensive structural alterations in the carbohydrate moieties of surface glycoproteins (see 14). Structural studies of the glycoproteins synthesized by Wga^{RVIII} cells are in progress.

Two Wga^{R} mutants selected from CHO cells have been partially characterized by Briles et al. (12). One of the Wga^{R} cell lines described by these authors is designated clone 1021 and it exhibits many properties similar to those of Wga^{RII} cells. The other mutant (clone 13) possesses certain properties similar to Wga^{RVIII} cells (for example, a very high degree of resistance to

WGA) but clone 13 cells appear to be blocked in the synthesis of GM$_3$ at a step beyond the synthesis of GC, in contrast to Wga^{RVIII} cells, which do not appear to synthesize increased amounts of GC. Clearly, more biochemical and genetic studies are required to determine whether these Wga^R cell lines arise from identical or different mutations. Suffice to say at present that the partial characterization of these mutants has revealed the complexity of their respective phenotypes. Since each of the mutants Wga^{RI}, Wga^{RII}, Wga^{RIII} and Wga^{RVIII} may be isolated in a single step selection (1; Stanley, manuscript in preparation), it is likely that they all arise from single mutational events. Thus these mutants should prove invaluable in defining the biosynthetic links between glycoprotein and glycolipid metabolism in animal cells.

Acknowledgements

The author wishes to thank Susan Kroener, Grace Vevona and Lisa Youkeles for excellent technical assistance. This work was supported by National Science Foundation grant No. PCM76-84293. The author is the recipient of an American Cancer Society Junior Faculty Award.

Literature Cited

1. Stanley, P.; Caillibot, V.; Siminovitch, L. Cell, 1975, 6, 121.
2. Stanley, P.; Siminovitch, L. Somatic Cell Genetics, 1977, 3, 391.
3. Stanley, P. "Surface Carbohydrate Alterations of Mutant Mammalian Cells Selected for Resistance to Plant Lectins". Lennarz, W.J. Ed. In "Biochemistry of Proteoglycans and Glycoproteins", Plenum Publishing Co., New York, (in press).
4. Robertson, M.A.; Etchison, J.R.; Robertson, J.S.; Summers, D.F.; Stanley, P. Cell, 1978, 13, 515.
5. Stanley, P.; Sudo, T.; Carver, J.P. J. Cell Biol. (in press).
6. Siakotos, A.N.; Rouser, G. J. Am. Oil Chem. Soc., 1965, 42, 913.
7. Svennerholm, L. Biochim. Biophys. Acta., 1957, 24, 604.
8. Aminoff, D. Biochem., 1961, 81, 384.
9. Yogeeswaran, G.; Stein, B.S.; Sebastian, H. Cancer Research, 1978, 38, 1336.
10. Kean, E.L.; Roseman, S. J. Biol. Chem., 1966, 241, 5643.
11. Yogeeswaran, G.; Murray, R.K.; Wright, J.A. B.B.R.C., 1974, 56, 1010.
12. Briles, E.B.; Li, E.; Kornfeld, S. J. Biol. Chem., 1977, 252, (3), 1107.
13. Li, E.; Kornfeld, S. J. Biol. Chem., 1978, 253, 6426.
14. Stanley, P.; Carver, J.P. Adv. in Exptl. Med. and Biol., 1977, 84, 265.
15. Yu, R.K.; Ledeen, R.W. J. of Lipid Research, 1970, 11, 506.

RECEIVED January 2, 1980.

Induction of Ganglioside Biosynthesis in Cultured Cells by Butyric Acid

PETER H. FISHMAN and RICHARD C. HENNEBERRY

National Institute of Neurological and Communicative Disorders and Stroke, The National Institutes of Health, Bethesda, MD 20205

The pleiotropic biochemical changes induced in mammalian cells in culture by butyric acid are many, varied and well documented (1). Although the initial reports on this subject appeared only six years ago, the number of published papers in this area has increased rapidly and is now approaching a hundred. The first reported observations of an effect of this fatty acid on cultured cells involved morphological changes. Ginsburg et al (2) noticed striking alterations in the shape of several lines of cultured cells including HeLa after exposure to butyrate. Independently, Wright (3) reported that butyrate caused morphological changes in Chinese hamster ovary (CHO) cells. Contrary to popular assumption, in neither of these studies was butyrate being used as a control for butyrylated derivatives of cyclic nucleotides.

In 1974, our laboratories reported that the activity of CMP-sialic acid:lactosylceramide sialyltransferase and amount of its biosynthetic product ganglioside GM3 (N-acetylneuraminylgalacto-sylglucosylceramide) increased dramatically in butyrate-treated HeLa cells (4). More recently, we have found that the ganglio-side GM1 (galactosyl-N-acetyl-galactosaminyl-[N-acetylneuraminyl]-galactosylglucosylceramide) is also increased in HeLa cells exposed to butyrate (5). GM1 has been implicated as the cell sur-face receptor for cholera toxin (6,7). In this article, we will review the effects of butyrate on cell morphology and ganglioside synthesis and provide conclusive evidence that GM1 is the cholera toxin receptor. In addition, we will describe some novel effects of cycloheximide on the turnover of membrane gangliosides.

Effects of Butyrate on Cell Morphology

In HeLa cells, the striking morphological alterations which follow exposure of the cells to butyrate are characterized by the extension of neurite-like processes (Fig. 1). No significant differences in the fine structure of the cell surface was observed by scanning electron micrography (Fig. 1). In addition to butyrate, propionate and pentanoate but not other homologous

fatty acids altered HeLa morphology (2,8). Cyclic AMP or its butyrylated derivatives did not induce shape changes in HeLa (2)
although they did effect the morphology of other cells such as
CHO (9). HeLa cells responded to butyrate in serum-free medium
which by itself had no effect on cell shape (8). In contrast,
neuroblastoma cells extended long processes when deprived of
serum (10) but also developed these neurites when exposed to
butyrate (11).

The formation of the neurite-like processes appears to be
dependent on assembly of microtubules as colchicine and Colcemid,
antimicrotubule drugs, prevented shape changes in the presence of
butyrate (2,8). The amount of tubulin per cell did not change
when HeLa were treated with butyrate (R.C.Henneberry, unpublished
observations). The role of microtubule assembly was further
explored with a calcium ionophore which alters intracellular
calcium levels and thus promotes microtubule depolymerization.
The ionophore both prevented and reversed butyrate-mediated process formation in HeLa (12).

Morphological changes were prevented in butyrate-treated HeLa
cells by actinomycin D and cycloheximide (2,8,13). After removal
of butyrate, the cells reverted to a normal morphology over a
24 h time course (2,8,12,13). When butyrate-treated cells were
detached from the culture dishes with trypsin, they assumed a
spherical shape; and, when replated in the absence of butyrate,
their neurite-like processes transiently re-extended (13). This
re-extension was blocked when cycloheximide but not the calcium
ionophore was included during the initial exposure of the cells
to butyrate (13). Process formation, however, did resume in the
presence of cycloheximide (13). These results were interpreted
as indicating that the fatty acid induces a protein(s) required
for process formation which can accumulate in the absence of processing and promote processing in the absence of inducer (13).

Induction of GM3 Biosynthesis by Butyrate

When HeLa cells were cultured in medium supplemented with 5 mM
sodium butyrate, their content of GM3 increased (Fig.2a). Increases
varied from 3.5 to 5-fold depending on the experiment (4,8,12,13).
When the butyrate was removed and the cells were cultured in normal
medium for 24 h, the GM3 levels returned to those found in untreated cells (Fig. 2a). Similar results were observed when N-[acetyl-
^{3}H]-D-mannosamine, a precursor of sialic acid, was also included
in the culture medium. In the butyrate-treated cells, radioactivity associated with GM3 increased 6.5-fold: and 24 h after butyrate was removed, the amount of labeling returned to control
values (Fig. 2b). We also were able to label the GM3 by means of
a cell surface labeling technique. Control and butyrate-treated
cells were exposed to 10 mM sodium periodate and the oxidized
sialyl residues were reduced with NaB^{3}H$_4$. There was 5.5-fold
more ^{3}H associated with the GM3 recovered from the butyrate-

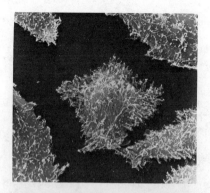

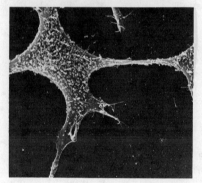

Figure 1. Scanning electron micrographs of untreated (top) and butyrate-treated HeLa cells (bottom). Photographs taken by Saleem Jahangeer at George Washington University (×1280).

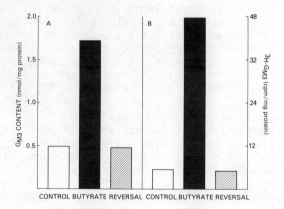

Figure 2.   Effect of butyrate treatment on GM3 content of HeLa cells

HeLa cells were cultured in medium containing N[acetyl-^{3}H]-D-mannosamine (50 μCi/mL) for 24 hr with (solid bars) and without (open bars) 5 mM sodium butyrate. In addition, butyrate-treated cells were cultured an additional 24 hr in fresh medium (without label and butyrate) (hatched bars). The cells were harvested and analyzed for GM3 content and radioactivity. (Data from Refs. 8, 13.)

treated cells than that from the control cells ($\underline{1}$). Butyrate treatment had no significant effect on any of the other major glycosphingolipids found in HeLa cells ($\underline{8}$; however, see below).

Exposure of HeLa cells to butyrate had no effect on the activity of GM3-sialidase when GM3 specifically labeled in the sialic acid residue was used as substrate (Fig. 3a). We were unable to detect any "ecto"-sialidase activity in either control or butyrate-treated cells ($\underline{14}$) although others have postulated that such an enzyme is important in regulating plasma membrane gangliosides ($\underline{15},\underline{16}$). In contrast, the activity of the specific sialyltransferase involved in GM3 biosynthesis increased over 20-fold following butyrate treatment (Fig. 3b). The effect was specific as activities of the other glycosphingolipid transferases that could be measured in HeLa cells were not altered in butyrate-treated cells ($\underline{4},\underline{8},\underline{17}$).

Increased sialyltransferase activity was dose and time dependent, and reversible ($\underline{8}$). Maximal activity was obtained by exposing the cells to 5 mM butyrate for 24 h. Following removal of butyrate, the enzyme had a half-life of 7 h and activity reached control levels by 24 h. Of the numerous short chain fatty acids and derivatives tested, only butyrate, pentanoate and propionate were effective ($\underline{8}$).

Butyrate appears to induce sialyltransferase activity as addition of actinomycin D or cycloheximide to the medium along with butyrate blocked the increase in activity ($\underline{4},\underline{8}$). Specific cell cycle inhibitors such as thymidine and colcemid did not cause an increase in activity in control cells or prevent induction in butyrate-treated cells ($\underline{8}$). Induction of sialyltransferase activity also occurred in serum-free medium ($\underline{8}$). When homogenates of control and butyrate-treated cells were admixed and assayed for sialyltransferase activity, there was no evidence of an inhibitor in the former or activator in the latter cells ($\underline{8}$).

In recent work on CHO cells, it had been suggested that the effects of butyrate are mediated by cyclic AMP ($\underline{18}$). We found, however, that cyclic AMP (2 mM), its mono- (1 mM) and dibutyryl (0.5 mM) derivatives, theophylline and prostaglandins did not cause an elevation in sialyltransferase activity ($\underline{1}$). Choleragen, which is a potent and persistant activator of adenylate cyclase (see below), also did not elevate sialyltransferase activity in HeLa cells (Table I) or alter cell morphology (unpublished observations). Thus, it is unlikely that these effects of butyrate are mediated by elevation of cyclic AMP levels.

Increased GM3 content was also observed in another strain of HeLa exposed to butyrate but not in butyrate-treated normal human fibroblasts (experiments in collaboration with E. Stanbridge, University of California at Irvine and R. O. Brady, NINCDS). Butyrate appeared to have similar effects on GM3 biosynthesis in KB cells, another human carcinoma-derived cell line ($\underline{20}$). Butyrate-treated KB cells had 9-fold elevated levels of sialyltransferase activity. In contrast, butyrate as well as dibutyryl-

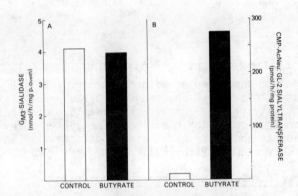

*Figure 3. Effect of butyrate treatment on GM3-sialidase and sialyltransferase
activities of HeLa cells*

HeLa cells were cultured for 24 hr with (open bars) and without (closed bars) 5 mM
sodium butyrate, harvested, and assayed for indicated enzyme activity. A: GM3-sialidase
activity assayed with GM3 specifically labeled with [^{14}C]-N-acetylneuraminic acid as substrate (14)). B: Sialyltrasferase activity assayed with CMP-[^{14}C]-N-acetylneuraminic acid
and lactosylceramide as substrates, and synthesis of labeled GM3 determined (17). Data
from Refs. 14, 17).

cyclic AMP had no (12) or only a small effect (21) on GM_3 content and sialyltransferase activity in CHO cells. Although the latter cells respond morphologically to both agents, they already have high levels of GM_3 and sialyltransferase activity comparable to levels found in butyrate-treated HeLa cells (12,21). These results suggest that only transformed cells with a low GM_3 content will exhibit an increase in GM_3 synthesis in response to butyrate.

Table I. Effect of Choleragen and Sodium Butyrate on Induction of Sialyltransferase Activity in HeLa Cells[a]

Experiment Number	Treatment Time (h)	Sialyltransferase Activity		
		Control	Butyrate	Choleragen
			(pmol/h/mg protein)	
1	4	0.7	4.8	0.9
2	8	- -	91.6	5.6
3	24	12	225	6.5

[a]In each experiment, HeLa cells were cultured for the indicated time in medium supplemented with no further additions, 5 mM sodium butyrate or 12 nM choleragen and assayed for sialyltransferase activity (17). Separate experiments demonstrated that this concentration of toxin maximally activated adenylate cyclase by 2 h whereas adenylate cyclase was not activated in cells exposed to butyrate for 24 h (19).

Induction of Choleragen Receptors by Butyrate

HeLa cells exposed to sodium butyrate bound more ^{125}I-choleragen than untreated cells (5). The increase in toxin receptors was time dependent; maximal levels were observed by 48 h (Fig. 4a). When butyrate was removed from the culture medium, the cells slowly lost their toxin receptors with a half-life of 32 h (Fig. 4b). The increase in choleragen binding depended on the butyrate concentration in the culture medium; maximal binding was observed with 5 mM butyrate (Fig. 4c). The increased binding of choleragen to butyrate-treated HeLa cells was due to an increase in receptors and not a change in affinity (Fig. 5). When ^{125}I-choleragen concentrations were varied over a 100-fold range and binding was plotted as percent saturation, the binding curves for control and treated cells were superimposable (Fig.5a). Half-saturation was observed at 6×10^{-10} M for both. Furthermore, unlabeled toxin was equally effective in inhibiting the binding of iodotoxin to control and treated cells (Fig. 5b); 50% inhibition occurred at 1.5×10^{-9} M for both. Of a number of fatty acids tested, butyrate and pentanoate were the most effective in inducing new choleragen receptors with smaller increases occurring in the presence of hexanoate and propionate (Table II).

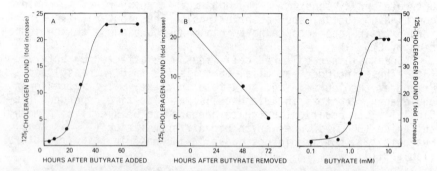

Figure 4.   *Effect of butyrate treatment on choleragen binding to HeLa cells*

A: Cells exposed to medium containing 5mM sodium butyrate for times indicated. B: After 48 hr, as in A, medium replaced with fresh medium without butyrate. C: Cells exposed for 48 hr to medium containing the indicated concentrations of butyrate. Specific binding of ^{125}I-choleragen determined as described in Ref. 5. (Data from Ref. 5.)

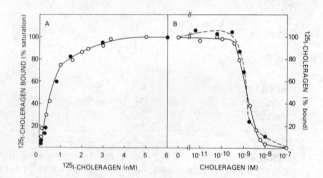

Figure 5.   *Effect of labeled and unlabeled choleragen concentrations on ^{125}I-choleragen binding to control and butyrate-treated HeLa cells*

HeLa cells cultured for 48 hr in medium with (○) and without (●) 5 mM sodium butyrate, washed, harvested, and assayed for specific ^{125}I-choleragen binding (5). A: effect of ^{125}I-choleragen concentration on iodotoxin binding. B: inhibition of ^{125}I-choleragen binding by increasing concentrations of unlabeled choleragen.

The induction of toxin receptors by butyrate also occurred in serum-free medium (5). Serum contains gangliosides including GM1 and GM1-deficient cells can absorb GM1 from serum and become responsive to choleragen (22). Cells exposed simultaneously to butyrate and cycloheximide did not exhibit an increase in choleragen binding (unpublished observations). Thus, butyrate appears to induce toxin receptors de novo in a manner analogous to the induction of GM3.

Several other cultured cell lines were affected by butyrate (5). These included rat glial C6 cells (12-fold increase) and Friend erythroleukemic cells (4-fold increase). The increase in choleragen receptors in Friend cells was also time dependent (Table III). In addition, butyrate appeared to be specific; dimethylsulfoxide (DMSO) induces erythroid differentiation (23) as does butyrate (24) but it did not cause an increase in toxin receptors (Table III).

Table II. Binding of Choleragen to HeLa Cells
Treated with Various Fatty Acids[a]

Fatty Acid (5 mM)	^{125}I-Choleragen Bound
	(fmol/mg protein)
None	17.3
Acetate	18.1
Propionate	67.8
Butyrate	1056
Pentanoate	631
Hexanoate	150
Isobutyrate	22.4

[a]Data from Fishman & Atikkan (5). HeLa cells were cultured for 48 h with the indicated fatty acid (as the sodium salt) and assayed for specific ^{125}I-choleragen binding.

Evidence that GM1 is the Choleragen Receptor

Previous studies have implicated GM1 as the cell surface receptor for choleragen. GM1 can precipitate the toxin, block its binding to cells and membranes and inhibit its biological effects (25-28). Cells and membranes treated with GM1 bind more iodotoxin and exhibit an enhanced response to the toxin (29-32). There is a correlation between GM1 content and choleragen binding and action in cultured mouse cells (33) and intestinal cells from different species (34). A line of transformed mouse fibroblasts that lacks GM1 and is unresponsive to choleragen can take up exogenous

GM1 from the culture medium. The GM1-treated cells now responded to the toxin (35) whereas cells that took up other gangliosides did not (36). Choleragen specifically protected the newly in-corporated GM1 from oxidation by galactose oxidase and sodium periodate (37); endogenous GM1 on cultured human fibroblasts was also protected by the toxin (37). Finally, choleragen interacted in a highly specific manner with liposomes containing GM1 but not other related glycosphingolipids (38-41).

Table III. Binding of Choleragen to Friend Erythroleukemic Cells Treated with Sodium Butyrate or Dimethylsulfoxide[a]

Treatment	^{125}I-Choleragen Bound		
	24 h	48 h	96 h
		(pmol/mg protein)	
None	3.89	3.89 (3.15)	2.29
2 mM Butyrate	9.74	14.2 (12.6)	- (10.7)
100 mM DMSO	4.22	3.97 (4.72)	2.09(1.87)

[a]Friend erythroleukemic cells were grown in suspension culture in Ham's F-12 medium containing 10% fetal calf serum as indicated and assayed for specific ^{125}I-choleragen binding as described elsewhere (5). Values are the mean of triplicate determinations; values in parenthesis are from a separate experiment.

If GM1 is the choleragen receptor, then butyrate-treated cells should have an increase in GM1 content. This is demonstrated in Table IV. Although GM1 could not be detected in control HeLa cells, they would contain less than 1 pmol per mg protein based on the limits of the sensitivity of the analytical procedure (5). GM1 was quantitated in the butyrate-treated cells (28.5 pmol per mg protein) and this increase is similar to the 32-fold increase in toxin binding observed in cells from the same experiment. The delipidated residue contained less than 1% of the toxin binding found in intact cells (Table IV). In addition, removal of the cells from the culture dishes with trypsin as opposed to the mechanical scraping routinely used had no effect on ^{125}I-choleragen binding to either control or butyrate-treated cells (5).

Finally, the ganglioside fraction isolated from butyrate-treated cells inhibited ^{125}I-choleragen binding to cells and membranes (Table V); and, when incorporated into liposomes, the liposomes bound iodotoxin (Table VI). Although GM3 is the major ganglioside in these cells, GM3 was a 1000-fold less effective inhibitor than GM1 and, when incorporated into liposomes, did not bind iodotoxin to any significant extent. From the data in Tables V and VI, we calculated that butyrate-treated HeLa cells which bound 1.1 and 2.6 pmol of toxin contained 15 and 40 pmol of GM1 per mg protein, respectively. Both of these estimates are in good agreement with the chemical analyses presented in Table IV.

Table IV. Effect of Sodium Butyrate on Choleragen
Receptors and GM1 Content of HeLa Cells[a]

Treatment	^{125}I-Choleragen Bound		GM1
	Intact Cells	Delipidated Residue	Content
	(fmol/mg protein)		(pmol/mg protein)
None	54.8	0.29	1
Butyrate	1770	0.45	28.5±3.73

[a]Data from Fishman & Atikkan (5). In a separate experiment,
butyrate-treated cells that bound 1575 fmol of toxin per mg protein contained 22 pmol of GM1 per mg protein.

Table V. Effect of Gangliosides on Binding of
Choleragen to Cells[a]

Ganglioside	Concentration	^{125}I-Choleragen Bound
	(nM)	(% inhibition)
GM1	60	87.8
	30	77.6
	10	51.2
	6	13.7
GM3	5000	14.2
	500	8.7
Gangliosides from butyrate-treated HeLa Cells		
	30 μl	66.1
	10 μl	36.2

[a]Butyrate-treated HeLa cells were incubated with 5 x 10-10 M [125]I-
choleragen in 0.2 ml of Tris-buffered saline containing 0.1% bovine serum albumin and the indicated inhibitor for 30 min and
assayed for bound iodotoxin (5). Binding was corrected for non-
specific binding as determined in the presence of 200 nM unlabeled
toxin. Gangliosides were isolated from butyrate-treated HeLa
cells (12.3 mg protein) which bound 1.1 pmol of toxin/mg protein
and dissolved in 1 ml of buffer. Separate experiments with [3]H-
GM1 indicated that 81.3-86.3% was recovered by the isolation
procedure.

The above results indicate that the increased choleragen
binding to butyrate-treated HeLa cells is associated with increased GM1 content. This was confirmed with Friend erythroleukemic cells (Table VII). Untreated Friend cells have measurable

amounts of GM1 and, following butyrate-treatment, the amount of GM1 increased 4-fold, which is identical to the increase in choleragen receptors. We believe that this is convincing evidence that the choleragen receptor is indeed GM1. As final proof, we conducted the following experiment.

HeLa cells were treated with 5 mM butyrate for 48 h or with 1 μM GM1 for 1 h. Both types of treated cells bound similar amounts of ^{125}I-choleragen and about 60-fold more than did control cells (Fig. 6a). When the cells were exposed to a saturating dose of toxin, the time course and extent of cyclic AMP accumulation were similar for both the butyrate-treated and the GM1-treated HeLa cells and more rapid than in the control cells (Fig. 6b). Thus, butyrate induces choleragen receptors in HeLa cells that are functionally equivalent to those created by incorporation of exogenous GM1.

Table VI. Binding of Choleragen to Liposomes Containing Gangliosides[a]

Liposomes Incubated With	^{125}I-Choleragen Bound (cpm)
125 pmol GM3	1,622
1.25 pmol GM1	66,615
gangliosides from butyrate-treated HeLa cells (23.7 μg protein)	49,911

[a]Liposomes prepared as described previously (41) were incubated with the indicated gangliosides dissolved in buffer for 24 h at 25°C. The suspensions were centrifuged at 10^5g for 10 min and the liposomes were suspended in buffer (41). Portions were assayed for specific binding of iodotoxin (41). Gangliosides were isolated from butyrate-treated HeLa which bound 2.59 pmol of toxin per mg protein. Separate experiments with labeled gangliosides indicated that ~20% of the gangliosides was taken up by the liposomes under these conditions.

Table VII. Effect of Sodium Butyrate on Choleragen Receptors and GM1 Content of Friend Erythroleukemic Cells

Treatment	^{125}I-Choleragen Bound	GM1 Content
	(pmol/mg protein)	
None	3.15	92
Butyrate	12.6	358

Data from Fishman & Atikkan (5).

Effects of Cycloheximide

Reversion of HeLa cells to normal morphology after removal of the butyrate was preceded by a decay of sialyltransferase activity

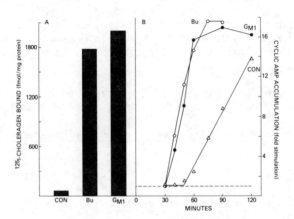

*Figure 6.   Comparison of butyrate and GM1 treatment of HeLa cells on choleragen
binding and activity*

*HeLa cells cultured for 48 hr in medium containing no additions (CON, △) or 5 mM
sodium butyrate (Bu, ○) in 35-mm wells. Medium removed and replaced with medium
199 buffered with 25 mM HEPES; 1 μM GM1 (GM1, ●) added to some of the wells
containing control cells. After 1 hr at 37°C, medium removed, and all cells washed
several times with phosphate-buffered saline. A: Cells incubated for 30 min in medium
199 containing 25 mM HEPES, 0.1% bovine serum albumin, and 10 nM ^{125}I-choleragen,
washed three times with cold PBS, and assayed for bound iodotoxin and protein; values
are mean of triplicate determinations and are corrected for non-specific binding as measured in the presence of 1 μM unlabeled choleragen. B: Cells incubated for the indicated
times in medium 199 containing 25 mM HEPES, 0.01% BSA, 0.5 mM methylisobutyl-
xanthine, and 10 nM choleragen and assayed for intracellular cyclic AMP (19); values are
mean of triplicate determinations and represent the fold increase in cyclic AMP levels
over cells not exposed to choleragen.*

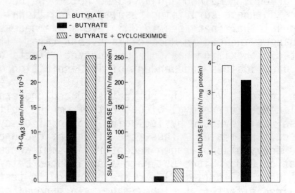

Figure 7.   Effect of butyrate and cycloheximide on GM3 metabolism in HeLa cells (data from Ref. 13)

A: HeLa cells cultured in medium containing 5 mM sodium butyrate and 50 µCi/mL N-[acetyl-³H]-D-mannosamine; after 24 hr, one set of cells harvested and medium replaced with fresh medium with and without 0.5 µg/mL cycloheximide on the other two sets, which were harvested after another 19 hr; cells then analyzed for radioactive GM3. B: Essentially same as A except the cells were cultured in unlabeled medium and assayed for sialyltransferase activity. C: same as B except the cells were assayed for GM3-sialidase activity.

236   CELL SURFACE GLYCOLIPIDS

and a decrease of GM3 content to basal levels (8,13). Morphological reversion was not blocked by puromycin but was blocked by
cycloheximide at drug concentrations that inhibited protein synthesis by about 50% (13). With puromycin, GM3 levels returned to
basal (13) whereas they remained elevated in the presence of cycloheximide (13). In addition, the specific radioactivity of the
GM3 labeled by culturing the cells' in medium containing N-[acetyl-
^{3}H]-D-mannosamine declined when the butyrate was removed but remained unchanged in the presence of 0.5 μg/ml of cycloheximide
(Fig. 7a). Sialyltransferase activity, however, did decay (Fig.
7b) and the drug had no effect on sialidase activity (Fig. 7c).
Thus, in the absence of inducer and in the presence of cycloheximide, the induced enzyme activity disappeared from the cell; but,
the induced product remained and appeared not to be degraded. We
have interpreted these surprising results as additional support
for our proposal that increases in GM3 levels are necessary for
the shape changes induced in HeLa cells by butyrate. It also
appears that cycloheximide has some unexpected effects on the
turnover of gangliosides independent of its inhibition of protein
synthesis.

As shown in Table VIII, the butyrate-induced choleragen receptors declined after removal of butyrate from the culture medium;
this decrease was prevented by including 0.5 μg/ml of cycloheximide in the culture medium. Thus, the effect of cycloheximide on
blocking the decay of choleragen receptors is similar to the
drug's effect on GM3. Since butyrate also induces beta-adrenergic
receptors in HeLa cells (19,42), we are planning to test whether
this unexplained effect of cycloheximide is a general one on plasma membrane components or is specific for gangliosides.

Table VIII. Effect of Cycloheximide on Butyrate-Induced
Choleragen Receptors In HeLa Cells[a]

| Additions to Medium | | Choleragen Receptors |
0-48 h	48-96 h	(fmol/mg protein)
None	None	37.8
Butyrate	Butyrate	1361
Butyrate	None	766
Butyrate	Cycloheximide	2224

[a]HeLa cells were cultured with and without 5 mM sodium butyrate
for 48 h. The medium was then replaced with fresh medium containing butyrate and cycloheximide (0.5 μg/ml) as indicated.
After an additional 48 h, the cells were washed, harvested and
assayed for specific ^{125}I-toxin binding as described elsewhere
(5). Values are the mean of triplicate determinations and there
was less than 10% variation amongst the three determinations.

Conclusions

Although there is a close association between induction of
ganglioside synthesis and morphological alterations in HeLa cells
by butyrate, the increase in ganglioside content by itself appears
not to be sufficient to elicit the changes in cell shape. Adding
GM3 to the culture medium did not alter the morphology of HeLa
cells (unpublished observations) or KB cells (20). An increase
in gangliosides may, however, be a necessary requirement as shape
changes were never observed in their absence. In addition, the
unusual ability of cycloheximide to maintain the cells in a morphological altered state in the absence of butyrate correlated
with a retention of induced levels of gangliosides. Undoubtedly,
additional components involved in the assembly of microtubules
must also be induced by butyrate.

Butyrate appears to have its most profound effects on neoplastic cells such as HeLa; in addition to morphological and biochemical differentiation, the fatty acid inhibits cell growth (2).
Previous studies have established a correlation between decreased
ganglioside synthesis and malignant transformation (43-46).
Transformed baby hamster kidney and newborn rat kidney cells exhibited a loss of GM3 and sialyltransferase activity (43,44).
Mouse cells transformed by various oncogenic agents had decreased
levels of GM1 and more complex gangliosides compared to their
normal counterparts (43,45). More recently, it has been reported
that transformed hamster cells exposed to butyrate lost many of
their transformed properties and acquired a more normal phenotype
(47). This reversal in phenotype was accompanied by the appearance of only a few new polypeptides in the butyrate-treated cells
(48).

Little is known about the function of gangliosides. Gangliosides may serve as cell surface receptors (45) and as biotrans-
ducers of membrane-mediated information (6). The ganglioside GM1
has been implicated as the receptor for choleragen. Our studies
clearly indicate that butyrate induces toxin receptors and GM1 in
parallel. In addition, the toxin receptors induced by butyrate
are functionally indistinguishable from exogenous GM1 that has
been absorbed by the cells. We believe that these observations
are the quintessential evidence that GM1 alone is the receptor
for choleragen.

LITERATURE CITED

1. Fishman, P.H.; Brady, R.O.; Henneberry, R.C.; Freese, E.
"Cell Surface Carbohydrate Chemistry"; Harmon, R., Ed.; Academic Press, Inc., New York, 1978; pp. 153-180.
2. Ginsburg, E.; Salomon, D.; Sreevalsan, T.; Freese, E. *Proc.*
Nat. Acad. Sci. USA, 1973, 70, 2457-2461.
3. Wright, J.A. *Exp. Cell Res.*, 1973, 78, 456-460.
4. Fishman, P.H.; Simmons, J.L.; Freese, E.; Brady, R.O. *Bio-*
chem. Biophys. Res. Commun., 1974, 59, 292-299.

5. Fishman, P.H.; Atikkan, E.A.   J. Biol. Chem., 1979, 254,
    4342-4344.
6. Brady, R.O.; Fishman, P.H.   Adv. Enzymol., 1979, 50, 303-323.
7. Svennerholm, L.; this volume.
8. Simmons, J.; Fishman, P.H.; Brady, R.O.; Freese, E. J. Cell
    Biol., 1975, 66, 414-424.
9. Hsie, A.W.; Puck, T.T. Proc. Nat. Acad. Sci. USA, 1971, 68,
    358-361.
10. Seeds, N.W.; Gilman, A.G.; Amano, T.; Nirenberg, M.W. Proc.
    Nat. Acad. Sci. USA, 1970, 66, 160-167.
11. Sheu, C.W.; Salomon, D.; Simmons, J.L.; Sreevalsan, T.;
    Freese, E. Antimicrob. Agents and Chemother., 1975, 7, 349-
    363.
12. Henneberry, R.C.; Fishman, P.H.; Freese, E. Cell, 1975, 5,
    1-9.
13. Henneberry, R.C.; Fishman, P.H. Exp. Cell Res., 1976, 103,
    55-62.
14. Tallman, J.F.; Henneberry, R.C.; Fishman, P.H. Arch. Bio-
    chem. Biophys., 1977, 182, 556-562.
15. Yogeeswaran, G.; Hakomori, S. Biochemistry, 1975, 14, 2151-
    2156.
16. Schengrund, C.L.; Rosenberg, A.; Repman, M.A. J. Cell Biol.,
    1976, 70, 555-561.
17. Fishman, P.H.; Bradley, R.M.; Henneberry, R.C. Arch. Biochem.
    Biophys., 1976, 172, 618-626.
18. Storrie, B.; Puck, T.T.; Wenger, L. J. Cell Physiol., 1978,
    94, 69-76.
19. Henneberry, R.C.; Smith, C.C.; Tallman, J.F. Nature, 1977,
    268, 252-254.
20. Macher, B.A.; Lockney, M.; Moskal, J.R.; Fung, Y.K.; Sweeley,
    C.C. Exp. Cell Res., 1978, 117, 95-102.
21. Briles, E.B.; Li, E.; Kornfeld, S. J. Biol. Chem., 1977, 252,
    1107-1116.
22. Fishman, P.H.; Bradley, R.M.; Moss, J.; Manganiello, V.C. J.
    Lipid Res., 1979, 19, 77-81.
23. Friend, C.; Scher, W.; Holland, J.G.; Sato, G. Proc. Nat.
    Acad. Sci. USA, 1971, 68, 378-382.
24. Leder, A.; Leder, P. Cell, 1975, 5, 319-322.
25. Cuatrecasas, P. Biochemistry, 1973, 12, 3547-3558.
26. Holmgren, J.; Lonnroth, I.; Svennerholm. L. Infect. Immun.,
    1973, 8, 208-214.
27. King, C.A.; van Heyningen, W.E. J. Infect. Dis., 1973, 127,
    639-647.
28. Staerk, J.; Ronneberger, H.J.; Wiegandt, H.; Ziegler, W.
    Eur. J. Biochem., 1974, 48, 103-110.
29. Cuatrecasas, P. Biochemistry, 1973, 12, 3558-3566.
30. Gill, D.M.; King, C.A. J. Biol. Chem., 1975, 250, 6424-6432.
31. King, C.A.; van Heyningen, W.E.; Gascoyne, N. J. Infect.
    Dis., 1976, 133, S75-81.

32. O'Keefe, E.; Cuatrecasas, P. J. Membrane Biol., 1978, 42, 61-79.
33. Hollenberg, M.D.; Fishman, P.H.; Bennett, V.; Cuatrecasas, P. Proc. Nat. Acad. Sci. USA, 1974, 71, 4224-4228.
34. Holmgren, J.; Lonnroth, I.; Mansson, J.E.; Svennerholm, L. Proc. Nat. Acad. Sci. USA, 1975, 72, 2520-2524.
35. Moss, J.; Fishman, P.H.; Manganiello, V.C.; Vaughan, M.; Brady, R.O. Proc. Nat. Acad. Sci. USA, 1976, 73, 1034-1037.
36. Fishman, P.H.; Moss, J.; Vaughan, M. J. Biol. Chem., 1976, 251, 4490-4494.
37. Moss, J.; Manganiello, V.C.; Fishman, P.H. Biochemistry, 1977, 16, 1876-1881.
38. Moss, J.; Fishman, P.H.; Richards, R.L.; Alving, C.R.; Brady, R.O. Proc. Nat. Acad. Sci. USA, 1976, 73, 3480-3483.
39. Moss, J.; Richards, R.L.; Alving, C.R.; Fishman, P.H. J. Biol. Chem., 1977, 252, 797-798.
40. Richards, R.L.; Moss, J.; Alving, C.R.; Fishman, P.H.; Brady, R.O. Proc. Nat. Acad, Sci. USA, 1979, 76, 1673-1676.
41. Fishman, P.H.; Moss, J.; Richards, R.L.; Brady, R.O.; Alving, C.R. Biochemistry, 1979, 18, 2562-2567.
42. Tallman, J.F.; Smith, C.C.; Henneberry, R.C. Proc. Nat. Acad. Sci. USA, 1977, 74, 873-877.
43. Fishman, P.H.; Brady, R.O. "Lipid Chemistry and Biochemistry"; Perkins, E.G.; Witting, L.A., Eds.; Academic Press, Inc., New York, 1975; pp. 105-126.
44. Hakomori, S. Biochim. Biophys. Acta, 1975, 417, 53-80.
45. Fishman, P.H.; Brady, R.O. Science, 1976, 194, 906-915.
46. Critchley, D.R.; Vicker, M.G. Cell Surf. Rev., 1977, 3, 308-370.
47. Leavitt, J.; Barrett, J.C.; Crawford, B.D.; Ts'o, P.O.P. Nature, 1978, 271, 262-265.
48. Leavitt, J.; Moyzis, R. J. Biol. Chem., 1978, 253, 2497-2500.

RECEIVED December 10, 1979.

Regulation of Glycoconjugate Metabolism in Normal and Transformed Cells

JOSEPH R. MOSKAL, MICHAEL W. LOCKNEY, CHRISTOPHER C. MARVEL, PEGGY A. MASON, and CHARLES C. SWEELEY

Department of Biochemistry, Michigan State University, East Lansing MI 48824

STEPHEN T. WARREN and JAMES E. TROSKO

Department of Pediatrics and Human Development, Michigan State University, East Lansing, MI 48824

Abstract. TPA and RA have significant effects on glycolipid and glycoprotein biosynthetic enzymes in several cultured cell systems. This suggests that these compounds as well as other "tumor promoters" will be useful in further studies on the regulation and control of glycoconjugate metabolism (metabolic perturbants). Butyrate, TPA and RA appear to exert their effects at different points in the cell cycle. These results could mean that tumor promotion, differentiation and virus infection occur at discrete points in the cell cycle. Membrane glycoconjugates may participate in these processes in a dynamic time-dependent way.

Introduction

A large body of literature has developed dealing with the chemistry and metabolism of glycosphingolipids and other glycoconjugates ($\underline{1}$, $\underline{2}$, $\underline{3}$). However, only recently have research efforts been addressed to the possible interrelationship between the regulation of glycoconjugate metabolism and the control of cell growth and transformation - whether it be the expression of various differentiated functions or transformation by viruses or carcinogens into a "tumorigenic state".

Bosmann and Winston ($\underline{4}$) were the first ones to examine the possible cell cycle dependence of glycolipid and glycoprotein synthesis. They concluded that glycolipid synthesis occurs almost exclusively in the G_2 and M phases while glycoprotein synthesis peaks during the S period. Wolfe and Robbins ($\underline{5}$), using radiolabeled palmitate and sugars followed by isolation and thin layer chromatographic characterization, found, however, that simple glycolipids (glucosylceramide, lactosylceramide and GM_3 ganglioside) were synthesized throughout the cell cycle in equal amounts, whereas triglycosylceramide and tetraglycosylceramide were labeled only in the G_1 and S phases. Forssman hapten was synthesized throughout the cell cycle but anti-Forssman antibody adhered to cells maximally in the G_1 and early S phases. They also reported that mitotic cells exposed all detectable antigens and that as cells moved through the cycle much of the Forssman antigen became cryptic.

0-8412-0556-6/80/47-128-241$5.75/0

At this time Chatterjee et al. (6) reported a maximal incorporation of galactose into the glycolipids of synchronized KB cells during late M and/or early G_1 phases. They also found a large increase in the levels of gangliosides and total neutral glycolipids during this period. Gahmberg and Hakomori (7), using the cell surface labeling technique of galactose oxidase:sodium borotritiide, reported that monoglycosylceramide through pentaglycosylceramide were labeled maximally during the G_1 phase and minimally during the S phase. Moreover, they found relatively constant amounts of these glycolipids throughout the cell cycle. It was concluded that while glycolipid synthesis occurs in the M-G_1 phase, as cells traverse the mitotic cycle, the exposure of glycolipids at the cell surface varies with the cell cycle. They suggest that a cell surface glycoprotein, "galactoprotein a" (found in confluent cultures of non-transformed cells but at lower levels in virally transformed cells), may be a key factor in the organization of membrane glycolipids and may explain why transformed cell glycolipids show a high rate of labeling throughout the cell cycle.

Using techniques such as galactose oxidase:sodium borotritiide reduction and lectin binding to study cell surface changes, Gahmberg and Hakomori (8) have observed that virus-transformed but not control hamster cells contained lacto-N-neotetraosylceramide on the cell surface, that this exposure was cell-cycle specific, and that normal and transformed cells interact with lectins via a different glycoprotein for each cell type. The authors concluded that the binding sites of the lectins were specific glycoproteins and that these interacting proteins are significantly different in normal compared to transformed cells (9).

Schnaar et al. (10) have synthesized polyacrylamide gels with covalently linked carbohydrates to study their interaction with cell membrane glycoconjugates. They report that chicken hepatocytes, in a temperature and calcium dependent way, interact strongly with only N-acetylglucosamine-linked polyacrylamide. Orosomucoid, minus sialic acid and galactose, was a potent inhibitor of this interaction. Lingwood et al. (11) have made monoclonal antibodies directed against various glycolipids and glycoproteins and have shown that when bound to temperature-sensitive virally transformed cells they could inhibit the expression of the "oncogenic" state at the permissive temperature. While these studies are not directly pertinent to the regulation of glycoconjugate metabolism, they clearly implicate glycoconjugates as playing a significant role in cell metabolism and growth control.

Since the first reports of in vitro glycosyltransferase activities (12, 13, 14) a number of papers have appeared which deal with elucidation of glycolipid biosynthetic pathways in several model systems (15-20). The earliest attempts to implicate glycolipid metabolism in cell transformation came from investigations of the effects of viral transformation. Studies by Brady and Mora (21), Grimes (22), Hakomori (23), Bosmann (24, 25) and Den et al. (26) have all shown significant differences in cellular glycolipid biosynthetic capability when comparing non-transformed with virally transformed cells. In general, irrespective of the cell line or the virus, a simplification of glycolipid patterns has been observed and often glycosyltransferase activities present in non-transformed cell lines were

absent in their transformed counterpart (27). Furthermore, it has been
suggested that this simplification of the oligosaccharide chain upon viral
transformation may have important implications with respect to a tumor
cell's ability to interact with molecules involved in growth control (28).

Studies dealing with cellular transformation to a non-proliferative
or "differentiated" state have been reported by a number of investigators. Yeung et al. (29) have shown a significant elevation in N-
acetylgalactosaminyltransferase activity in clones of adrenal tumor cells
after dibutyryl-cyclic adenosine monophosphate (dBcAMP) treatment.
Moskal et al. (30) and Basu et al. (31) have also reported significant
changes in glycolipid sialyltransferase and galactosyltransferase activities in cultured murine neuroblastoma cells after dBcAMP treatment, as
well as between spinner cultured cells and T-flask grown cells. Fishman
and coworkers have reported a dramatic morphological change and an
elevation in CMP-NeuAc: lactosylceramide sialyltransferase activity
after treatment of HeLa cells with butyric acid (32, 33, 34). They found
that these effects were dependent on protein synthesis and suggested
that the increased levels of II^3-α-N-acetylneuraminosyllactosylceramide
(GM_3) were necessary for expression of the morphological changes.
Macher et al. (35) also reported a significant elevation in this
sialyltransferase in human epithelial carcinoma cells (KB) and observed
changes in cell surface labeling patterns and galactosyltransferase (UDP-
Gal-lactosylceramide galactosyltransferase) activity (36), suggesting that
butyrate-treated cells could be used to study many aspects of glycoconjugate metabolism. Recently, Presper et al. (37) have reported two
fucosyltransferase activities from human (IMR-32) neuroblastoma cells.
They also found that 6-thioguanine but not bromodeoxyuridine-induced
differentiation caused a marked elevation in fucose containing glycoconjugates. On the other hand, Dawson et al. (38) have reported that
enkephalins caused a dose-dependent decrease in the incorporation of
radiolabeled glucosamine or galactose into glycolipids and glycoproteins
in cultured neuroblastoma cells. These investigators suggested that their
results may be interpreted in terms of a cyclic AMP mediated process
(5).

Recently, a class of compounds called "tumor promoters" have
emerged that, unlike "differentiating" agents such as butyrate or
dBcAMP, stimulate cell proliferation and tumorigenesis. Since tumor
promoters alter cell growth patterns in an opposing manner to differentiating agents, they may also alter glycoconjugate metabolism. If this
were the case, glycoconjugates could be involved in the regulation of cell
growth, and studies with tumor promoters might be a useful approach to
elucidate the mechanisms of regulation of glycoconjugate metabolism.
The following is a brief review of the "two-stage theory of carcinogenesis" in which tumor promoters play an integral part, and some of the
effects tumor promoters have been reported to have on cells.

Two-Stage Theory of Carcinogenesis. Historically, the induction of
skin tumors was accomplished by repeated applications of a potent
carcinogen (39). Berenblum (40), however, found that croton oil, when
administered together with a carcinogen, led to more tumors than the
carcinogen alone. Mottram (41) then reported that after multiple

applications of croton oil, only one treatment with a carcinogen was necessary to cause tumors. It was later shown that diesters of the diterpene alcohol, phorbol, are the active components of croton oil (42, 43, 44) (Figure 1 shows the general phorbol structure with a list of substituted phorbols and their carcinogenic efficacy). These findings led Boutwell and coworkers to establish the 2-stage protocol for tumor induction in mouse skin and a model system for the 2-stage theory (45, 46). Briefly, it was found that 1) a small single dose of a carcinogen caused no tumor formation in mouse skin (however, the mouse skin is said to be initiated), 2) multiple applications of tumor promotors alone caused no tumor formation, but 3) if a tumor promoter was applied to initiated mouse skin, tumors did arise. Furthermore, if the order of treatments was reversed no tumors were seen (47). Initiation appears to be permanent, since tumors formed when promoter was added as long as one year after initiation (48). More recently, O'Brien et al. (49) have reported that a single application of tumor promoter (croton oil or 12-O-tetradecanoyl-phorbol-13-acetate:TPA) caused a rapid stimulation of ornithine decarboxylase (ODC) activity (2-300-fold induction, reaching maximal levels 4-5 hours after treatment) in mouse skin. Verma and Boutwell (50) later reported that retinoic acid (RA) (the general structure is shown in Figure 2), when applied with TPA, could completely inhibit the formation of tumors.

At about this time O'Brien (51) and Boutwell (52) proposed the two-stage theory of carcinogensis. Stated simply, the induction of tumors requires first the "initiation" of a cell by a carcinogen. This process is irreversible and is believed to occur at the genetic level. Following initiation a tumor promoter must be introduced. Promotion is believed to be a reversible phenomenon accompanied by an induction in ODC activity as a key step. The model implicates ODC induction as an essential feature of tumor promotion, based on the following evidence: 1) the degree of induction of enzyme activity (ODC) correlates well with the promoting ability of various concentrations of TPA and other phorbol esters of varying promoter efficacy, 2) retinoic acid inhibits the ability of TPA to induce tumors and also inhibits ODC induction, and 3) tumors produced by TPA treatment have high levels of ODC activity, with malignant tumors possessing higher levels than benign tumors. O'Brien and Diamond (53) have recently reported a bioassay system, based on ODC induction by tumor promoters, to analyze the metabolism of the phorbol diester tumor promoters.

Research on the various biochemical systems affected by tumor promoters, in particular TPA, has been recently reviewed by Diamond et al. (54) and Werner and coworkers (55) have reviewed the early effects of phorbol esters on the membranes of cultured cells. The latter group reports that TPA causes permeability changes in 3T3 cell membranes and experimental evidence is cited that phorbol esters interact specifically with a membrane-specific macromolecule rather than passive adsorption by the membrane lipid matrix. One of the earliest observed effects of TPA is a significant modification in the transport of potassium, sodium and phosphate. Lee and Weinstein (56) have found that the addition of phorbol esters immediately stimulated the uptake of 2-deoxyglucose in

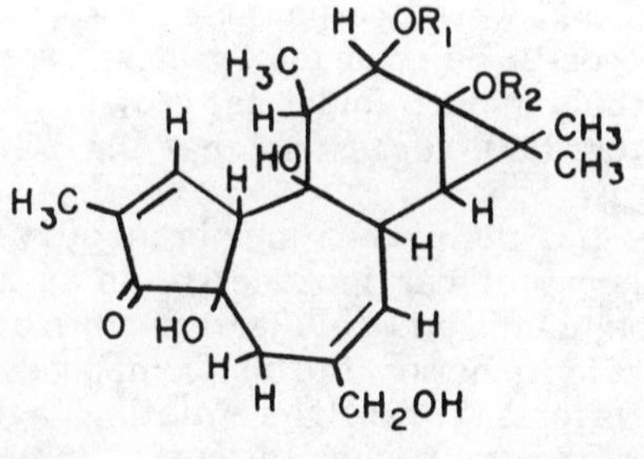

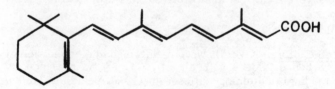

In vivo Tumor
Promoting Activity

$R_1 = R_2 = H$ —

$R_1 = CO(CH_2)_{12}CH_3; R_2 = COCH_3$ ++++

$R_1 = R_2 = CO(CH_2)_8CH_3$ +++

$R_1 = R_2 = COCH_3$ +

$R_1 = R_2 = COC_6H_5$ ++

Figure 1. General structure of the tumor-promoting component, phorbol, of croton oil. 12-O-tetradecanoyl-phorbol-13-acetate (TPA) is the most potent promoter of the phorbol diesters. Phorbol didecanoate and phorbol dibenzoate, among others, have promoting ability but to a lesser extent than TPA. Phorbol alone has been reported to have no capacity to induce tumors as ODC.

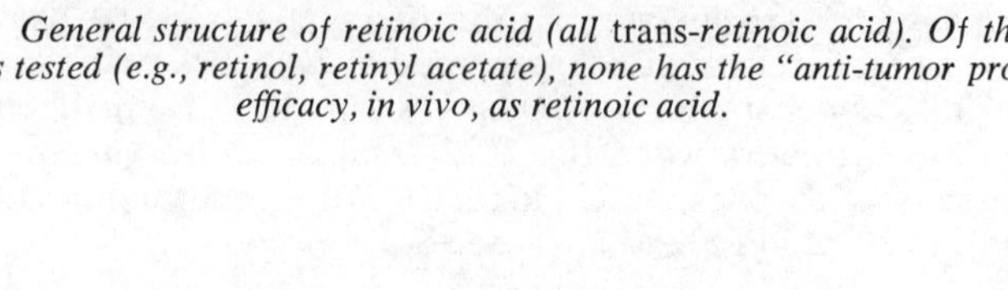

Figure 2. General structure of retinoic acid (all trans-retinoic acid). Of the many derivatives tested (e.g., retinol, retinyl acetate), none has the "anti-tumor promoter" efficacy, in vivo, as retinoic acid.

cultured cells. This enhancement peaked after 90 minutes, persisted as long as three hours and was temperature dependent. These results support the idea that transport mechanisms rather than effects on intracellular metabolism were responsible for these observations. Another interesting observation reported by these investigators was that tumor promoting phorbols inhibit epidermal growth factor (EGF) binding to cell surface receptors (57), suggesting that the EGF receptor may be the binding site for TPA.

Thus, it appears that the plasma membrane plays a very important part in the 2-stage theory of carcinogenesis. In an attempt to further elucidate how glycoconjugates and cellular transformation are linked, we began a study of the effects of some of the compounds involved in tumor promotion. Previous efforts with "differentiating" agents set the stage for similar studies with compounds directly implicated in cellular proliferation and tumorigenesis. In these studies, then, the effects of tumor promoters on cell morphology, glycoconjugate metabolism and composition, and cell cycling activity are reported, and the results are discussed in terms of how tumor promoters might affect cell metabolism.

Materials

All materials were obtained from the following sources: human epidermoid carcinoma (KB) cells from The American Type Culture Collection (Rockville, MD.); nontransformed (NIL 8) and virally-transformed (NIL 8HSV) cells were a gift from Dr. P.W. Robbins (Department of Biology and Center for Cancer Research, Massachusetts Institute of Technology); modified Eagle's medium, calf serum, trypsin, and fetuin from Grand Island Biological Company (Detroit, MI.); radiolabeled CMP-sialic acid, UDP-galactose, sialic acid and DL-ornithine from New England Nuclear (Boston, MA.); unlabeled UDP-galactose from Sigma (St. Louis, MO.). Lactosylceramide was purified from canine intestine. II^3-α-N-acetylneuraminosyl-lactosylceramide (GM_3) and mixed gangliosides were from Supelco, Inc., (Bellefonte, PA.) and high performance thin layer chromatography (HPTLC) plates (silica gel 60; without fluorescent indicator) were from EM laboratories (Elmsford, N.Y.).

Methods

Cell Culture. KB cells were maintained in a humidified atmosphere of 5% carbon dioxide - 95% air at 37°C in the presence of modified Eagle's medium containing calf serum (10%), penicillin (100 μg/ml) and streptomycin (100 units/ml). Cells were routinely subcultured with 0.25% trypsin and stocks were discarded after twenty passages. All drugs were administered with fresh media 24 hours after subculture in the following concentrations: TPA, 1.6 μM; RA, 1.6 μM; butyric acid, 2mM. Drug treatments were for 20-24 hours. Cells were harvested for enzyme assays with phosphate-buffered saline containing 0.05% EDTA and stored at -20°C in 0.32 M sucrose.

Glycolipid Sialyltranferase Assays. Complete incubation mixtures contained the following components (in micromoles), in a final volume of 0.05 ml: lactosylceramide, 0.05; Triton CF-54-Tween 80 (2:1 w/w), 200µg; Cacodylate-HCl buffer, pH 6.5, 10; magnesium chloride, 0.1; CMP-NeuAc (1.5 x 10^6 cpm/µmole), 0.05; and 0.2 to 0.6 mg of protein. Enzyme reactions were incubated for 60 min at 37°C, terminated by the addition of 0.6 µmole of EDTA (pH 7.0) and assayed by the double chromatographic procedure of Basu (16).

Glycoprotein Galactosyltransferase Assays. Complete incubation mixtures contained the following components (in micromoles, unless otherwise stated) in a final volume of 0.05 ml: MES buffer, pH 6.7, 12.5; manganese chloride, 0.5; 0.5% Triton X-100; desialized (mild acid hydrolysis) and degalactosylated (58) fetuin, 125 µg; UDP-galactose (specific activity, 8.8 x 10^6 cpm/µmole), 0.0025; protein, 1-50 µg. Incubations were carried out for 60 min at 37°C and were terminated by the addition of 5 ml of cold 5% phosphotungstic acid in 0.5 M HCl. Precipitates were collected on millipore filters (0.45 micron pore size), washed twice in the acid mixture, and dissolved in 1% SDS-0.1N NaOH. After neutralization with 1N HCl, samples were counted by liquid scintillation spectrometry using 10 ml of a Toluene Triton X-100 based liquid scintillation cocktail.

Ornithine Decarboxylase Assays. The double-chamber assay system of Moskal and Basu (59) was used to measure enzyme activity in the form of [^{14}C] carbon dioxide evolution. The assay conditions of O'Brien and Diamond (60) were used and consisted of the following components (in micromoles, unless otherwise stated) in a total volume of 100 µl: sodium phosphate buffer, pH 7.2, 5.0; EDTA, 1.0; dithiothreitol, 5.0; pyridoxal-5'-monophosphate, 0.2; L-ornithine (specific activity 0.5 x 10^6 cpm/µmole), 0.1 and protein, 0.1-0.5 mg. Incubations were carried out at 37°C for 60 min, and the reactions were terminated by the addition of 200 µl of 2M sodium citrate followed by a post-incubation period of 3 hours at 37°C to insure maximal release of radiolabeled carbon dioxide.

Scanning Electron Microscopy. KB cells were synchronized by the following procedure: 2mM thymidine was added for 20 hr followed by release for 8 hr and shaking for mitotic cells. After mitotic selection, cells were briefly suspended in and triturated with PBS-EDTA (0.05%) before transfer to flasks with fresh media. Synchronized or drug-treated cells were washed three times with PBS (calcium and magnesium free) and fixed for 30 min in 3% glutaraldehyde (EM grade, Polysciences, Warrington, PA.) in 50 mM cacodylate-HCl buffer, pH 7.2. The glutaraldehyde was removed by washing the cells three times with cacodylate buffer and dehydration with graded ethanol-water solutions was performed followed by several washes with absolute ethanol. Critical point drying of the samples was carried out in a Bomar SPC-900/EX apparatus (Bomar Co., Tacoma, WA.) using carbon dioxide as the carrier gas. Gold coating of the samples was done using a Film-Vac Mini Coater (Englewood, N.J.) to a density of 200 A. An ISI super III scanning electron microscope was used (International Scientific Instruments, Santa Clara, CA.).

Light Microscopy. Cells were photographed using an Olympus (model IMT) inverted microscope.

Thymidine Uptake Studies. Tritiated thymidine (52 mCi/μmole; 0.1 μCi/ml) was added to cells for 60 min at 37°C. Cells were then washed with PBS, incubated at 4°C for 15 min in the presence of ice cold 5% TCA, rinsed with TCA and scraped from the flasks with a rubber policeman. Cells were again washed with PBS, and solubilized in 0.1N NaOH overnight. Aliquots were assayed for protein (62) and radioactivity (scintillation fluid: 100 ml Biosolve (Beckman, Fullerton, CA.), 7g of PPO and 0.6 g of POPOP per liter of toluene).

Results and Discussion

Glycoconjugate Metabolism Studies. CMP-NeuAc: lactosylceramide sialyltransferase activity: This particular sialyltransferase was assayed because of the dramatic induction in activity reported by Fishman et al. (32) and Macher et al. (35) after treatment with butyrate, a drug with putative anti-tumor properties (63). Thus, gangliosides (in particular GM₂) may play an important role in growth control and differentiation and sialyltransferase could be a pivotal enzyme in the regulation of such processes. Table I gives the results of the incubation of KB cells in log- phase growth with butyrate, TPA or RA as described in Methods. Butyrate caused an approximately 5-fold increase in KB sialyltransferase, as expected. TPA treatment also resulted in a 5-fold elevation of sialyltransferase activity. The most dramatic elevation in enzyme activity, however, was seen when cells were treated with RA, in which case a 10 to 15 fold increase in activity was observed. We have also seen similar increases in sialyltransferase activity after RA treatment of both non-transformed and virally transformed hamster embryo cells (NIL). However, sialyltransferase activity in the non-transformed NIL cells was slightly decreased after TPA treatment (64).

Table I. The Effect of Various Compounds Implicated in the 2-Stage Theory of Carcinogenesis on CMP-NeuAc:lactosylceramide Sialyltrans- ferase Activity in Human Epithelial Carcinoma with (KB) Cells

CONDITION	SIALYLTRANSFERASE ACTIVITY* (pmoles/mg protein/hr)	% CONTROL
CONTROL	477	
BUTYRATE	2549	534
TPA	2605	546
RA	6726	1410

*Minus endogenous values CMP-NeuAc:lactosylceramide sialyltransferase

Glycoprotein Galactosyltransferase Assays. In order to further investigate the effects of TPA and RA on glycoconjugate metabolism, UDP-galactose:DSG-fetuin galactosyltransferase activity was assayed. Table II shows the results of this study. The effects of TPA and RA on

Table II. The Effect of Various Compounds Implicated in the 2-Stage Theory of Carcinogenesis on UDP-galactose:DSG-fetuin Galactosyltransferase Activity in Human Epithelial Carcinoma (KB) Cells

CONDITION	GALACTOSYLTRANSFERASE ACTIVITY* (nmoles/mg protein/hr)	% CONTROL
CONTROL	61	
TPA	95	155
RA	122	200

*Minus endogenous values UDP-Gal:DSG-fetuin galactosyltransferase

this enzyme, relative to controls, were quite similar to those with sialyltransferase in that TPA clearly led to an elevation in activity but RA gave almost twice as much activity as that observed with TPA. The magnitude of the changes in galactosyltransferase activity with TPA and RA is significantly lower than that seen with the sialyltransferase (Table I). It is possible that the differences observed with these in vitro assay systems, while indicative of total enzyme activity present, do not accurately reflect the activities in vivo. This could be due to compartmentalization of substrates, enzymes, or both in the cell. Alternatively, the alterations in sialyltransferase activities may actually be of greater magnitude than galactosyltransferase activities, suggesting a more marked change in glycolipid composition than that in the glycoprotein fraction. DeLuca and coworkers (65) have recently reported that retinal can be phosphorylated to retinyl phosphate in vivo and that 10% of the total mannolipid synthesized in rat liver is in the form of mannosylretinyl phosphate (the rest being in the form of dolicholmannosyl phosphate (66)). DeLuca et al. (67) also reported that in cultured epidermal cells the addition of exogenous retinyl acetate leads to a significant increase of both galactose and mannose in glycoconjugates. They further suggest that retinoids may act as carriers for different monosaccharides in different tissues. Perhaps retinoic acid facilitates glycoprotein glycosyltransferase activity, as suggested above. Nevertheless, the differences between the glycosyltransferases involved in glycoprotein and glycolipid biosynthesis is unclear at the present time, making it difficult to explain the changes in glycolipid sialyltransferase activity after RA treatment.

Based on data cited previously it appears that TPA's primary effect is at the cell surface (55, 56). This does not exclude the possibility that TPA is transported and metabolized to an "active" intermediate, which then affects glycosyltransferase activity. However, it is possible (particularly in light of the structural dissimilarity between TPA and RA) that these compounds affect glycoconjugate metabolism in different ways.

Morphological Studies. Scanning Electron Microscopy. We undertook an analysis of the morphological changes induced by TPA, RA, and butyrate to further examine the possibility that TPA and RA exert their effects on glycoconjugate metabolism by different mechanisms. We reasoned that differences in morphology produced by TPA and RA could be related to the changes in glycosyltransferase activities. In addition, we investigated the effects of TPA and RA on butyrate-induced morphological changes. Figure 3 shows a series of micrographs of synchronized KB cells at mitosis (3a), early G_1 (3b), late G_1-S (3c) and S-phase (3d). This experiment was performed in order to document the rich variety of morphologies possible in "control" cell populations. It can be seen that the smooth, rounded mitotic cells quickly attach, spread out and become covered with microvilli. By late G_1 the microvilli have already begun to disappear and, as cells enter the S-phase, the plasma membrane appears to be very smooth and spread out. Porter et al. (68) have previously reported such morphological changes for synchronized CHO cells. They also reported that non-transformed cells did not show the appearance of microvilli during the cell cycle and speculated that these membrane components may be involved in facilitating the transport of nutrients in rapidly growing cells.

Figure 4 shows micrographs of butyrate-treated KB cells (4a) and TPA-treated KB cells (4b). Retinoic acid treated cells (results not presented) show the relatively normal morphology of a middle to late G_1-phase cell (see Figure 3c). Figure 5 is a series of micrographs taken after treatment of KB cells with butyrate and TPA for 24 hr (5a), after pretreatment of cells with butyrate for 24 hr followed by TPA treatment for 24 hr (5b), and after pretreatment with TPA for 24 hr followed by butyrate treatment for 24 hr (5c). There is a dramatic difference between TPA- and butyrate-treated cells. Moreover, it appears that TPA was able to reverse butyrate-induced morphological effects but not the converse. When TPA and butyrate were added together a "mixed" morphology appeared. When similar experiments were performed with butyrate and RA, a more pronounced "butyrate-like" morphology was observed in all cases.

Although these results are difficult to interpret in molecular terms, they provide a basis for further research. The morphological changes brought about from these mixing experiments suggest that the TPA-induced increases in glycosyltransferase activity are the result of a different process than those induced by RA or butyrate (manuscript in preparation).

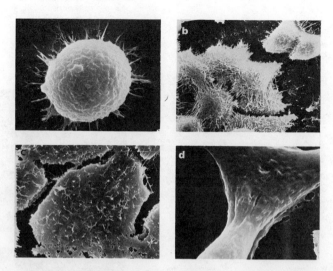

Figure 3.   Scanning electron micrographs of synchronized KB cells at various stages of cell cycle. Cells pretreated for 20 hr with 2 mM thymidine, released for 8 hr, and mitotically selected by shaking. (a): Mitotic cells ($\times$200); (b): early G_1 phase ($\times$800); (c): late G_1-early S phase ($\times$1200); (d): S phase ($\times$2000).

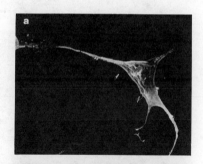

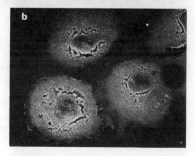

Figure 4.   Scanning electron micrographs of KB cells after butyrate treatment (a) ($\times$400) and after TPA treatment (b) ($\times$280). Cells prepared for microscopy simultaneously according to procedures described in text. The "membrane tearing" (Figure 5b) was consistently found only in cells treated with TPA and somewhat in synchronized, late G_1-early S phase cells.

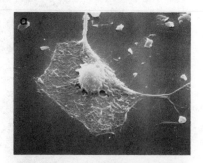

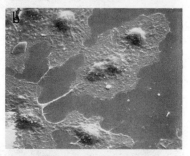

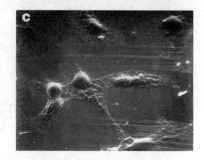

Figure 5.   Scanning electron micrographs of KB cells after treatment for 24 hr with butyrate and TPA (a) (×1000), butyrate for 24 hr followed by TPA for 24 hr (b) (×500), and TPA for 24 hr followed by butyrate (c) (×400).

Ornithine Decarboxylase Activity Versus Sialyltransferase Activity. Preliminary experiments were performed to implicate the glycoconjugates, functionally, in tumor promotion. As discussed earlier, the rate limiting enzyme in polyamine biosynthesis, ODC, plays an important role in the 2-stage theory of carcinogenesis as well as in the regulation of cell growth in general (69, 70). Furthermore, as shown in Figure 6, it was found that the addition of fresh serum-containing medium to growing (or confluent) KB cells resulted in a significant increase in ODC activity, reaching a maximum 4-5 hrs after treatment (O'Brien and Diamond (59) have reported that fresh serum added with TPA to hamster cells in culture led to a higher elevation of ODC than TPA alone). It was also found (Figure 7) that at least one of the factors involved in this serum-induced increase in ODC activity was a dializable, heat-labile component (manuscript in preparation). The role of serum factors and small molecular weight nutrients in the regulation of cell growth have been reported by Holley and coworkers (71, 72), Paul et al. (73), and Chen and Canellakis (74).

Fresh serum-containing medium was added to several cultured cell lines after 24 hr drug treatments and the changes in ODC activity measured. Table III compares the results of this experiment with the CMP-NeuAc:lactosylceramide sialyltransferase activity. This activity was determined after exposure of the cells to RA or TPA plus RA for 24 hr. The ability of fresh serum to induce ODC levels under similar conditions was determined by treating cells for 24 hr with the same drug combinations and then adding fresh serum-containing media. After 4 hr of serum stimulation, ODC activities were determined. Retinoic acid treatment of KB, NIL and NIL-HSV cells resulted in a significant elevation in serum-induced ODC activity. In each case, ODC activity reached a maximum value (at 4 hr) almost twice as great as that of controls. Sialyltransferse activity (as reported above) was also significantly higher. When NIL and NIL-HSV cells were treated with TPA and RA for 24 hr an even more dramatic increase in ODC by serum stimulation was observed. Likewise, sialyltransferase activities were significantly higher than those for RA-treated cells. The important point is that, in the few cases examined thus far, there was a correlation between elevated glycosyltransferase activity (and elevated GM_3 levels in the case of KB cells) and the "inducability" of ODC by serum.

Table III. Comparison of Serum-induced ODC Activity with CMP-NeuAc:
lactosylceramide Sialyltransferase Activity in Various Cell Lines

CONDITION	KB	NIL-8	NIL-HSV
CONTROL	6.77(477)	4.1(1116)	18.21(205)
RA	9.12(6726)	11.6(3800)	29.5(3175)
TPA + RA		179.52(7382)	194.22(3750)

Non-bracketed numbers: ODC activity; nmoles/mg protein/hr
Bracketed numbers: Sialyltransferase Activity; pmoles/mg protein/hr

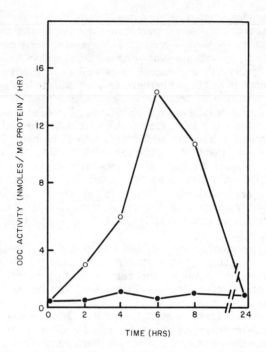

Figure 6. *Effect of fresh serum-containing media on ODC activity in logarithmically growing KB cells. At* t = 0 *fresh media added to one set of cultures with no addition to the other set. ODC activity then measured every 2 hr after media change.*

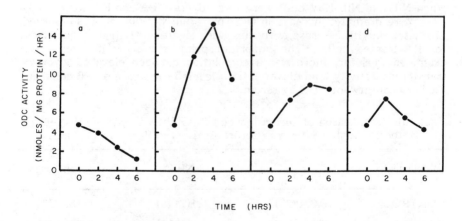

Figure 7. *Effects of various sera on ODC activity in KB cells. To log-phase KB cells, fresh media containing (a) no serum, (b) 10% calf serum, (c) 10% heat-inactivated calf serum, or (d) 10% dialyzed calf serum was added (t = 0). ODC activity then measured every 2 hr under each condition listed.*

These results support the possibility that membrane glycoconjugates play an important role in the regulation of ODC activity by extracellular effectors. Further support for this possibility comes from Boynton et al. (75), who reported that cultured cells could not reenter the cell cycle from G_o by the addition of serum if cells were pretreated with α-methylornithine, a competitive inhibitor of ODC, and Natraj and Datta (76), who have reported that glycosylation of the cell surface glycoconjugates of cells arrested in G_1 by serum depletion restored their ability to enter S-phase. They suggested that quiescent cells have underglycosylated membrane glycoconjugates, preventing the normal transport of nutrients necessary for continued cell growth and proliferation. Further experiments using serum stimulation of ODC and modifications of cell surface glycoconjugates are in progress.

Cell Cycle Specificity of Drug Action. One possible reason for the variations in glycosyltransferace activity between RA and TPA treated cells and the morphological results described above could be related to a cell cycle specific phenomenon. Perhaps butyrate, TPA, and RA act at different points in the cell cycle. Figures 8-10 (manuscript in preparation) give the results of experiments designed to answer this question. KB cells were mitotically selected (as described in methods) and approxmately 10^5 cells per data point were transferred into 25 cm^2 T-flasks. Butyrate, RA, or TPA were then added at various times after transfer. The incorporation of ^{3}H-thymidine (60 min pulses) was used to study the effects of these drugs on cell cycle behavior. The results of butyrate treatment are shown in Figure 8. If butyrate was added at mitosis or at 2, 4 or 6 hr after mitosis a significant suppression in ^{3}H-thymidine uptake was observed, suggesting that butyrate exerts its effects in the M or early G_1 phase. By late G_1, butyrate could no longer appreciably affect KB cell entry into the S phase.

In Figure 9 the results of the same experiment with TPA are shown. It appears that TPA has a pronounced effect on ^{3}H-thymidine uptake, irrespective of where in the G_1 phase the drug was added. Interestingly, this suppressive effect seems to reach a maximum approximately 2 hr after the drug was administered. When added early enough in G_1 the cells seem to recover to some extent and enter the S phase. Figure 10 shows that RA treatment affected ^{3}H-thymidine uptake and incorporation differently than either butyrate or TPA. Not until drug addition early in the S phase was any appreciable effect on ^{3}H-thymidine uptake noticed. Nowhere in the region between M and late G_1 was RA able to arrest thymidine uptake relative to controls in these cells.

Thus it appears that these drugs act at different points to modify cell cycle behavior. The implications of these results in the regulation of glycoconjugate metabolism and their involvement in tumor promotion remains to be seen. Further studies on changes in glycoconjugate metabolism and cell surface patterns as a function of cell cycle should prove very fruitful.

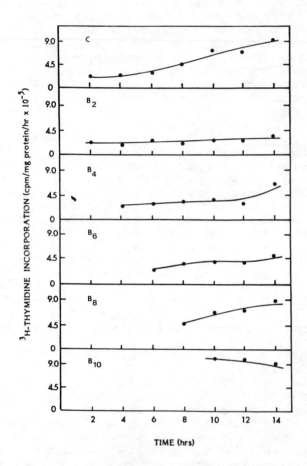

*Figure 8.   Effect of adding butyrate at different points in the cell cycle of mitotically
selected KB cells on [³H]-thymidine incorporation.*

Mitotically selected KB cells were plated into culture flasks and allowed 2 hr to attach to
the substrate. At 2-hr intervals after attachment 2 mM sodium butyrate (B) was added to
a set of flasks, and incorporation of [³H]-thymidine was measured every 2 hr after the
addition of the drug by a 1-hr [³H]-thymidine pulse, (see Materials and Methods). C: control cells, no butyrate added; B_2: B added 2 hr after plating; B_4: B added 4 hr after plating; B_6: B added 6 hr after plating; B_8: B added 8 hr after plating; B_{10}: B added 10 hr
after plating.

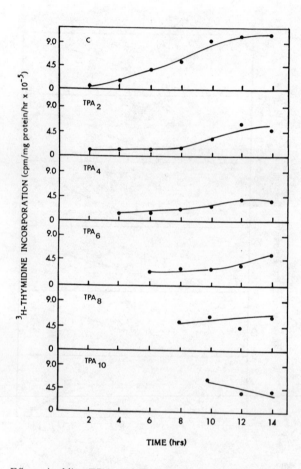

Figure 9. Effect of adding TPA at different points in the cell cycle of mitotically selected KB cells on [³H]-thymidine incorporation. Same protocol as Figure 8. TPA was added instead of B.

258   CELL SURFACE GLYCOLIPIDS

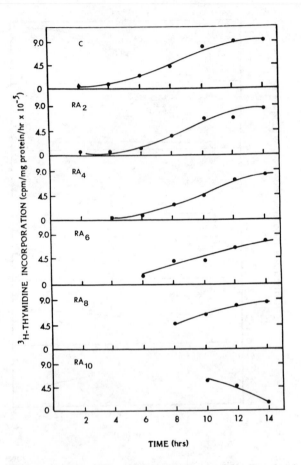

Figure 10. Effect of adding RA at different points in the cell cycle of mitotically selected KB cells on [³H]-thymidine incorporation. Same protocol as Figure 8. RA was added instead of butyrate.

Conclusions

At least some aspects of glycoconjugate metabolism were significantly affected by TPA and RA. The change in sialyltransferase activities was quite significant and was reflected by an increase in GM_3 levels in KB cells. The galactosyltransferase activity changes involved in glycoprotein biosynthesis were also significantly altered. Perhaps RA, TPA, and related phorbol esters will be valuable tools in further studies of glycoconjugate metabolism, affording the possibility of systematically looking at the enzymes involved in glycoconjugate metabolism and the possible interrelationships between glycolipid and glycoprotein biosynthesis. Clearly, other classes of tumor promoters should be investigated. In vivo studies with tumor-forming systems (e.g. mouse skin) could also shed light on the role of glycoconjugates in carcinogenesis.

Based on the results with non-transformed and virally transformed NIL cells, together with the morphological results presented here, it seems that RA and TPA induce changes in glycosyltransferase activities by separate mechanisms. Since TPA has been reported to have an important and rapid effect on the plasma membrane, it may be possible to regulate some aspects of glycoconjugate metabolism at the membrane level. Cell surface glycoconjugates have certainly been implicated in the regulation of cellular growth and differentiation. For example, Lingwood and Hakomori (11) have concluded from the results of their recent experiments with monovalent antibodies and temperature sensitive virally transformed cells that cell surface structures can influence gene expression. Moreover, it has been shown that the receptor for cholera toxin is the ganglioside, GM_1, and ligand-receptor binding leads to an elevation of cyclic AMP (77, 78, 79). Cuatrecasas and coworkers have shown that cholera toxin can interfere with the binding of epidermal growth factor to cell surfaces (80) and affects differentiation of melanoma cells (81). These results also suggest a role for cell surface glycoconjugates in the regulation of growth control and cellular differentiation (28).

The comparisons of changes in ODC activity with changes in sialyltransferase activity after various drug treatments, together with the other effects of RA and TPA reported here, begin to imply that membrane glycoconjugates may be involved in the regulation of ODC activity. Because TPA, epidermal growth factor and cholera toxin (57, 80) all appear to compete with each other for a membrane "receptor site", glycoconjugates also appear likely to participate in at least some receptor-related function for the phorbols. Studies with other structural classes of tumor promoters should prove interesting.

Recent reviews pertaining to such topics as the role of calcium in growth control (82), lectins as natural membrane components (83), fibronectins (84), and biochemical studies of the cell cycle (85, 86, 87, 88), together with many of the works cited earlier, suggest a dynamic, time-dependent relationship between extracellular factors, membrane components and the intracellular metabolic machinery as a cell goes from division to division. Currently available evidence strongly favors an important role for glycoconjugates in transformation. However, more experiments must be performed taking these relationships into consideration before a complete picture will emerge.

Acknowledgements

The authors are grateful to Ms. Dorothy Byrne for her assistance in the preparation of this manuscript. This research was supported in part by a grant (AM 12434) from the National Institutes of Health.

Literature Cited

1.   Whittig, L.A., Ed. "Glycolipid Methodology"; American Oil Chemists Society, Champaign, 1976.
2.   Horowitz, M.I. and Pigman, W., Ed. "The Glycoconjugates", Academic Press, New York, 1977.
3.   Harmon, R.E., Ed. "Cell Surface Carbohydrate Chemistry"; Academic Press, New York, 1978.
4.   Bosmann, H.B. and Winston, R.A. J. Cell Biol., 1970, 45, 23.
5.   Wolfe, B.A. and Robbins, P.W. J. Cell Biol., 1974, 61, 676.
6.   Chatterjee, S., Sweeley, C.C., Velicer, L.F. Biochem. Biophys. Res. Commun., 1973, 54, 585.
7.   Gahmberg, C.G. and Hakomori, S. Biochem. Biophys. Res. Commun. 1974, 59, 283.
8.   Gahmberg, C.G. and Hakomori, S. J. Biol. Chem., 1975, 250, 2438.
9.   Gahmberg, C.G. and Hakomori, S. J. Biol. Chem., 1975, 250, 2447.
10. Schnaar, R.L., Weigel, P.H., Kuhlenschmidt, M.S., Lee, Y.C. and Roseman, S. J. Biol. Chem., 1978, 253, 7940.
11. Lingwood, C.A., Ng, A. and Hakomori, S. Proc. Natl. Acad. Sci. USA, 1978, 75, 6049.
12. Basu, S., Kaufman, B. and Roseman, S. J. Biol. Chem., 1965, 240, PC4115.
13. Kaufman, B., Basu, S. and Roseman, S. Methods Enzymol., 1966, 8, 365.
14. Basu, S., Kaufman, B. and Roseman, S. J. Biol Chem., 1968, 243, 5802.
15. Hildebrand, J. and Hauser, G. J. Biol. Chem., 1969, 244, 5170.
16. Chien, J.L., Williams, T. and Basu, S. J. Biol. Chem., 1973, 248, 1778.
17. Kijimoto, S., Ishibashi, T. and Makita, A. Biochem. Biophys. Res. Commun., 1974, 56, 177.
18. Yip, M.C.M. and Dain, J.A. Biochem. J. 1970, 118, 247.
19. Basu, M. and Basu, S. J. Biol. Chem., 1972, 247, 1489.
20. Basu, S., Basu, M. and Chien, J.L. J. Biol. Chem., 1975, 250, 2956.
21. Brady, R.O. and Mora, P.T. Biochim. Biophys. Acta, 1970, 218, 308.
22. Grimes, W.J. Biochemistry, 1970, 9, 5083.
23. Hakomori, S. Proc. Natl. Acad. Sci. USA, 1970, 67, 1741.
24. Bosmann, H.B. Biochem. Biophys. Res. Commun., 1972, 48, 523.
25. Bosmann, H.B. Biochem. Biophys. Res. Commun., 1972, 49, 1256.

26. Den, H., Schultz, A.M., Basu, M. and Roseman, S. J. Biol. Chem., 1971, 246, 2721.
27. Fishman, P.H. and Brady, R.O., in "Biological Roles of Sialic Acid", Rosenberg, A. and Schengrund, C-L., Ed., Plenum Press, New York, 1976, p. 239.
28. Bennett, V., O'Keefe, E., and Cuatrecasas, P. Proc. Natl. Acad. Sci. USA, 1975, 72, 33.
29. Yeung, K.K., Moskal, J.R., Chien, J.L., Gardner, D.A. and Basu, S. Biochem. Biophys. Res. Commun., 1974, 59, 252.
30. Moskal, J.R., Gardner, D.A. and Basu, S. Biochem. Biophys. Res. Commun., 1974, 61, 751.
31. Basu, S., Moskal, J.R. and Gardner, D.A., in "Ganglioside Function: Biochemical and Pharmacological Implications", Porcellati, G., Ceccarelli, B., and Tettamanti, G., Ed., Plenum Press, New York, 1975, p. 45.
32. Fishman, P.H., Simmons, J.L., Brady, R.O. and Freese, E. Biochem. Biophys.Res. Commun., 1974, 59, 292.
33. Henneberry, R.C. and Fishman, P.H. Exp. Cell Res., 1976, 103, 55.
34. Fishman, P.H., Brady, R.O., Henneberry, R.C. and Freese, E., in "Cell Surface Carbohydrate Chemistry", Harmon, R.E., Ed., Academic Press, New York, 1978, p. 153.
35. Macher, B.A., Lockney, M., Moskal, J.R., Fung, Y.K. and Sweeley, C.C. Exp. Cell Res., 1978, 117, 95.
36. Sweeley, C.C., Fung, Y.K., Macher, B.A., Moskal, J.R. and Nunez, H. in "Glycoproteins and Glycolipids in Disease Processes", American Chemical Society, 1978, 80, p. 47.
37. Presper, K.A., Basu, M. and Basu, S. Proc. Natl. Acad. Sci. USA, 1978, 75, 289.
38. Dawson, G., McLawhon, R., Miller, R.J. Proc. Natl. Acad. Sci. USA, 1979, 76, 605.
39. Boutwell, R.K., in "Carcinogenesis", Vol. 2, 'Mechanisms of Tumor Promotion and Cocarcinogenesis", Slaga, T.J., Sivak, A.; Boutwell, R.K., Ed., Raven Press, New York, 1978, p. 49.
40. Barenblum, I. Cancer Res., 1942, 1, 44.
41. Mottram, J.C. J. Path. Bacteriol., 1944, 56, 181.
42. Van Duuren, B.L. Prog. Ex. Tumor Res., 1969, 11, 31.
43. Hecker, E. Methods Cancer Res., 1971, 6, 439.
44. Hecker, E. Cancer Res., 1968, 28, 2338.
45. Boutwell, R.K. Prog. Exp. Tumor Res., 1964, 4, 207.
46. Boutwell, R.K. CRC Crit. Rev. Toxicol., 1974, 2, 419.
47. Roe, F.J.C. Br. J. Cancer, 1959, 13, 87.
48. Van Duuren, B.L., Sivak, A., Katz, C., Seidman, I. and Melchionne, S. Cancer Res., 1975, 35, 502.
49. O'Brien, T.G., Simsiman, R.C. and Boutwell, R.K. Cancer Res., 1975 35, 1662.
50. Verma, A.K. and Boutwell, R.K. Cancer Res., 1977, 37, 2196.
51. O'Brien, T.G. Cancer Res., 1976, 36, 2644.
52. Boutwell, R.K., in "Origins of Human Cancer", Ed., 1977, Cold Spring Harbor Laboratory, p. 773.
53. O'Brien, T.G. and Diamond, L. Cancer Res., 1978, 38, 2567.

54. Diamond, L., O'Brien, T.G. and Rovera, G. Life Sci., 1978, 23, 1979.
55. Wenner, C.E., Moroney, J. and Porter, C.W., in "Carcinogenesis, Vol. 2, 'Mechanisms of Tumor Promotion and Cocarcinogenesis', Slaga, T.J., Sivak, A., and Boutwell, R.K., Ed., Raven Press, New York, 1978, p. 363.
56. Lee, L.S. and Weinstein, I.B. J. Cell Physiol., 1979, 99, 451.
57. Lee, L.S. and Weinstein, I.B. Science, 1978, 202, 313.
58. Spiro, R.G. J. Biol. Chem., 1964, 239, 567.
59. Moskal, J.R. and Basu, S. Anal. Biochem., 1975, 65, 449.
60. O'Brien, T.G. and Diamond, L. Cancer Res., 1977, 37, 3895.
61. Hirschberg, C.B., Goodman, S.R. and Green, C. Biochemistry, 1976, 15, 3591.
62. Bradford, M.M. Anal. Biochem., 1976, 72, 248.
63. Leavitt, J., Barrett, J.C., Crawford, B.D. and Ts'o, P.O.P. Nature, 1978, 271, 262.
64. Moskal, J.R., Lockney, M.W., Mason, P.A. and Sweeley, C.C. Fed. Proc., 1979.
65. Frot-Coutaz, J.P., Silverman-Jones, C.S. and DeLuca, L.M. J. Lipid Res., 1976, 17, 220.
66. Rosso, G.C., DeLuca, L.M., Warren, C.D. and Wolfe, G. J Lipid. Res., 1975, 16, 235.
67. DeLuca, L.M., Frot-Coutaz, J.P., Silverman-Jones, C.S., Roller, P.R. J. Biol. Chem., 1977, 252, 2575.
68. Porter, K., Prescott, D. and Frye, J. J. Cell Biol., 1973, 57, 815.
69. Bachrach, V. "Function of Naturally Occurring Polyamines", Academic Press, New York, 1973.
70. Russell, D.H. Ed. "Polyamines in Normal and Neoplastic Growth", Raven Press, New York, 1973.
71. Holley, R.W. and Kiernan, J.A. Proc. Natl. Acad. Sci. USA, 1974, 71, 2908.
72. Holley, R.W., Armour, R. and Baldwin, J.H. Proc. Natl. Acad. Sci. USA, 1978, 75, 339.
73. Paul, D., Brown, K.D., Rupniak, B.H. and Ristow, H.J. In Vitro, 1978, 14, 76.
74. Chen, K.Y. and Canellakis, E.S. Proc. Natl. Acad. Sci. USA, 1977, 74, 3791.
75. Boynton, A.L., Whitfield, J.F. and Isaacs, R.J. J. Cell Physiol., 1976, 89, 481.
76. Natraj, C.V. and Datta, P. Proc. Natl. Acad. Sci. USA, 1978, 75, 3859.
77. Cuatrecasas, P. Biochemistry, 1973, 12, 3547.
78. Cuatrecasas, P. Biochemistry, 1973, 12, 3558.
79. Gill, D.M. and King, C.A. J. Biol. C   1975, 250, 6424.
80. Hollenberg, M.D. and Cuatrecasas, P. Proc. Natl. Acad. Sci. USA, 1973, 70, 2964.
81. O'Keefe, E., and Cuatrecasas, P. Proc. Natl. Acad. Sci. USA, 1974, 71, 2500.
82. Berridge, M.J. J. Cyclic Neucleotide Res., 1975, 1, 305.
83. Bowles, D.J. FEBS, 1979, 102, 1.
84. Yamada, K.M. and Olden, K. Nature, 1978, 278, 179.

85. Pardee, A.B., Dubrow, R., Hamlin, J.L. and Kletzien, R.F. Ann. Rev. Biochem., 1978, 47, 715.
86. Potter, V.R. Br. J. Cancer, 1978, 38, 1.
87. Gelfant, S. Cancer Res., 1977, 37, 3845.
88. Smith, J.A. and Martin, L. Proc. Natl. Acad. Sci. USA, 1973, 70, 1263.

RECEIVED January 2, 1980.

Alterations in Cell Surface Glycosphingolipids and Their Metabolism in Familial Hypercholesterolemic Fibroblasts

S. CHATTERJEE[1], P.O. KWITEROVICH, and C. S. SEKERKE

Department of Pediatrics, Johns Hopkins University, Baltimore MD 21205

Relatively little is known about the possible interrelation-ships of the metabolism of the complex sugar-containing lipids, the glycosphingolipids[2] (GSL_s) and the plasma lipoproteins. These interrelationships may occur in the plasma compartment, on the surface of cells, or within the cell. Our purpose here will be to review briefly some of the previous work in the above area and to present some of our recent, preliminary data on GSL and lipoprotein metabolism. Our approach has been to study simul-taneously cultured human fibroblasts derived from both normal subjects and those heterozygous or homozygous for familial hyper-cholesterolemia (FH), a relatively common disorder of cholesterol and low density lipoprotein (LDL) metabolism.

Plasma Glycosphingolipids and Lipoproteins

Plasma Glycosphingolipids. Svennerholm and Svenner-holm (1) initially reported that the four major neutral GSL_s, GlcCer, LacCer, GbOse3Cer and GbOse4Cer were present in human plasma. Tao and Sweeley (2) next demonstrated that ganglioside G_{M3} was also present in human plasma. Increased amounts of cer-tain GSL_s have been found in the plasma of patients with lipid storage disorders; for example, elevated amounts of GlcCer in patients with Gaucher's disease and increased levels of GbOse3Cer in those with Fabry's disease (3,4). The plasma pool of GSL is perhaps synthesized in part in the liver, probably during the assembly of plasma lipoproteins.

Plasma Lipoproteins. The plasma lipids are trans-ported by four major lipoprotein classes. The plasma lipoproteins are synthesized and secreted only in the intestine and liver. Chylomicrons, the richest in triglyceride, are synthesized in the small intestine and transport dietary (exogenous) triglyceride and cholesterol. Very low density (prebeta) lipoproteins (VLDL)

primarily transport triglyceride and cholesterol of hepatic
(endogenous) origin and about 90% of VLDL are synthesized in the
liver. High density (alpha) lipoproteins (HDL) are also made in
the liver. However, both VLDL and HDL are secreted by the liver
as "nascent" lipoproteins, which rapidly undergo modification in
the plasma compartment during lipoprotein metabolism (see also
below). Low density (beta) lipoproteins (LDL) are derived primarily from the catabolism of VLDL. LDL are the major carriers
of plasma cholesterol. The HDL that result from the transfer of
lipids and apoproteins from chylomicrons and VLDL to the "nascent"
HDL contain about 50% by weight protein with the remainder about
equally divided between cholesterol and phospolipid.

The plasma lipoproteins contain eight major apoproteins, the
structure and function of which have recently been reviewed (5).
Briefly, the primary amino acid sequence is known for five of
these apoproteins. ApoB, a highly hydrophobic protein, is found
in chylomicrons, VLDL and LDL. It is the major polypeptide in
LDL and has been shown to be responsible, in part, for the
recognition of LDL by its receptor in cultured human fibroblasts
(7,10). The major polypeptides of HDL are apoA-I and apoA-II;
apoA-I activates lecithin cholesterol acyl transferase. In
addition, studies on the cellular level suggest that apoA-I may
regulate the content of the lipids in the cell membrane (8).

The major plasma lipoproteins are metabolized through a series
of complex and incompletely understood processes that have been
recently reviewed (9). We shall be primarily concerned here with
LDL metabolism in cultured cells. The metabolism of LDL, the end-
product of VLDL catabolism, has been studied both in vivo, using
human and porcine models, and in vitro in a number of cell types,
including cultured fibroblasts, smooth muscle cells, leukocytes
and isolated hepatocytes. The LDL pathway has largely been
worked out by the work of Goldstein and Brown (6) in cultured
human fibroblasts (see also below). Briefly, LDL is recognized
specifically by a cell surface receptor. The arginine and lysine
residues of LDL have been shown to be involved in the specific
recognition of LDL by human fibroblasts. LDL is then internalized by adsorptive endocytosis and catabolized within lysosomes.
For example, the apoB moiety of LDL is hydrolyzed into amino
acids and small peptides and the cholesteryl ester component
(primarily linoleate) is broken down into free cholesterol and
fatty acids. The cholesterol is either reesterified into cholesteryl ester (primarily oleate), or used for membrane biosynthesis. Cholesterol also performs several regulatory functions
through presently undefined mechanisms: 1) feedback inhibition
of the rate limiting enzyme of cholesterol biosynthesis, hydroxy-
methylglutaryl (HMG) CoA reductase (NADPH) [mevalonate: NADP$^+$
oxidoreductase (CoA acylating), E.C. 1.1.1.34]; 2) prevention of
the further biosynthesis of the LDL receptor thereby prohibiting
further entry of LDL into the cell. In patients homozygous for
FH, these aspects of the LDL pathway do not operate because LDL

is unable to enter the cell through the specific receptor which is deficient or defective in FH (6).

The metabolism of HDL probably involves interaction with both hepatic and peripheral cells, as well as with other lipoproteins. HDL may remove cholesterol from tissues, the "scavenger hypothesis" (11,12). The cholesterol may then be esterifed by the action of lecithin cholesterol acyl transferase. HDL may provide cholesterol to the liver for bile acid synthesis (13) and some HDL may be catabolized by the liver in the process. HDL has not been found to interfere with the binding of LDL in cultured human fibroblasts (6). However, in cultured human arterial cells, porcine or rat hepatocytes, and rat adrenal gland, there appears to be some competition of HDL with LDL binding sites, suggesting the presence of a "lipoprotein-binding" site (14).

Glycosphingolipids and Lipoproteins

Skipski (15) first reported that HDL_3 and lipoprotein deficient serum (LPDS) contained most of the plasma GSL. Subsequently, our laboratory (16), and several others (17,18), found that most of the plasma GSL_S were associated with the major lipoprotein classes, whereas LPDS did not contain significant amounts of GSL_S. We found that LDL carried about 60% plasma GSL with much of the remainder (20%) on HDL (Table I). A recent report by Danishefsky and co-workers (19) found a significant amount of plasma GlcCer bound to antithrombin III. However, antithrombin activity measurements in LDL and LPDS employing immunochemical (20) and biological assays (21), revealed that relatively more activity was associated with LPDS (22). These observations suggest that GlcCer on LDL presumably does not contribute to antithrombin activity.

The distribution of the plasma GSL_S on the lipoproteins has also been studied in several disorders of lipid and lipoprotein metabolism. Three different laboratories (16,18,23) showed that the total amount of plasma GSL was increased 3-5 fold in homozygotes for FH. Dawson et al (18) and Auran et al (23) found that the total increase in the plasma GSL_S in these patients was directly proportional to the increased amount of total plasma cholesterol in these FH patients. We studied as well the individual major lipoprotein classes isolated from a patient (D.D.) homozygous for FH. We found an absolute increase of GSL_S on the LDL of this patient (µmoles glucose/mg protein) (Table I). In patients with Fabry's disease, Clarke et al (17) reported a 3-5 fold increase in $GbOse_3Cer$ (per mg lipoprotein cholesterol) in both LDL and HDL. In several patients with abetalipoproteinemia, a disorder characterized by an absence of plasma LDL, these workers found that the total plasma GSL_S were slightly reduced when compared with normal plasma; HDL contained a 4-6 fold molar increase in GSL_S in these patients compared with plasma HDL from normal patients (18). Similar shifts of plasma GSL_S to other

TABLE 1

Glycosphingolipid (GSLs) Content[a] of Human Plasma and Plasma Lipoproteins Derived from a Representative Normal Male and Female Subject and One Patient with Homozygous Familial Hypercholesterolemia (FH)[b]

GSL[c]	Normal female (R.S.)[e]			Normal male (R.B.)					Homozygous FH[d] (D.D.)	
	Whole Plasma	LDL	HDL2	Whole Plasma	VLDL	LDL	HDL2	HDL3	Whole Plasma	LDL
GlcCer	4.77 (0.73)	4.60 (0.35)	2.70 (0.20)	6.90 (0.90)	5.20 (0.10)	4.20 (0.45)	2.2 (0.25)	0.55 (0.05)	9.12 (6.00)	23.30 (7.87)
LacCer	4.44 (0.68)	3.81 (0.29)	2.90 (0.21)	5.5 (0.72)	5.2 (0.10)	2.8 (0.30)	2.2 (0.20)	0.55 (0.05)	3.37 (1.14)	4.99 (1.68)
GbOse3Cer	1.96 (0.30)	2.71 (0.21)	0.55 (0.04)	3.80 (0.50)	2.60 (0.05)	2.30 (0.25)	1.80 (0.15)	1.10 (0.10)	7.0 (4.5)	9.29 (3.14)
GbOse4Cer	1.96 (0.30)	1.30 (0.10)	1.38 (0.10)	3.00 (0.40)	2.60 (0.05)	1.40 (0.15)	1.80 (0.15)	0.55 (0.05)	6.0 (3.9)	4.65 (1.57)
GM3	1.76 (0.27)	1.70 (0.13)	0.97 (0.07)	2.60 (0.35)	1.40 (0.02)	1.00 (0.10)	1.10 (0.10)	0.37 (0.03)	7.6 (4.9)	6.80 (2.29)

[a] The data are expressed as nmol glucose/mg of lipoprotein cholesterol and within brackets as μmole glucose/100 ml plasma.

[b] VLDL = very low density lipoproteins (d<1.006); LDL = low density lipoproteins (d 1.022-1.055); HDL2 = high density lipoproteins (d 1.063-1.12); HDL3 = high density lipoproteins (d 1.12-1.21).

[c] Different structural moieties of GSLs.

[d] No GSLs were detected in VLDL, HDL2, or HDL3 from D.D.

[e] Trace amounts of GSLs were found in the VLDL and HDL3 fractions of this subject.

Reprinted from (16) with permission of the publishers.

plasma lipoproteins were found in patients with hypobetalipoproteinemia and Tangier disease (deficiency of HDL) (18). Several tentative conclusions may be drawn from the above studies on GSL_S and lipoproteins. First, most of the plasma GSL_S are associated with the plasma lipoproteins; however, a small amount of GSL_S may also be associated with other plasma proteins. Second, in normal plasma the LDL appears to carry the majority of plasma GSL_S. However, in dyslipoproteinemic states the GSL_S can be primarily carried by other major lipoprotein classes. This suggests that the association of the plasma GSL_S with the lipoproteins is not covalent and is secondary to some other interaction of the GSL_S with lipids or proteins. Third, in some patients with the homozygous FH, but apparently not in all, there is an absolute increase per mg lipoprotein cholesterol in the GSL_S content of LDL. Similar findings have also been reported for $GbOse_3Cer$ of LDL and HDL in patients with Fabry's disease.

Metabolism of Glycosphingolipids in Cultured Human Fibroblasts

The metabolism of GSL_S has been studied in cultured human fibroblasts from normal subjects, patients with lipid storage diseases, and those with FH. The content of the GSL_S, as well as activities of the biosynthetic enzymes, the glycosyltransferases and the lysosomal GSL hydrolases, have been studied. Complex gangliosides, such as G_{M1}, G_{D1a}, have been found in this cell system to serve as receptors for cholera toxin and thyrotropin, respectively (24-26). More recently, G_{T1} and G_{D1a} have been postulated to be receptors for fibronectin in cultured fibroblasts (27).

Glycosphingolipid Content. Matalon and co-workers (28) demonstrated initially that GSL were present in normal human cultured fibroblasts. Subsequently, the content of GSL_S in normal cultured fibroblasts has been studied by others (29,30). The increased content of GSL_S in human skin fibroblasts has been shown in a number of lipid storage diseases (Table II). Increased levels of G_{M3} and G_{M2} in fibroblasts from subjects with Hurler Syndrome, a mucopolysaccharidosis, have been shown by Matalon et al (28). Although GalCer, I^3SO_3-GalCer, G_{M2} and G_{M1} are present in insignificant amounts in normal human fibroblasts, the recent availability of microdensitometric methods have made possible their quantification (26,31).

We have also studied the content and cell surface GSL_S in a family with FH (30). The neutral GSL_S and G_{M3} were increased 3 to 5 fold in the cultured fibroblasts from the homozygous FH proband (30); levels of GSL_S in both obligate heterozygous FH parents were intermediate between those of normal cells and of the homozygote. In a later study by Fishman and co-workers (32), normal amounts of gangliosides and phospholipids were found in

two unrelated FH homozygotes; these workers attributed our previous findings to normal variability. However, the increased
amounts of GSL_S in the obligate heterozygous FH parents that we
studied were outside the normal range of Fishman and co-workers
(30,32). Data will be presented here that indicate that there is
some alteration in the metabolism of the GSL_S in this family with
FH.

Glycosphingolipid Glycosyltransferases. Most work
on the biosynthesis of GSL_S has been attempted in brain tissue extracts from young animals, because of the large quantities of GSL
that appear during myelination (33,34). In vitro experiments
have indicated the existence of two major pathways for neutral GSL
biosynthesis. These are: a) acylation of a long chain base
followed by transfer of Gal from UDP-Gal (35,36); and b) transfer
of Gal to a long chain base to form psychosine (37) followed by
acylation (36,38). The conversion of ceramides containing 2-
hydroxy fatty acids to other neutral GSL was studied by the use of
deuterium-labeled substrates and mass spectrometric analysis of
the products (39). It was suggested that ceramides are converted
to neutral GSL_S without hydrolysis of the amide bond. Furthermore,
ceramides containing hydroxy fatty acids accept predominantly
galactose to form cerebrosides (40). In contrast, ceramides containing nonhydroxy fatty acids serve as precursors for gangliosides (41). The conversion of ceramides to GSL occurs by a step-
wise transfer of sugars from nucleotide sugars in the presence of
specific glycosyl transferases (42). Detergents and manganese
ions are required for optimal activity (42). Individual enzymes
involved in the biosynthesis of the major GSL_S have been studied in
considerable detail, and the sequence of glycosyl transferase
steps has been established, as described in several reviews (42,
43).

The activities of some of the glycosyltransferases can be
stimulated by lysolecithin (presumably because of its detergent-
like action) (44) and butyric acid (45,46) presumably by exposing
potential enzymic activity that was inaccessible to substrate (45).
In contrast, dibutyryl cyclic adenosine monophosphate has been
shown to decrease glycosyl transferase activity in SV40 mouse
fibroblasts without significantly affecting the activity of the
normal mouse fibroblasts (47). Cells approaching confluency have
higher levels of galactosyl transferase activity than the corresponding sparsely confluent or densely confluent cultures (48).
Maximum biosynthesis of GSL occurs during the $M-G_1$ phase of the
cell cycle (46,49). Recently, glioblastoma cells treated with
estradiol were found to reversibly decrease the cholesterol content of the membranes to 60% of normal; concomitantly, the cerebroside sulfotransferase activity increased to a value of 200% of
normal (50). These studies suggest that cerebroside sulfotransferase activity is modulated by the changing cholesterol/phospo-
lipid ratio in cells involved in myelination. In view of these

findings, particular care must be taken in the interpretation of glycosyltransferase data of cells grown in tissue culture. Burton et al (51) and Brady et al (52) showed that a wide variety of radioactive sugars, particularly [^{14}C]glucosamine [^{3}H] N-acetyl mannosamine, serve as specific precursors of the sialic acid moiety of GSL. [^{14}C], radioactive galactose and glucose were also shown to serve as precursors for almost all the sugar moieties of GSL$_S$ of mouse fibroblasts (53) and rat brain gangliosides (51). Whereas, [^{14}C] glucose was shown to be incorporated into the sugar, sphingosine and fatty acid moieties of GSL$_S$ of human diploid fibroblasts (54). These studies provided an opportunity to investigate the biosynthesis and turnover of GSL and to assist in the elucidation of their structure. For example, using D-[U-^{14}C] glucose, Dawson et al (54) showed that the de novo synthesis of the six major GSL$_S$ occured in human fibroblasts. The half-life for individual GSL was 2-3 days, a value similar to the turnover of membrane-bound phospholipids. Similar half-life values of individual GSL$_S$ of mouse fibroblasts were reported in another study in which [^{14}C] galactose was employed as a radioactive sugar precursor (55). No alteration in the biosynthesis of GbOse$_3$Cer, as measured by the incorporation of label into sugar, fatty acid and sphingosine moieties of fibroblasts with Fabry's disease, was reported (54). The absence of LacCer suggested the inability of Fabry cells to catabolize GbOse$_3$Cer.

Glycosphingolipid Hydrolases. The deficiency of lysosomal glycosylhydrolases has been shown in a number of lipid storage diseases (as summarized in Table II) using cultured fibroblasts (29).

Glycosphingolipid and Lipoprotein Metabolism in Cultured Human Fibroblasts

Although the metabolism of LDL in cultured fibroblasts has been studied extensively (6), relatively little is known about the interrelationship of GSL and lipoprotein metabolism in this system. For example, it is not known whether GSL$_S$ carried on LDL are involved in the recognition of LDL by the cell surface receptor, or in internalization and fusion with the lysosomal membrane. The GSL$_S$ are probably not involved as "recognition sites" on LDL, since partial delipidation of LDL does not abolish the recognition of the macromolecule by the cell (56). The metabolic fate of the GSL carried on LDL is not known. The GSL on LDL may be hydrolyzed within the lysosomes, and their constituents utilized by the cell. Further, it is not known whether GSL$_S$, one of the components of GSL or some other moiety of LDL, such as cholesterol or glycopeptide, exerts an influence on GSL biosynthesis, specifically the glycosyl transferase system.

We have been interested in the contribution made by plasma GSL towards cell membrane-lipoprotein interaction with regard to:

Table II

Lysosomal Glycosyl Hydrolase Deficiency and the Accumulation of Glycosphingolipids in Fibroblasts in "Glycolipidosis"

Disease	Deficient Enzyme	Tentative Structure of Glycosphingolipid Accumulating in Human Skin Fibroblasts	Abbreviated form of GSL
Farber's	ceramidase	Ceramide	Cer
Gaucher's	β-Glucosidase	Glc-Cer	Glc-Cer
[a]Globoid cell leukodystrophy	β-Galactosidase	Gal-Cer	Gal-Cer
[a]Metachromatic leukodystrophy	Arylsulfatase A	SO_4-Gal-Cer	I^3SO_3Gal Cer
Fabry's	α-Galactosidase	Gal-Gal-Glc-Cer	$GbOse_3Cer$
Sandhoff Jatzkewitz, G_{M2} gangliosidosis type II	β-N-Acetylhexosaminidase A and B	GalNAc-Gal-Gal-Glc-Cer	$GbOse_4Cer$
I-Cell	generalized lysosomal hydrolase	Gal-Glc-Cer NeuAc NeuAc-NeuAc-Gal-Glc-Cer	G_{M3} G_{D3}
Tay Sach's, G_{M2} gangliosidosis type I	β-N-Acetylhexosaminidase A	Gal-Nac-Gal-Glc-Cer NeuAc	G_{M2}
G_{M1} gangliosidosis type I	β-Galactosidase	Gal-GalNAc-Gal-Glc-Cer NeuAc	G_{M1}

[a]Fibroblasts from these patients did not accumulate GalCer or sulfatide although the specific enzyme deficiencies could be readily demonstrated (29).

a) the normal homeostasis of GSL in normal human fibroblasts; and
b) the pathological consequences in FH. We have previously shown
(30) that fibroblasts from the proband (T.B.,a FH receptor negative
homozygote) and their two obligate heterozygote parents (S.B. and
T.B.) contained 2-5 fold higher than normal levels of GSL.
    In this chapter we shall review the previously obtained data
on normal and FH fibroblasts, as well as present our most recent
findings on the biosynthesis, degradation and egress of cellular
GSL.

Materials

    Galactose oxidase was from Worthington (800 units/mg),
$D[1-^3H]$ galactose (12.4 Ci/m mol) 2-deoxy-$[1-^3H]$ glucose (15 Ci/
mM) L-$[U-^{14}C]$ leucine (18.8 Ci/mM) $[6-^3H]$thymidine, 2 Ci/mM and
potassium borotritiide (18.5 Ci/m Mol; 97% purity were purchased
from Amersham/Searle. $Na^{125}I$ (13 to 17 mCi/µg in 0.05 N NaOH) was
from New England Nuclear.

Patient Population. The proband of the B family, T.B.,
was referred to the Lipid Research Clinic at The Johns Hopkins
Hospital at the age of five years because of hypercholesterolemia
of 900 mg/100 ml. She had multiple planar xanthomas that had
first appeared at three years of age. The patient was free of
symptoms of ischemic heart disease. The index lipoprotein pattern
was type IIb (57), with marked hypercholesterolemia, hyperbeta-
lipoproteinemia, a mild hyperprebetalipoproteinemia and hyper-
triglyceridemia. None of the relatives of T.B. had xanthomas or
corneal arcus; one (J.S.) developed signs of premature coronary
atherosclerosis at the age of 43 years. Increased total plasma
and LDL cholesterol levels were transmitted over three generations on both maternal and paternal sides of the family (Fig. I).
The parents of the proband, S.B. and K.B., had endogenous hyper-
triglyceridemia as well. Two normolipidemic members of this
family (S.B., Jr. and E.B.), were also studied.
    The unrelated controls in this study were healthy, normo-
lipidemic male (R.W.) and female (L.W.) adults (58), and two newborns (H.S.F. and B.P.). Fibroblasts from the unrelated patients
(GM 2000, M.C.; GM 3040, D.S.), previously characterized in detail
as receptor-negative FH homozygotes (6,59), were obtained from
the cell repository, Camden, New Jersey, and from Dr. A.B.
Khachadurian (59), respectively.

Methods

Biochemical Characterization of Fibroblasts

    Fibroblasts were classified as normal FH heterozygous
or receptor negative FH homozygous following characterization
according to the four biochemical assays described by Goldstein

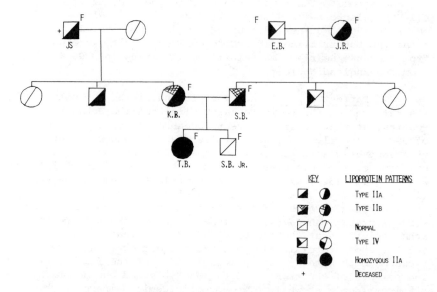

Figure 1.   Pedigree pattern of the "B" family.   Relationship of the proband (T.B.) with the clinical phenotype of homozygous familial hypercholesterolemia to her other relatives whom we studied is shown.   Lipoprotein patterns were determined after ultracentrifugation using NIH cutpoints (57).   F indicates that fibroblast cell lines were established from skin biopsies.   Males, ☐; Females, ○.

and Brown (6). These studies to be reported elsewhere (58) included the measurements of: 1) ^{125}I LDL binding; 2) LDL proteolysis; 3) cholesterol esterification; and 4) hydroxy-methyl-glutaryl CoA reductase activity.

Cell Culture

Depending upon the design of an experiment, cells were grown in 60 mm petri dishes, 150 cm^2 plastic flasks (Falcon no. 3024), or roller bottles in MEM supplemented with 12% fetal bovine serum and antibiotics in 95% air, 5% CO$_2$ water humidified incubator at 37°. All experiments were initiated with confluent monolayer of cells grown for about 5-15 generations.

Human Plasma Lipoproteins

Human plasma lipoproteins and lipoprotein deficient plasma (LPDS) were prepared from normal plasma of healthy individuals by differential ultracentrifugation on KBr gradients (16).

Cell Surface Labeling of Glycosphingolipids

The galactose moiety of the cell surface GSL and glycoproteins was labeled with [^{3}H] following treatment of the cells with galactose oxidase, followed by reduction with KB[^{3}H]$_4$, as previously described (30,60,61).

Metabolic Labeling of Cells with [^{3}H] Galactose

Approximately 5 x 10^6 cells were seeded in glass roller bottles and grown for 12 days in minimum essential medium (MEM) containing 12% fetal calf serum. Subsequently, medium was removed and the monolayers washed 3 times with PBS maintained at 37°. To each bottle was added 20 ml of fresh MEM containing 1% fetal calf serum (FCS) and 5% lipoprotein deficient serum (LPDS) and 10 µCi of [^{3}H] galactose. Following incubation for 48 hrs at 37°, the medium was removed and frozen. The monolayers were washed 5 times with ice cold PBS, harvested with a rubber policeperson and centrifuged (500 x g, 5 min, 4°). The pellets were resuspended in PBS, washed and centrifuged. Finally, the cell pellets were resuspended in water and sonicated for 30 seconds in a Bransonic sonifier. Suitable aliquots of the cell suspension were withdrawn for isolation of individual GSL$_5$ (see below) and measurement of protein and radioactivity.

Incubation of Co-cultured Cells with [^{3}H] Galactose

Approximately 2.5 x 10^6 each of normal (B.P.) and FH receptor negative homozygous fibroblasts (T.B.) were seeded into roller bottles and grown as described above.

On the 12th day, medium was removed, monolayers washed and cells incubated in fresh MEM plus 1% FCS and radioactive galactose for 48 hrs. Subsequently, medium was removed and the cells processed as described above.

Incubation of Cells with [14C] leucine, and [3H] thymidine, and [3H] 2-deoxyglucose

Approximately 2 x 10^5 normal and FH homozygous fibroblasts were seeded in 60 mm petri dishes and grown for 6 days in MEM containing 10% fetal bovine serum. Subsequently, the medium was removed, monolayers washed and 2 ml of fresh MEM containing 5% LPDS, 1% FCS and 2 µCi each of the above isotopes added and incubated for 48 hrs. Subsequently, medium was removed, the monolayers washed twice with 5 ml of PBS and 2% BSA and thrice with PBS. The entire monolayer was solubilized in 1 ml of 1 N NaOH. Suitable aliquots of the cell extracts were used for protein and radioactivity measurements.

Measurement of 125I Low Density Lipoprotein Binding and Degradation in Normal, Familial Hypercholesterolemic, Homozygous Fibroblasts and Co-cultured Fibroblasts

Approximately 2 x 10^5 normal and FH homozygous fibroblasts and 1 x 10^5 cells each of the normal and FH lines were seeded in 60 mm Falcon petri dishes and cultured in MEM containing 10% fetal calf serum. On the sixth day medium was removed, the monolayers washed and replenished with 2 ml of fresh MEM containing 5% LPDS. Following incubation for 4 hrs at 37°, medium was removed and further processed for the measurement of 125I LDL degradation outlined in detail elsewhere (6,58). The monolayers were washed twice with ice cold PBS - 2% BSA and thrice with PBS in a cold room and solubilized overnight in 1 N NaOH. Suitable aliquots of the cell extracts were processed for protein and radioactivity measurements. All assays were carried out in duplicate in two separate sets of dishes. The data is expressed as ng LDL bound or degraded/mg protein.

Measurement of Egress of Glycosphingolipids into the Culture Medium

Monolayers of normal and FH cultured fibroblasts were incubated in the presence of [3H] - galactose in either LPDS medium or medium containing FCS. The medium was then collected and dialyzed at 4°C for 24 hrs against 4 changes of 4L water. Antibody against human serum was then added to the culture medium. Following incubation for 48 hrs at 4° the samples were centrifuged at 30,000 x g for 30 min at 4°. The supernatants were carefully withdrawn and the pellets rinsed with ice cold PBS and frozen until further use. The supernatants were freeze-dried.

Suitable aliquots of human kidney GSL_S were added to the antibody-precipitated pellets and GSL isolated as described below.

Isolation and Quantitation of Glycosphingolipids from Cultured Fibroblasts, Plasma Lipoproteins and Tissue Culture Medium

GSL_S were extracted from unlabeled and radiolabeled normal and mutant FH fibroblasts essentially as described earlier (30, 62). GSL_S in plasma or plasma lipoproteins were isolated after Vance and Sweeley (3). The methodology for the further isolation, purification and quantitation of GSL_S by gas liquid chromatography was followed essentially as described earlier (30,62).

Measurement of Radioactivity Incorporated into Individual Glycosphingolipids

Following the separation of GSL by thin-layer chromatography, the GSL_S were stained with iodine vapour and gel zones that corresponded in chromatographic migration with authentic human GSL of known chemical structure were scraped, transferred into scintillation vials and counted in 10 ml of "Liquiscint" (National Diagnostics, Parsippany, New Jersey).

Measurement of UDP-Gal:lactosylceramide galactosyltransferase Activity in Normal and Mutant Cells

Galactosyl transferase activity of normal and mutant cells was pursued following modifications of the procedure of Hildebrandt and Hauser (63). Product characterization was carried out by thin layer chromatography.

Other Methods

Protein was measured by the method of Lowry et al (64). Crystalline bovine serum albumin (Sigma Chemical Company) served as a standard.

Results

Glycosphingolipid Composition of Normal and Familial Hypercholesterolemic Cells

When the cells were grown in the presence of fetal calf serum, GlcCer, LacCer, $GbOse_3Cer$ and $GbOse_4Cer$ were the major neutral GSL_S and G_{M_3} and G_{D_3} the major gangliosides in both normal and FH cells (Table IIIA). The two heterozygote FH cell lines, S.B. and K.B., had a 2 to 3 fold increase in these GSL_S, except for LacCer, which was not elevated. The FH homozygote line, T.B., contained approximately 5 fold higher levels of GSL compared to normal

Table III

Glycosphingolipid levels (nmol glucose/mg of protein) in cultured fibroblasts from normal and familial hypercholesterolemic subjects

Glycosphingolipids (GSL)	A. Medium supplemented with 12% fetal calf serum				B. Medium supplemented with 5% lipoprotein-deficient serum			
	Normal	Heterozygote male	Heterozygote female	Homozygote	Normal 24 hr	Normal 5 days	Homozygote 24 hr	Homozygote 5 days
Neutral GSL								
GlcCer	0.37	0.52	0.87	1.51	0.35	0.30	0.90	0.65
LacCer	0.10	0.08	0.11	0.08	0.07	0.07	0.10	0.07
GbOse$_3$Cer	0.43	1.23	1.01	2.37	0.45	0.70	0.99	0.73
GbOse$_4$Cer	0.38	1.03	1.00	2.00	0.40	0.59	0.99	0.84
Subtotal	1.28	2.86	2.99	5.96	1.27	1.66	2.98	2.29
Gangliosides								
G$_{M3}$	0.88	3.50	3.00	4.62	1.00	1.25	2.30	1.60
G$_{D3}$	0.11	0.32	0.16	0.47	0.11	0.15	0.27	0.20
Subtotal	0.99	3.82	3.16	5.09	1.11	1.40	2.57	1.80
Total	2.77	6.68	6.15	11.05	2.38	3.06	5.55	4.09

Glycosphingolipids (GSL$_s$) were isolated from fibroblasts grown in tissue culture medium supplemented with 12% fetal calf serum or 5% lipoprotein-deficient serum (see Materials and Methods). The quantities of the purified GSLs were determined following acid-catalyzed methanolysis and gas-liquid chromatography of trimethylsilyl-methyl glycosides, using mannitol as an internal standard (56). All analyses were performed in triplicate.

Reprinted from (30) with permission of the publishers.

cells. When the cells were transferred to LPDS medium, and incubated for 24 hr, the cellular content of the GSL_S, except LacCer, were reduced to about one half in the homozygous FH fibroblasts (Table IIIB). These reduced levels were still 2-3 fold higher than those found in the normal fibroblasts, whose GSL content was unchanged after 24 hrs in LPDS medium. After 5 days, the GSL in the homozygous line T.B. were reduced an additional 25%, except LacCer, which remained unchanged; the total content of GSL was then only 1.3 times that of normal, and the slight elevation primarily resided in GlcCer, $GbOse_4Cer$ and $^GM3$. After 5 days in LPDS medium, both $GbOse_3Cer$ and $GbOse_4Cer$ increased in normal fibroblasts by 50% and $^GM3$ by 25%; the total GSL increased 34%. The levels of GlcCer, LacCer and $^GD3$ were essentially unchanged in the normal cells regardless of whether they were grown in medium supplemented with LPDS for 24 hr or 5 days.

Cell Surface Labeling Pattern of Glycosphingolipids of Normal and Familial Hypercholesterolemic Fibroblasts

The exposure of the GSL_S on the cell surface of normal and FH cells was studied employing galactose oxidase (GAO), followed by reduction with KB $[^3H]_4$ (60,61). The incorporation of ^3H into the hexose, sphingosine and fatty acid moieties of the GSL_S of normal fibroblasts, before and after treatment with galactose oxidase is presented in Table IV. In normal cells, there was a marked increase in the incorporation of tritium in the hexose moiety of $GbOse_3Cer$ (30 fold), $GbOse_4Cer$ (87 fold), $^GM3$ (82 fold) and $^GD3$ (40 fold) following treatment with GAO. There was little incorporation of ^3H into the galactose moiety of LacCer. Most of the GlcCer was not susceptible to oxidation with galactose oxidase since more than 90% of its hexose moiety was glucose rather than galactose. Some incorporation of ^3H into the fatty acid and sphingosine moieties of the GSL_S occurred through the reduction of their unsaturated bonds. The homozygous FH cell line, T.B., had an increased uptake of label into the hexose moiety of $GbOse_4Cer$ (4 fold), $^GM3$ (2 to 3 fold), and $GbOse_3Cer$ and $^GD3$ (1 to 2 fold), compared to normal (Table V). However, when the amount of radioactivity incorporated into each GSL was divided by its cellular content (Table IIIA), there was a moderate decrease in the total specific radioactivity for each GSL except LacCer.

Incorporation of Radioactive Leucine, Thymidine and 2-Deoxyglucose in Normal and Familial Hypercholesterolemic Fibroblasts

Confluent cells were incubated with a wide variety of radioactive isotopes for 48 hrs in medium containing either FCS or LPDS. Subsequently, the medium was removed and the monolayer washed twice with ice cold PBS-2% BSA and five times with ice cold PBS. The entire monolayer was solubilized in 1 n NaOH and suitable aliquots used for radioactivity and protein measurements.

Table IV

Incorporation of tritium (cpm/mg of protein)
into individual moieties of glycosphingolipids of normal
human skin fibroblasts in the presence and absence of
galactose oxidase

Glycosphingolipid moiety		Galactose oxidase	
		Treated	Untreated
GlcCer:	Hexose	25	21
	Sphingosine	151	24
	Fatty acid	24	6
LacCer:	Hexose	37	25
	Sphingosine	71	28
	Fatty acid	24	11
GbOse$_3$Cer:	Hexose	2995	100
	Sphingosine	1001	300
	Fatty acid	602	55
GbOse$_4$Cer:	Hexose	5655	65
	Sphingosine	1006	250
	Fatty acid	635	31
G$_{M3}$:	Hexose	1225	15
	Sphingosine	252	56
	Fatty acid	209	10
G$_{D3}$:	Hexose	395	10
	Sphingosine	319	25
	Fatty acid	76	12

Normal human fibroblasts grown in media with 12% fetal calf serum were treated with or without galactose oxidase and then subjected to reduction with potassium borotritiide. GSL$_s$ were isolated and subjected to acid-catalyzed methanolysis (62) (see Materials and Methods). Methyl glycosides, methyl sphingosine, and fatty acid methyl esters were isolated by extraction of the total hydrolysate with solvents as described previously (62). Radioactivity was measured in triplicate in aliquots of these extracts.

Reprinted from (30) with the permission of the publishers.

Table V

Distribution of radioactivity in individual
glycosphingolipids (GSLs) of human fibroblasts after
treatment with galactose oxidase*

GSL	Normal	Homozygote
GlcCer	200	367
	(540)	(243)
LacCer	132	167
	(1,320)	(2,087)
GbOse₃Cer	4,598	6,850
	(10,693)	(2,890)
GbOse₄Cer	7,296	25,050
	(19,200)	(12,525)
G_M3	1,686	4,250
	(1,916)	(919)
G_D3	790	1,450
	(7,182)	(3,085)

Fibroblasts grown in media with 12% fetal calf serum were treated with galactose oxidase, followed by reduction with potassium borotritiide. The individual GSLs were isolated and the radioactivity was measured in triplicate (see Materials and Methods).
*The radioactivity is given as cpm/mg of protein (approximately 2×10^6 cells). Specific radioactivity (cpm/nmol of glycosphingolipid) is provided within the parentheses.
Reprinted from (30) with the permission of the publishers.

The specific activity data is shown in Table VI. The protein
content per dish is presented in parenthesis. The rates of protein and DNA synthesis and the transport of hexose in normal cells
grown in FCS was somewhat higher than the FH homozygous cells.
However, the protein content per dish among normal and mutant
cells was similar. When cells were incubated in LPDS medium there
was a decrease in the rate of [^{14}C] leucine incorporation into
both normal and mutant cells. Apparently, this decrease was not
due to decreased transport and uptake of this amino acid into
cells but rather due to a 10-30% increase in protein content.
Furthermore, there was a 25% increase in the rates of DNA synthesis in FH homozygous cells incubated in LPDS medium. In contrast,
normal cells did not exhibit such   phenomena. While a 25-30%
decrease in the transport of 2-deoxyglucose was observed in normal
cells incubated in LPDS medium compared with cells grown in FCS
medium, no such differences were observed in the FH homozygous
cells. These observations suggest that there is little difference
in the protein turnover or in transport of hexoses among normal
and FH mutant cells.

Incorporation of Radiolabled Galactose into Glycosphingolipids of Normal and FH Homozygous Cells

Radioactive galactose has been previously shown to be
suitable for labeling the GSL of cultured mouse fibroblasts (55).
Confluent cultures were incubated for 48 hrs in the presence of
[^{3}H] galactose in 1% FCS. The cells were then harvested, the
lipids extracted and the fraction containing GSL isolated as described in Methods. The GSL containing fraction was subjected to
thin layer chromatography and the zones corresponding to the major
GSL scraped, eluted and counted. In 4 normal cell lines, each of
the major neutral GSL$_s$ and ^{G}M3 and ^{G}D3 incorporated [^{3}H] galactose (Table VII). In the homozygote from the B family, there was
a 2 to 5 fold increase in the incorporation of [^{3}H] galactose into
the individual neutral GSL$_s$ and major gangliosides, compared to
the average 4 normal cell lines. In distinct contrast, the incorporation of [^{3}H] galactose into an unrelated FH receptor negative homozygote line (GM 2000) was several fold below the average
normal values for each of the major GSL (Table VII).

Metabolic Labeling Pattern of Glycosphingolipids in
Familial Hypercholesterolemic Heterozygous Fibroblasts
and Co-cultured Normal and Familial Hypercholesterolemic
Homozygous Fibroblasts

One possible explanation for the alteration of GSL metabolism in T.B. is that the abnormality is related to or dependent
upon the characteristics of cell growth in culture. To address
this possibility, we co-cultured the cells from T.B. along with
normal human fibroblasts. For comparative purposes, we also

Table VI

Incorporation of radioactivity derived from [14C] Leucine, [3H] Thymidine and [3H]2-deoxy glucose into normal and familial hypercholesterolemic homozygous fibroblasts incubated in fetal calf serum and lipoprotein deficient serum containing medium

| | Normal | | | | Homozygous | | | |
| | L.W. | | R.W. | | T.B. | | G.M.3040 | |
Isotope	FCS	LPDS	FCS	LPDS	FCS	LPDS	FCS	LPDS
[14C]-Leucine	539 (310)	502 (335)	522 (250)	516 (285)	414 (330)	392 (445)	402 (325)	409 (405)
[3H]-Thymidine	30 (310)	29 (345)	22 (205)	27 (225)	15 (320)	20 (400)	15 (355)	20 (400)
[3H]-2-deoxy glucose	39 (310)	27 (340)	27 (210)	20 (255)	18 (310)	17 (410)	10 (362)	8 (422)

(c.p.m./mg protein)

In these experiments, normal and mutant cells (2 x 10^5 cells/dish) were seeded and grown for 6 days in 5 ml of medium supplemented with 10% fetal calf serum and antibiotics. On the 6th day, medium was removed and the monolayers washed twice with PBS/2% BSA and 5 times with PBS at room temperature. Two ml of fresh medium containing 1 μ Ci each of L-[U14C] leucine, [6-3H] thymidine, 2-deoxy [1-3H] glucose antibiotics, and 1% fetal calf serum or 5% lipoprotein deficient serum was added to each dish and the cells incubated for 48 hrs. Subsequently, the medium was removed and the monolayers washed as described above, but at 4° C. The entire monolayer was solubilized in 1 N NaOH. Radioactivity and protein were subsequently measured in the alkali solubilized cell extracts. The data are expressed as c.p.m. per mg protein. The data within parentheses represent total protein (ug) in individual petri dishes.

Table VII

Incorporation of Radioactivity Derived from [3H] Galactose into the
Individual Glycosphingolipids of Normal and Familial Hypercholesterolemic
Homozygous (FH) Cells Cultured in Fetal Bovine Serum

| Glycosphingolipid | Normal | | | | | Homozygote | |
| | H.S.F. | S.B. Jr. | L.W. | B.P. | Average Normal | G.M. 2000 | T.B. |
	(c.p.m./mg protein)						
GlcCer	110	377	329	145	240	92	1161
LacCer	82	73	84	266	126	34	253
GbOse$_3$Cer	449	1231	2030	450	1040	395	3754
GbOse$_4$Cer	323	290	287	209	277	171	741
G$_{M3}$	336	369	480	226	352	109	996
G$_{D3}$	83	98	122	43	86	50	192

Table VII. In a typical experiment, normal and mutant fibroblasts (one 150 cm^2 flask, each containing approximately 2 x 10^6 cells) were incubated for 6 days in 20 ml medium containing 10% fetal calf serum and antibiotics. On the sixth day, medium was removed, the confluent monolayers washed five times with warm PBS, and further incubated for 48 hrs in 10 ml medium containing 1% fetal calf serum, antibiotics, and 5 µ Ci of [^{3}H]-D-galactose. Subsequently, the medium was removed, the monolayers washed with PBS, harvested, and centrifuged at 500 x g for 5 min at 4°. The cell pellet was washed thrice with 40 vol of ice cold PBS. The washed cell pellets were suspended in a small volume of water, sonicated and a suitable aliquot withdrawn for total protein and radioactivity measurements. The remainder of the cell pellets were subjected to solvent extraction for the purposes of isolation of GSL$_S$ according to previously described procedures (30). The lower phase lipids were fractionated by silicic acid column chromatography (62). The neutral GSL$_S$ in the glycolipid fraction were then separated by thin layer chromatography (62). The chromatoplates were then air-dried and zones corresponding in chromatographic migration with mono, di, tri and tetra glycosyl ceramides were scraped off and their radioactivity determined. The upper phase was dialyzed at 4° for 24 hr against 3 changes 4 L of water. The dialysates were dried *in vacuo*, resuspended, and individual gangliosides separated by thin layer chromatography on plates coated with silica gel G. The chromatograms were developed in chloroform-methanol-ammonium hydroxide-water (60:35:1:7 v/v). The chromatoplates were then air-dried and zones corresponding in chromatographic migrations to ^{G}M3 and ^{G}D3 were scraped off and their radioactivity determined.

studied an obligate FH heterozygous fibroblast line (K.B.).
Following incubation with 1% FCS and [^{3}H] galactose for 48 hrs,
the incorporation of radioactivity into the individual GSL was
measured. The data are presented in Table VIII. The similarity
in the incorporation of [^{3}H] galactose into the individual GSL of
co-cultured cells and FH heterozygous cells is clearly evident.
With the exception of LacCer, the amounts of [^{3}H] galactose incorporated into the co-cultured and FH heterozygous cells were intermediate between our homozygous FH cell line (T.B.) and the average
of four normal fibroblast lines (Tables VII, VIII). There was
some overlap between the values in the co-cultured or heterozygous
FH cells with those from a given normal cell line (Tables VII,
VIII). This suggests that a considerable amount of variation can
occur in normal cell lines in GSL metabolism and that some overlap may be expected between normal and FH heterozygote populations. There was also a 4 fold increase in the incorporation of
[^{3}H] galactose into ^{G}D3 in the heterozygote K.B., compared to the
average normal, and a 2 fold increase over T.B. We are unable to
explain this observation at the present time.

^{125}I Low Density Lipoprotein Binding and Degradation in Normal, Familial Hypercholesterolemic Homozygous, and Co-cultured Cells

Normal human fibroblasts (line L.W.) bound and degraded
[^{125}I] LDL by two mechanisms: the high affinity, saturable recep-
or mediated uptake, and the low affinity non-saturable bulk phase
pinocytosis (Fig. 2). In contrast, the FH homozygous fibroblasts
(T.B.) took up [^{125}I] LDL only through the nonspecific mechanism.
Co-cultured normal and FH homozygous cells carried out LDL binding and degradation, as predicted, at one half the capacity seen
in normal fibroblasts. These data suggest that at least half of
the confluent co-cultured fibroblasts had the normal complement
of LDL receptors and the other half, presumably representative of
the FH cells, had no functioning LDL receptors and contributed
nothing towards total LDL binding activity.

Effects of Incubation of Fibroblasts with Lipoprotein Deficient Medium on the Incorporation of Radioactivity Derived from [^{3}H] Galactose into Cellular Glycosphingo-lipids and Cell Culture Medium

Confluent monolayers of a normal cell line L.W., and our
FH homozygous cell line (T.B.), were incubated for 48 hrs in
medium containing lipoprotein deficient serum and [^{3}H] galactose.
The medium was then collected, dialyzed against water, and pre-
cipitated with an antibody against whole human serum. GSL$_S$ were
isolated from both the cells and the precipitate from the culture
medium. Since both the normal and FH fibroblasts incorporated 3
to 4 fold more radioactivity into GbOse$_3$Cer and GbOse$_4$Cer when

Table VIII

Incorporation of radioactivity derived from
[^{3}H] Galactose into the individual glycosphingolipids of
co-cultured normal and familial hypercholesterolemic
homozygous cells and a familial
hypercholesterolemic heterozygous fibroblast line

Glycosphingolipid	Co-cultured cells (c.p.m/mg protein)	Heterozygote (K.B.)
GlcCer	620	450
LacCer	259	147
GbOse$_3$Cer	1684	1898
GbOse$_4$Cer	402	537
G$_{M3}$	498	497
G$_{D3}$	84	332

In a typical experiment, 1×10^6 normal cells (B.P.) and familial hypercholesterolemic homozygous cells (T.B.) were co-cultured in medium containing 10% fetal calf serum and antibiotics. In another flask 2×10^6 familial hypercholesterolemic heterozygous cells (K.B.) were grown in the same medium. On the sixth day, medium was removed, the confluent monolayers washed five times with warm PBS and further incubated for 48 hrs with [^{3}H]-D-galactose as described in the legend for Table VII. GSL$_s$ were isolated and their radioactivity measured as described in the legend for Table VII.

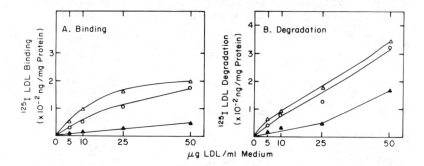

Figure 2. Measurement of [¹²⁵I] low density lipoprotein binding and degradation in normal, mutant, and co-cultured cells. Experimental details of these assays are described in the text.

A: low-density lipoprotein binding in normal (△), co-cultured (○), and familial hyper-cholesterolemic homozygous (▲) cells. B: low density lipoprotein degradation in normal (△), co-cultured (○), and familial hypercholesterolemic homozygous (▲) cells incubated in fetal calf serum (Table VII), it was of interest to study if the cells egressed these GSL_S into the culture medium when incubated with LPDS. The data shown in Table IX indicate that neither the normal or mutant FH fibroblasts egressed significant amounts of GSL_S labeled with [^{3}H] galactose into the medium. Moreover, when the GSL fraction derived from culture medium alone (without mixing with carrier GSL) were separated by thin layer chromatography, no visible (iodine positive) bands could be observed. These observations suggest that little, if any membrane GSL, particularly $GbOse_3Cer$ or $GbOse_4Cer$ of the normal and mutant FH fibroblasts, are egressed into the culture medium.

UDP-Gal:Lactosylceramide Galactosyltransferase Activity of Normal and Mutant Cells

One possible explanation for the marked increase in both the levels of, and the incorporation of, [^{3}H] galactose into GL_{3a} in the FH homozygote line T.B., is that the activity of GL_{2a} - galactosyl transferase is increased. The assay for lactosyl-ceramide - uridine diphosphate galactosyltransferase activity was therefore performed. There was a 2 fold variation in the galac-tosyltransferase activity among four normal human fibroblast lines (Table X). However, the homozygous FH line, T.B., had almost a 3 fold higher enzyme activity than the average for the normal cell lines (Table X). In separate experiments, microsomal preparations from the normal fibroblast lines, L.W. and S.B. Jr. (sibling of T.B.), were mixed separately with equal amounts of a microsomal preparation derived from T.B. fibroblasts. The LacCer galactosyl transferase activity was measured and the actual values obtained were very close to those predicted, based upon the activities of individual microsomal preparations of the normal and FH homozygous fibroblasts. The results of the mixing experiments also suggests that the FH homozygote line T.B. does not contain any endogenous factor that activates the LacCer galactosyltrans-ferase activity. Finally, another FH homozygote line, GM 2000, contained only 18.5% of normal fibroblast transferase activity. The activity in GM 2000 was 17 fold lower than that in the FH homozygous line, T.B. (Table X). There is therefore a clear difference in the capacity to synthesize a major fibroblast sur-face GSL, i.e., $GbOse_3Cer$, in the two homozygote lines. These differences do not appear to be due to normal cell variation.

Discussion

We report here further studies on GSL metabolism in the cultured fibroblasts from a family with FH. Evidence from four different kinds of biochemical experiments supports the hypothesis that there are alterations in GSL metabolism in affected members of this family. First, there were marked increases in the cellu-lar content of GSL_S. Second, employing the galactose oxidase-

Table IX

Effects of incubation of normal and familial hypercholesterolemic homozygous
fibroblasts in lipoprotein deficient medium on the incorporation of
radioactivity derived from [^{3}H] galactose into
cellular glycosphingolipids and culture medium

Glycosphingolipid	Normal cells	Normal culture medium	Homozygote cells	Homozygote culture medium
		(c.p.m./mg protein)		
GlcCer	427	10	209	56
LacCer	81	15	65	30
GbOse$_3$Cer	564	20	910	28
GbOse$_4$Cer	107	20	245	25
G$_{M3}$	667	15	400	23
G$_{D3}$	80	10	100	21

In a typical experiment, normal and FH mutant cells were grown to con-
fluence as described in the legend for Table VII. On the sixth day, medium
was removed, the monolayers washed twice with PBS-2% BSA and five times with
PBS. Ten ml of fresh medium containing 5% lipoprotein deficient serum, antibiotics and 5 μ Ci of [^{3}H]-D-galactose was added to each flask and cells
incubated for another 48 hrs. Subsequently, the cells were harvested, GSL$_S$
isolated and their radioactivity measured as described in the legend for Table
VII.

KB^3H_4 reduction procedure of labeling cell surface components, we found significant increases in the incorporation of [^{3}H] by the GSL_S on the surface membrane. Third, there was a marked increase in the incorporation of [^{3}H] galactose into the GSL_S. Fourth, there was an almost three-fold increase in the activity of lactosyl-ceramide galactosyltransferase in the proband. The strengths of this study include the following. The hypercholesterolemic members of this kindred have been well-characterized as having the "receptor-negative" form of FH, as judged by the four biochemical criteria of Goldstein and Brown (6). Detailed studies of members from a single kindred minimize (but do not eliminate) problems of genetic heterogeneity that are present in less extensive studies of a number of patients from many different families. Finally, our studies have been expanded to include four normolipidemic controls and an unrelated receptor-negative FH homozygote.

The conclusions that may be drawn from the current data are limited. For example, it is not possible to generalize the findings of alterations in GSL metabolism in one family with FH to other families with phenotypic receptor-negative FH. It is unlikely, however, that these alterations are confined to this one family. We have also found an increased GSL_S on LDL from another unrelated patient (D.D.) with the clinical phenotype of homozygous FH (16). The precise relation of these alterations in GSL_S to the LDL receptor defect is, however, unclear. The altera-tions in the GSL metabolism may be completely unrelated to the LDL receptor defect and simply reflect the expression of another gene that regulates activity of lactosylceramide galactosyl transferase. Conversely, one or more abnormalities in GSL metabo-lism may be involved in the pathogenesis of a "functionless" LDL receptor defect. Little is known about the nature of the LDL receptor except that it is protein or glycoprotein in nature (65). It is conceivable that the GSLs, or some other lipid class such as phospholipids, are associated with the cell surface LDL recep-tor, and may have a role in LDL recognition, binding and inter-nalization. Or, the alterations in GSL metabolism may be "second-ary" to the LDL receptor defect. The expression of the gene(s) at the "LDL receptor locus," (66) may be superimposed on the action of gene(s) at another locus (epistasy). The molecular basis of LDL receptor defect in this family may be different than in other families with receptor-negative FH, thereby producing secondary alterations in GSL metabolism that are not seen in all patients classified as receptor negative FH. For example, the GSL levels in fibroblasts of GM 2000 are normal or low (32) and the activity of lactosylceramide galactosyltransferase is depressed. In this FH homozygote, the inability to bind LDL and internalize the LDL-LDL receptor complex may have resulted in the failure to suppress glycoprotein synthesis, the glycosyltransferases and consequently GSL synthesis. Since the classification of these hypercholes-terolemic patients is currently made on the basis of functional assays, the degree of possible genetic heterogeneity on a molecular level is not known (67).

Table X

Uridine diphosphate galactose: lactosylceramide
galactosyltransferase activity in cultured normal and
familial hypercholesterolemic human fibroblasts

Cell Strain	Glycosyltransferase activity (c.p.m./mg protein/hr)		
	Anticipated Value	Actual Value Obtained	
1. Normal			
R.W.			2,486
H.S.F.			2,944
L.W.			3,389
S.B., Jr.			5,017
2. Mixed*			
L.W. + T.B.	6,515	4,100	
S.B., Jr. + T.B.	7,329	5,626	
3. Familial Hypercholes-terolemic Homozygous			
T.B.			9,643
GM 2000			640

*One half the amount of microsomal protein from normal and mutant cells
indicated in items 1 and 3 were mixed and glycosyl transferase activity
measured as described in the text.

Table X. In a typical experiment, normal and mutant cells
(2 x 10^6/ flask) were cultured as described in the legend for
Table VI except that during the 48 hr incubation in 1% fetal calf
serum, no isotope was added. The cells were harvested and microsomes isolated. The assay system contained 100 µg of microsomal
protein, 100 µg cutscum, 50 µg triton x-100, 0.5 µmol $MnCl_2$, 0.15
µmole lactosylceramide, 0.15 µmole UDP-[^{3}H] Galactose (20 µCi/
µmole) and 2.5 µM Mes buffer (pH 6.4) in a total volume of 50 µl.
Following incubation for 2 hr at 37°, the reaction was terminated
with 2 ml of chloroform-methanol (2:1 v/v). GbOse3Cer (1 n mole)
was added as carrier. The lipid extract was subject to Folch
partitioning and the lower phase was chromatographed on silica
gel H coated thin layer plates. The chromatoplates were developed in chloroform-methanol-water (100:42:6 v/v) dried, and the
GSL_S scored following exposure of the chromatogram in iodine
vapour. Gel zones corresponding in chromatographic migration
with $GbOse_3Cer$ were scraped and radioactivity measured by
scintillation spectrometry. Appropriate boiled enzyme blanks
served as control. The control data was deduced from the experimental data and the activity of the enzyme expressed as c.p.m.
incorporated into trihexosyl ceramide/mg protein. The mixed
enzyme assays consisted of one half the amount of microsomal
protein from normal and mutant cells indicated above plus the
regular amounts of substrate, ions, detergents, buffer and glycolipids.

Since most of the GSL$_S$ in serum are associated with lipoproteins (16-18), these complex molecules are the major exogenous
source of GSL in cultured cells. Besides some de novo biosynthesis of GSL$_S$, serum supplemented in the culture medium is the major
source of GSL$_S$ in cultured fibroblasts (29). Indeed, cultured
fibroblasts derive most of their lipids, including cholesterol,
from LDL (6,68). However, cultured cells may also take up purified GSL$_S$ added to the culture medium (26,69). GSL$_S$ have also
been shown to be taken up from lipoproteins in other systems. For
example, several erythrocyte blood group antigens, p^k (GbOse$_3$Cer),
P (GbOse$_4$Cer) and Le^a and Le^b (both of which are GSL) (70), are
not synthesized by the red cell but are acquired from plasma lipoproteins (71,72). Dawson et al (29) analyzed the culture medium
before and after incubation with human fibroblasts and found very
little change in the concentration of GSL$_S$ in the medium. We
found that very little of the cellular GSLs prelabeled with [^{3}H]
galactose were released into the culture medium, even after 48
hours of incubation in lipoprotein deficient medium. These observations suggest that the mechanism of egress of GSL into the culture medium is not favored in normal human fibroblasts; in contrast, significant amounts of cholesterol is egressed in these
cells.

Most investigators use fetal calf serum in their cell culture
medium. In contrast to human serum, calf serum contains very low
levels of GSL$_S$ and fetal calf serum contains virtually none (29).
Medium (100 ml), supplemented with either 10% calf or fetal calf
serum, contains 0.1 and 0.02 u mole GSL$_S$, respectively (29).
Increasing the serum supplement from 10 to 30% had little effect
on the intracellular GSL content (29). In this study, the cells
were incubated in medium containing 12% fetal bovine serum. The
total GSL content of our normal fibroblasts was 2.77 n mole
glucose/mg cell protein, in agreement with values reported by
others (Dawson et al (29)). The amounts of the individual major
neutral GSL$_S$, namely GlcCer, LacCer, GbOse$_3$Cer, GbOse$_4$Cer, and of
the major gangliosides, ^{G}M3 and ^{G}D3, agreed well with values
reported by others in normal cultured fibroblasts (29,31). Fishman et al (32) reported a 3 fold variation in the content of ^{G}M3
in normal human fibroblasts. Fibroblasts from newborn foreskin
had a 3 fold lower level of ^{G}M3 than those from normal adult
fibroblasts. Such variability in ^{G}M3 among normal human fibroblasts may be due to difference in cell volume or size, in vitro
cellular aging accompanied by a decline in cell replication, in
vitro passage, or incubation temperature (73). Finally, cells
from newborns appear to divide rapidly, encompass less area, and
are smaller than adult fibroblasts (73).

The total amounts of GSL$_S$ in the cells of the obligate heterozygous FH parents of T.B. were two to three fold higher than
normal, while those of the FH homozygote, T.B. were four to five
fold higher. Each of the major neutral GSL$_S$ and major gangliosides was increased to about the same extent, except for the

content of LacCer which was similar to normal. Recently, Fishman et al (32) studied the cellular content of gangliosides and phospholipids in two other FH homozygotes and four lines of normal cells. They found no consistent differences between the FH and normal cells and attributed our previous findings in the B family to normal variation. A close inspection of the data in Table XI, however, indicates that the differences in the cellular gangliosides and phospholipid content were not confined to our FH homozygous lines. The cellular lipid levels of our two FH hetero-zygous lines were also elevated and there was no overlap between the lipid levels in these heterozygous cells and the homozygous FH subjects studied by Fishman and co-workers (32). The different results for gangliosides in the two studies may be, in part, related to the methods used, since we employed gas-liquid chroma-tography and Fishman et al (32) used densitometric scans of thin-layer plates charred with resorcinol. We believe, however, that the apparent discrepancy is probably due to differences in the phenotypic presentation of "homozygous" FH (see also above). This conclusion is supported by two additional lines of evidence. First, there was a 17-fold difference in the activity of lactosyl-ceramide galactosyltransferases between T.B. and GM 2000, one of the homozygotes studied by Fishman et al (32). Second, recent studies in our laboratory have indicated that the protein profile of isolated plasma membranes from this FH homozygote (T.B.) is distinct from that of two unrelated homozygotes (including GM 2000) (74). The protein profiles of all three homozygotes differ from those of normal fibroblasts as well (74). These biochemical differences may indeed reflect more than one mutation that affects the LDL receptor, and that are not distinguishable by the physiological assays currently used.

After 24 hr in LPDS medium, there was a marked decrease in the GSL content of T.B. cells (except for LacCer) and by 5 days the GSL levels in these FH homozygous cells approached those of normal. This finding may be related to: 1) egress of GSL into the culture medium in the presence of LPDS medium; 2) decreased biosynthesis of GSL_S via the glycosyltransferase system; 3) de-creased uptake of GSL from the LPDS medium, compared to the serum supplemented medium; and, 4) after 5 days, cells cultivated in the LPDS medium for more than two cell doublings may have under-gone marked changes in cell growth behavior, transport and repli-cation. We found no egress into the culture medium of GSL labeled with [^{3}H] galactose in normal or T.B. cells after 24 hours of incubation in LPDS medium. Therefore, in contrast to our earlier suggestion (30), egress of GSL does not appear to explain the decrease of GSL in T.B.'s cells in LPDS medium. The most likely explanation is that there is a decrease in the biosynthesis of the GSL in LPDS medium. This would suggest that there are factor(s) in serum that stimulate the biosynthesis of GSL in these FH cells. This factor is unlikely to be LDL since LDL is not bound or internalized by T.B.'s cells. However, nonspecific

Table XI

Differences in the lipid content of cultured
fibroblasts in familial hypercholesterolemia[+]

	Fishman et al (32)		Chatterjee et al (30)		
	N*	HMZ	N	HTZ	HMZ
GANGLIOSIDES (nmole/mg)	1.91 (1.22-2.81)	2.89;1.91	0.99	3.82;3.16	5.09
PHOSPHOLIPIDS (ug/mg)	161	160; 180	145	260; 213	415

N - normal; HTZ - heterozygotes; HMZ - homozygotes.

[+]In both studies, lipids were isolated from fibroblasts grown in the presence of media supplemented with fetal calf serum.

*Values represent the mean of four normal controls (range).

interaction of these cells with LDL may occur. Moreover, since
the fluidity of FH plasma membranes has been shown to be altered
by the presence and absence of cholesterol in the medium (75),
and since some glycosyltransferases are localized on the plasma
membrane, alteration in membrane fluidity may have had an effect
on the availability and activity of the glycosyltransferases.
It is also unlikely that the increased GSL in these cells is
related to the uptake of GSL from serum supplemented medium,
since the medium has previously been shown not to be the primary
source of cellular GSL in cultured fibroblasts (29) (see also
above). The effect of serum and lipoproteins on GSL biosynthesis
will be the subject of future studies in this laboratory.

At least four GSL_S were exposed on the cell surface of normal
and FH fibroblasts. LacCer took up very little label, in agreement with data obtained from BHK and NIL cells (60) and red blood
cells (61). Erythrocytes also do not react with an antibody
against LacCer (76). LacCer probably does not extend enough to
be accessible to galactose oxidase or it may primarily have an
intracellular distribution. Most of the GlcCer was not susceptible
to oxidation with galactose oxidase since more than 90% of its
hexose moiety was glucose rather than galactose. Although the
mutant cells had an increase in labeling of surface GSL_S, the
specific radioactivities of the GSL_S were lower than those of the
normal cells. The pool of unlabeled GSL in the mutant cells
might be explained by abnormal increases in both the surface and
cellular GSL_S, in which the accumulation of cellular GSL_S is dis-
proportionately larger than that on the surface. Alternatively,
surface GSL might account for most of the increase in GSL, leading to an absolute increase in labeling, but with significant
amounts buried in the cell membrane and therefore less exposed to
GAO.

The accumulation of GSL_S in the cultured fibroblasts from
this family was studied further by examining the incorporation of
[^{3}H] galactose into the individual GSL_S. T.B. had a three to
four fold increase in the total incorporation of [^{3}H] galactose
into GSL_S, compared with the average of four normolipidemic control cell lines. The most striking difference was the incorporation of galactose into GalCer of T.B. cells, because it suggested
that: a) relatively extensive interconversion (epimerization) of
this precursor to Glc was occurring; and b) the presence of trace
amounts of galactocerebroside might be contributing towards
increased incorporation of galactose. Also of interest was the
below normal incorporation of [^{3}H] galactose into the GSL_S of GM
2000, a receptor negative FH homozygote previously found to have
"barely detectable" neutral GSL_S on TLC (32). The results of
these incorporation studies must be interpreted with caution.
The amount of radiolabeled precursor incorporated into a molecule
or series of molecules may be affected by the rate of uptake of
the precursor by the cell, the amount (pool) of unlabeled precursor and unlabeled nucleotide sugars available inside the cell,

the intracellular rates of the biosynthesis and subsequent incorporation of the precursor itself, and the rates of biosynthesis, degradation exchange and metabolic interconversion of the molecule(s) into which the precursor is being incorporated. We have shown here that the uptake of 2-deoxyglucose, as a general measure of the uptake of carbohydrate precursors, was not abnormal in T.B. Further, the rates of protein and DNA synthesis in these cells were normal, suggesting that the cells were not being rapidly turned over and using up excessive labeled precursor in the process. These preliminary data also suggest that the cells were not arrested in mitotic or S (DNA synthetic) phase of the cell cycle in which maximum GSL synthesis and glycoprotein synthesis have been shown to occur (62). Considered together, our observations suggest that there is relatively more synthesis and insertion of GSL_s into T.B.'s cell membrane than normal cell membrane, whereas cellular protein turnovers were similar.

The possibility that the cells of T.B. were producing a factor that increased the incorporation of [3H] galactose into GSL was pursued by performing a "mixing" experiment in which cells from T.B. were co-cultured with normal cells. That about equal amounts of each cell line were present in the culture was indicated by the results of the LDL binding and degradation experiments. The co-cultured cells bound and degraded LDL to an extent that was intermediate between that of the homozyous FH cell line and the normal cells. Using this system, the amount of [3H] galactose incorporated into GSL_s was close to that predicted on the basis of the incorporated values of each cell line alone and was also close to that found in the FH heterozygous cell line of the mother of T.B. It is therefore unlikely that the cells of T.B. produce endogenous factor(s) that accelerate the incorporation of [3H] galactose into GSL_s.

The possibility that the above changes in the cellular content and metabolism of GSL in this family may be related to increased synthesis of GSL was next tested by assaying for the activity of the enzyme, lactosylceramide galactosyltransferase. LacCer serves as the major precursor for the synthesis of neutral GSL as well as gangliosides (33,34,42,54). The conversion of LacCer to $GbOse_3Cer$ and $^GM3$ occurs via specific glycosyltransferases and nucleotide sugars (42). The similarity in fatty acid composition between individual GSL_s in the BHK cells (77) and human fibroblasts (29) also supports the concept of direct metabolic interconversion. This system was chosen because LacCer appears to play a pivotal role in the GSL biosynthetic pathway in a wide variety of tissues including brain, kidney, and cells grown in tissue culture (33,42,54). Further, LacCer levels, in contrast to all the other major GSL_s was not increased in the cells of the B family. This suggested that LacCer may be used up rapidly and converted into $GbOse_3Cer$ and other GSL_s. In contrast, the levels of $GbOse_3Cer$ in T.B.'s cells were 5 fold higher than normal. We found that there was a 2 fold difference in the activity of this enzyme among normal fibroblasts; however, the activity of this enzyme in T.B. was almost 3 fold higher than the average normal value. Furthermore, when we mixed normal and T.B. microsomes, the galactosyltransferase activities were intermediate between normal and mutant cells. This observation suggested that: a) there was no specific induction of the synthesis of $GbOse_3Cer$ in mixed microsomal preparations presumably due to the absence of endogenous factors in T.B. microsomes; b) these observations are compatible with metabolic experiments employing radioactive galactose; and c) the increased levels of $GbOse_3Cer$ in T.B. could at least, in part, be explained on the basis of increased activity of galactosyltransferase.

One of the most striking aspects of the transferase data was that there was a 17 fold difference in the activities of lactosylceramide galactosyltransferase in T.B. and GM 2000, two patients with "receptor negative" homozygous FH. We are not able to explain this discrepancy at present. Further studies of this enzyme are indicated in a number of other FH homozygotes to determine the basis for this difference, whether a decreased activity is the rule in most FH homozygotes, and what the relevance of these findings is with respect to the basic molecular defect(s) underlying receptor-negative FH patients. One may speculate that the presence of marked depression of this transferase may be secondary to the inability of LDL to enter these cells (i.e., GM 2000) or may be indicative of some primary genetic abnormality in the transferase system(s).

Literature Cited

1. Svennerholm, E.; Svennerholm, L. Biochim. Biophys. Acta, 1963, 70, 432-438.
2. Tao, R.V.P.; Sweeley, C.C. Biochim. Biophys. Acta, 1970 218, 372-375.
3. Vance, D.E.; Sweeley, C.C. J. Lipid Res., 1967, 8, 621-630.
4. Vance, D.E.; Krivit, W.; Sweeley, C.C. J. Lipid Res., 1969, 10, 188-192.
5. Smith, L.C.; Pownall, H.J.; Gotto, A.M. Ann. Rev. Biochem., 1978, 47, 751.
6. Goldstein, J.L.; Brown, M.S.; Ann. Rev. Biochem., 1977, 46, 897.
7. Mahley, R.W.; Innerarity, T.L.; Pitak, R.E.; Weisgraber, K.H.; Brown, J.H.; Gron, E. J. Biol. Chem., 1977, 252, 7279.
8. Stein, Y.; Glangeand, M.C.; Stein, O. Biochim. Biophys. Acta, 1975, 380, 106.
9. Schaefer, E.J.; Eisenberg, S.; Levy, R.I. J. Lipid Res., 1978, 19, 667.
10. Weisgraber, K.H.; Innerarity, T.L.; Mahley, R.W. J. Biol. Chem., 1978, 253, 9053-9062.
11. Glomset, J.A.; Norum, K.R. Adv. Lipid Res., 1973, 11, 1.
12. Miller, G.J.; Miller, N.E. Lancet, 1975, 1, 16.

13. Schwartz, C.C.; Halloran, C.G.; Vlahcevic, F.R.; Gregory,
     D.H.; Swell, L. Science, 1978, 200, 62.
14. Bachorik, P.S.; Kwiterovich, P.O.; Cooke, J.C.; Biochemistry,
     1978, 17, 5287-5299.
15. Skipski, V.P.; Barclay, M.; Barclay, R.K.; Fetzer, V.A.;
     Good, J.J.; Archibald, F.M. Biochem. J., 1967, 103, 340-352.
16. Chatterjee, S.; Kwiterovich, P.O. Lipids, 1976, 11, 462-466.
17. Clarke, J.T.R.; Stoltz, J.M.; Mulcahey, M.R. Biochim. Bio-
     phys. Acta, 1976, 431, 317-325.
18. Dawson, G.; Kruski, A.W.; Scanu, A.M. J. Lipid Res., 1976,
     17, 125-131.
19. Danishefsky, I.; Zweben, A.; Slomiany, B.L. J. Biol. Chem.,
     1978, 253, 32-37.
20. Mannan, E.F. Determination of Antithrombin III. In:
     Thrombosis and Bleeding Disorders in Coagulation Disorders,
     Band, N.U.; Bieller, F.K.; Deutsch, E.; and Mannan, E.F.
     (eds) Academic Press Inc., New York, 1971, 268-277.
21. Goitel, S.N.; Wessler, S. Thrombos. Res., 1975, 7, 5.
22. Bell, W., Chatterjee, S.; Kwiterovich, P.O. (unpublished
     observations).
23. Auran, T.B.; Zavoral, J.H.; Krivit, W. Thromb. Res., 1974,
     5, 173-184.
24. Cuatrecasas, P. Biochemistry, 1973, 12, 3558-3566.
25. Mullin, B.R.; Fishman, P.H.; Lee, G.; Aloj, S.M.; Ledley,
     F.D.; Kohn, L.D.; Brady, R.O. Proc. Nat. Acad. Sci. USA,
     1976, 73, 842-846.
26. Fishman, P.H.; Bradley, R.M.; Moss, J.; Manganiello, V.C.
     J. Lipid Res., 1978, 19, 77-81.
27. Kleinman, H.K.; Martin, G.R.; Fishman, P.H. Proc. Nat. Acad.
     Sci. USA, 1979, 76, 3367-3371.
28. Matalon, R.; Dorfman, A. Proc. Nat. Acad. Sci. USA, 1966,
     56, 1310-1316.
29. Dawson, G.; Matalon, R.; Dorfman, A. J. Biol. Chem., 1972,
     247, 5944-5950.
30. Chatterjee, S.; Sekerke, C.S.; Kwiterovich, P.O. Proc. Nat.
     Acad. Sci. USA, 1976, 73, 4339-4343.
31. Fishman, P.H.; Moss, J.; Manganiello, V.C. Biochemistry,
     1977, 16, 1871-1875.
32. Fishman, P.H.; Bradley, R.M.; Brown, M.S.; Faust, J.R.;
     Goldstein, J.L. J. Lipid Res., 1978, 19, 304-308.
33. Basu, S.; Kaufman, B.; Roseman, S. J. Biol. Chem., 1968,
     243, 5802.
34. Chien, J.L.; William, T.J.; Basu, S. J. Biol. Chem., 1973,
     248, 1778.
35. Basu, S.; Schultz, A.M.; Roseman, S. Fed. Proc., 1969, 28,
     540.
36. Basu, S.; Schultz, A.M.; Basu, M.; Roseman, S. J. Biol.
     Chem., 1971, 246, 4272-4279.
37. Cleland, W.W.; Kennedy, E.P.; J. Biol. Chem., 1960, 235, 45.
38. Sribney, M. Biochim. Biophys. Acta, 1966, 125, 542.

39. Hammarstrom, S.; Sammuelsson, B. Biochem. Biophys. Res. Comm., 1970, 41, 1027.
40. Morell, P.; Radin, N.S. J. Biol. Chem., 1970, 245, 342.
41. Basu, S.; Kaufman, B.; Roseman, S. J. Biol. Chem., 1973, 248, 1388-1394.
42. Roseman, S. Chem. Phys. Lipids, 1970, 5, 270.
43. Fishman, P.H., Chem. Phys. Lipids, 1974, 13, 305.
44. Shier, W.T.; Trotter, J.T. Fed. Eur. Biochem. Soc., 1976, 62, 165.
45. Fishman, P.M.; Bradley, R.M.; Henneberry, R.C. Arch. Biochem. Biophys., 1976, 172, 618.
46. Lockney, M.; Moskal, J.R.; Fung, J.K.; Macher, B.A. Fed. Proc., 1978, 37, 2722.
47. Painter, R.G.; White, A. Cell Biol., 1976, 73, 837.
48. Chandrabose, K.A., Graham, J.M., MacPherson, L.A. Biochim. Biophys. Acta, 1976, 429, 112.
49. Chatterjee, S.; Velicer, L.F.; Sweeley, C.C. Biochem. Biochys. Res. Commun., 1973, 54, 585-592.
50. Siegrist, H.P.; Burkart, T.: Hoffman, K.; Wiesmann, U.; Herschkowitz, N. Biochim. Biophys. Acta, 1979, 572, 160-166.
51. Burton, R.M.; Garcia, L.B.; Golden, M.; Balfour, Y.M. Biochemistry, 1963, 2, 580.
52. Brady, R.O.; Borek, C.; Bradley, R.M. J. Biol. Chem., 1969, 244, 6552.
53. Chatterjee, S.; Murray, R.K. Proc. Can. Fed. Biol. Sci., 1972, 15, 156.
54. Dawson, G.; Matalon, R.; Dorfman, A. J. Biol. Chem., 1972, 247, 5951-5958.
55. Chatterjee, S. (Ph.D. Thesis, University of Toronto, 1972).
56. Shireman, R.B.; Fisher, W.R. Biochim. Biophys. Acta, 1979, 572, 537-540.
57. Fredrickson, D.S.; Levy, R.I. In, Metabolic Basis of Inherited Diseases, Stanbury, J.B.; Wyngaarden, J.B.; Fredrickson, D.S. (eds), McGraw-Hill, New York, p. 545.
58. Chatterjee, S.; Ose, L.; Kwiterovich, P.O.; Sekerke, C.S., manuscript in preparation.
59. Fung, C.H.; Khachadurian, A.K. Fed. Proc., 1978, 37, 295.
60. Gamberg, C.G.; Hakomori, S.I. J. Biol. Chem., 1973, 248, 9311-9317.
61. Steck, T.L.; Dawson, G. J. Biol. Chem., 1974, 249, 2135-2142.
62. Chatterjee, S.; Sweeley, C.C.; Velicer, L.F. J. Biol. Chem. 1975, 250, 61-66.
63. Hildebrandt, J.; Hauser, G. J. Biol. Chem., 1969, 244, 5170-5180.
64. Lowry, O.H.; Rosebrough, N.J.; Farr, A.L.; Randall, R.J. J. Biol. Chem., 1959, 193, 265-275.
65. Chatterjee, S.; Kwiterovich, P.O.; Sekerke, C.S. J. Biol. Chem., 1979, 254, 3704-3707.

66. Goldstein, J.L.; Brown, M.S.; Stone, N.J. Cell, 1977, 12, 629.

67. Kwiterovich, P.O.; Bachorik, P.S.; Chatterjee, S. In: The Genetic Analysis of Common Diseases: Applications to Predictive Factors in Coronary Heart Disease, Sing, C. and Skolnick, M. (eds), Alan Liss, in press.

68. Howard, B.; Kritchevsky, D. Biochim. Biophys. Acta, 1969, 187, 293-301.

69. Sourander, P.; Hansson, H.A.; Olsson, Y.; Svennerholm, L. Acta Neuropathol., 1971, 6, 231-242.

70. Hakomori, S.; Strycharz, G.D. Biochemistry, 1968, 7, 1279-1285.

71. Marcus, D.M.; Case, L.E. Science, 1969, 164, 553-555.

72. Naiki, M.; Marcus, D.M. Biochem. Biophys. Res. Commun., 1974, 60, 1105-1111.

73. Mitsui, Y.; Schneider, E.L. In: Mechanism of Aging and Development, 1976, 5, 45-56.

74. Chatterjee, S.; Sekerke, C.S.; Kwiterovich, P.O. Circulation 1978 (abstract).

75. Haggerty, D.F.; Kalra, V.K.; Popjak, G.; Reynolds, E.E.; Chiappelli, F. Arch. Biochem. Biophys., 1978, 189, 51.

76. Hakomori, S.J. Vox Sang., 1964, 16, 478-482.

77. Hakomori, S.J.; Teather, C.; Andrews, H. Biochem. Biophys. Res. Commun., 1968, 33, 563-568.

Footnotes

[1] To whom correspondence should be addressed.

[2] Abbreviations: The nomenclature of glycosphingolipids recommended by the IUPAC-IUB Commission will be adopted in this review. Described below are the previous abbreviations followed by the proposed abbreviations and the detailed structure of individual glycosphingolipids. GL-1a (GlcCer) = Glc-β-(1$\rightarrow$1')-Cer; GL-1b (GalCer) = Gal-β-(1$\rightarrow$1')-Cer; GL-2a (LacCer) = Gal-β-(1$\rightarrow$4)-Glc-β-(1$\rightarrow$1')-Cer; GL-3a (GbOse$_3$Cer)=Gal-α-(1$\rightarrow$4)-Gal-β-(1$\rightarrow$4)-Glc-β-(1$\rightarrow$1') Cer; GL-4a (GbOse$_4$Cer)=GalNAc-β-(1$\rightarrow$4)-Gal-α-(1$\rightarrow$4)-Gal-β-(1$\rightarrow$4)-Glc-β-(1$\rightarrow$1')-Cer; ^{G}M3 (II$^3\alpha$ NeuAc-LacCer)=NeuAc-α-(2$\rightarrow$3) Gal-β-(1$\rightarrow$4) Glc-β-(1$\rightarrow$1') Cer; ^{G}D3 (II^3 (NeuAc)$_2$-LacCer)=NeuAc-α-(2$\rightarrow$8)-NeuAc-α-(2$\rightarrow$3)-Gal-β-(1$\rightarrow$4)-Glc-β-(1$\rightarrow$1')-Cer.

RECEIVED January 2, 1980.

Biochemical, Morphological, and Regulatory Aspects of Myelination in Cultures of Dissociated Brain Cells from Embryonic Mice

RONALD A. PIERINGER, GRAHAM Lem. CAMPBELL, NARAYAN R. BHAT, G. SUBBA RAO, and LOUIS L. SARLIEVE[1]

Department of Biochemistry, Temple University School of Medicine, Philadelphia, PA 19140

Myelination is an essential step in the development of the nervous system of higher animals, which occurs early in life, and is completed within a relatively short period. The central nervous system of the rat and mouse, for example, is myelinated most actively between the 10th and 20th day after birth (1). During this period dramatic morphological and biochemical changes have been observed. The biochemical parameters which best measure this temporal change are the enzymes and compounds most closely associated with myelination. Cerebrosides (2), galactosyl diglycerides (3), sulfatides (2), sulfogalactosyl diglycerides (4) and the enzymes catalyzing their synthesis (3, 4, 5, 6), the myelin basic protein (7) and cyclic nucleotide phosphohydrolase (8) are very useful molecular markers of myelination.

The identification of the regulators of myelination and their mechanism of action at the molecular level has only been partially uncovered by studies on the intact animal. Although in vivo studies have implicated thyroid hormone (9, 10, 11, 12) as a potentially important regulator, these studies using whole animals can not demonstrate whether thyroxine is acting directly or indirectly on the myelin producing cells. Manipulation of one hormone in vivo invariably affects the availability and concentration of many other hormones. In contrast to the intact animal the growth of animal cells in culture offers the possibility of measuring a direct interaction between hormone, such as thyroxine and a myelin-producing cell such as the oligodendrocyte.

Several types of in vitro culture systems have evolved for the study of myelination. Cultures of cerebellar explants produced myelin-like processes, which can be stimulated to grow by thyroid hormone (13). However, the explant system produces only small amounts of tissue which may be sufficient for morphological and histochemical studies but are usually insufficient

---

[1]Current address: Centre de Neurochimie 67085 Strasbourg France

for biochemical measurements. On the other hand, primary cultures of cells dissociated from embryonic mouse or rat brain produce significant quantities of myelin-like tissue when grown as aggregates either on polylysine coated plastic surfaces (14, 15) or in suspension (16 17). Recent biochemical investigation by Sheppard et al. (18) and Matthieu and coworkers (19 20) have demonstrated the suitability of the suspension culture system for the study of myelination in vitro. In independently initiated studies we have adapted the surface adhering primary culture system of Sensenbrenner (14) and Yavin and Yavin (15) to a form suitable for studying the regulation of myelination by thyroid hormone and other effectors in vitro. In this paper we will demonstrate the efficacy   of studying myelination in these cultures; and present data showing the direct influence of L-triiodothyronine (T_3) on myelination in these cultures as monitored by the synthesis of some of the lipids closely associated with myelin. Parts of these investigations have been published (21 22).

Materials and Methods

Dulbecco's modified Eagle's medium, calf serum (heat-inactivated), and antibiotic mixture were obtained from GIBCO, Grand Island, NY. Serum from thyroidectomized calf was purchased from Rockland Farms, Gilbertsville, Pa. Sterile culture dishes and flasks were supplied by Fisher Scientific Co., Pittsburgh, Pa. Polylysine (M_r = 80,000) and 3, 5, 3'-triiodothyronine were from Sigma Chemical Co., St. Louis. $H_2/^{35}S/0_4$ (3.6 Ci/mmol) and $/^3H/$ galactose (2.1 Ci/mmol) were obtained from New England Nuclear, Boston, Ma. Pregnant mice (ICR) were supplied by Charles River Laboratories, Boston, Ma.

Culture Condition. Fifteen-day-old embryos numbering 11 to 14 per litter were removed by Caesarean section (23) and placed in Dulbecco's modified Eagle's medium augmented with glucose (600 mg %), 0.23% sodium bicarbonate, 90 units/ml penicillin, 90 µg/ml of fungizone, adjusted to pH 7.0. Cerebral hemispheres (cerebra) were dissected and temporarily placed in the medium. Cells were then dissociated mechanically (24) by passing through a nylon mesh (82 µ). The cells were collected in a small volume (~1 ml for three cerebra) of the above medium supplemented with 20% heat-inactivated calf serum. Aliquots (1 ml) of the cell suspension containing, on an average, 10 to 15 x 10^6 cells/ml were added to 250 ml polylysine-coated (25) plastic tissue culture flasks containing 9 ml of growth medium. Cultures were incubated at 37° under an atmosphere of 90% air, 10% CO_2 and 90% relative humidity. The medium in the flasks containing non-attached cells was carefully removed on the 4th day and replaced by 10 ml of the fresh medium containing 20% calf serum. Thereafter, the medium was changed once a week.

Incorporation of Radioactive Precursors into Lipids. The
cultures of surface-adhering cells were exposed for 16 h to
either 400 µCi of $H_2[^{35}SO_4$ (final specific activity, 50 µCi/
µmol) or 10 µCi of $[^3H]$, galactose (the rate of incorporation
was linear during this period). After 16 hr. the radioactive
medium was removed and the cultures were washed four times with
0.9 NaCl. The cells were removed from the surface with a rubber
policeman and suspended in physiological saline. Lipids were
extracted by Bligh and Dyer procedure (26) and analyzed for
various lipids according to Neskovic, et al. (27)

Enzymes and Protein Assays. Aliquots of sonicated cells were
adjusted with 0.32 M sucrose to give 10% homogenates which were
they lyophilized, and stored over $CaCl_2$ desiccant at -20%. When
needed, portions of the powder were reconstituted with water to
the original concentration of about 6 to 10 mg protein per ml.
The assays for the enzyme synthesis of sulfogalactosyl-
glycerolipid, sulfatides and galactocerebrosides were carried
out as previously described respectively by Subba Rao, et al. (28)
Sarlieve, et al. (5, 29), and Neskovic, et al. (30). The assay
for 2', 3' cyclic nucleotide phosphohydrolase was performed according to the method of Prohaska, et al. (31). E. coli. alkaline
phosphatase type III-S, 2', 3' -cAMP, and sodium deoxycholate were
obtained from Sigma (St. Louis, Mo.). Protein was determined by
the method of Lowry, et al. (32) with crystalline bovine serum
albumin as the standard.

Preparation of coverslips for scanning electronmicroscopy.
Coverslips were pretreated as described by Campbell and Williams
(32). These coverslips were placed in a petri dish (35 mm) onto
which the cell suspension was plated out. Dulbeco's minimal
essential medium plus 20% fetal calf serum was added to the dish.
Each dish was then incubated at 37° in 10% CO_2. Each coverslip
was recovered, washed in Hanks B.S.S. (Gibco) and then immersed in
Hanks B.S.S. containing 4% glutaraldehyde for at least 4 hrs.
The coverslips were then prepared for scanning electron micro-
copy using the critical drying technique (34).

Results
    Our initial studies were carried out on cultures of dissocia-
ted cells from brains of one day old mice. Unfortunately these
preparations had lower than hoped for activities of certain biochemical parameters of myelination. For example, both the incorporation of $^{35}SO_4$ into sulfolipids and the activity of the cyclic
nucleotide phosphohydrolase attained relatively small activities
even at the age (8 to 11 days) of optimum activity in culture
(fig. 1 and 2). The dissociated cells from the 1 day old mouse
produced parallel developmental patterns of both myelin parameters
with peak activities at 8 to 11 days in culture. However cultures
prepared from 15 day fetal brains proved to have relatively

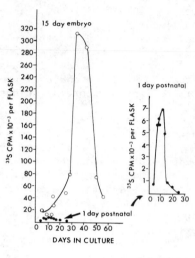

Figure 1. Incorporation of $^{35}SO_4^{2-}$ into lipids of dissociated cells from 15-day embryonic and from 1-day postnatal mouse brain grown for different days in vitro. The $^{35}SO_4^{2-}$ was exposed to the cells for 16 hr (a time in which the incorporation of ^{35}S still proceeded linearly). The data are expressed as activity from $^{35}SO_4^{2-}$ incorporated into the total number of cells (or total proteins) per flask (21).

Brain Research

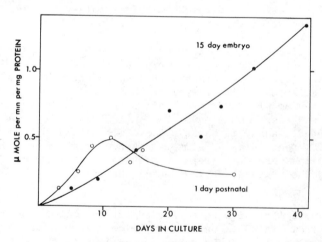

Brain Research

Figure 2. 2', 3'-cyclic AMP phosphohydrolase activity of dissociated cells derived from brains of 15-day embryonic or 1-day postnatal mice and grown for various days in culture (21).

high activities after about 33 to 41 days in culture in both the incorporation of $^{35}SO_4^=$ into sulfolipids and the cyclic nucleotide phosphohydrolase (fig. 1 and 2)

The remainder of our studies were all carried out on cells derived from the 15 day old fetus. The incorporation of $^{35}SO_4^=$ and $/^{-3}H(G)/$-galactose from the media into the sulfolipids of cells proceeded at a linear rate up to 16 hours (fig. 3). Two $/^{-35}S/$-sulfolipids were produced by these cells and these were identified by cochromatography with appropriate standards in several solvent systems as sulfogalactosylceramide and sulfogalactosyl diacyl- and alkylacylglycerol (21). The sulfogalactosylceramide contained 80 to 85% of the total radioactivity and the sulfogalactosyl glycerol lipid had 15 to 20%. The amount of ^{35}S in the sphingolipid shifted from 80 to 85% as the age of the culture changed from 20 to 41 days. (21). The quantity of radioactivity recovered in sulfogalactosylglycerol and sulfogalactosylmonoalkylglycerol, was 73% and 27% respectively. These ratios of sulfosphingolipid to sulfoglycerol lipid and of the diacyl to alkylacyl forms of the latter lipid are similar to the amounts found in rat brain by Pieringer, et al.

The incorporation of D/$^{-3}H(G)/$-galactose into the lipids of these growing cell cultures was also determined. The ability of the cells to incorporate the radioactivity of $/^{-3}H/$-galactose into four myelin-associated lipids (cerebroside, galactosyl glycerol lipid, sulfogalactosyl ceramide, and sulfogalactosyl glycerol lipid) over a 16 hr period in general increased with increasing age of the culture up to 41 days (Table I). These results parallel the data obtained with ^{35}S in the media. (figs. 1). Cells grown for certain days in culture incorporate more radioactivity into sulfatide than into cerebroside during the 16 hr exposure to $/^{-3}H/$-galactose. Some caution must be exercised in attempting to correlate the data of Table I with the actual concentration of each lipid in the culture. The data reflect the balance between the synthesis of the lipid from exogenous $/^{-3}H/$-galactose and the conversion of the synthesized lipid to some other metabolite(s) or degradation products(s) during a 16 hr. period. They do not necessarily indicate concentration especially since the amount of each lipid accumulated for a number of days in culture prior to the exposure to $/^{-3}H/$-galactose is unknown. Also it has not been established that the externally derived $/^{-3}H/$-galactose will be metabolized the same as or equilibrated with endogenously synthesized galactose.

Very little of $/^{-3}H/$-galactose precursor was converted to a glucocerbroside or to a lactosylceramide under these culture conditions (21). The activities of the sulfotransferase and galactosyltransferase, enzymes responsible for the synthesis of the myelin-associated sulfo- and glycolipids, were also measured at different days in culture (fig. 4). The activities of these enzymes in homogenates of cells derived from the 15 day embryo were relatively low at 8 days but increased until reaching

308   CELL SURFACE GLYCOLIPIDS

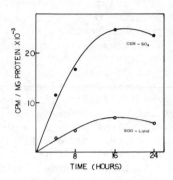

Figure 3. Time course of incorporation of $^{35}SO_4{}^{2-}$ into cerebroside sulfate (cer-SO₄) and sulfogalactosyl glycerol lipid (SGG-lipid) in dissociated brain cells from 15-day mouse embryos grown 19 days in culture

TABLE I

Incorporation of [³H]-galactose into lipids in cultures of dissociated brain cells from 15 day embryo at different days in culture.

Days in Culture	Cer	MGD	SG-Ceram	SGG-Lipid
	CPM/mg Protein/16 hrs.			
5	275	58	222	69
9	373	59	279	87
15	321	73	484	125
20	1441	322	2537	650
25	515	269	3662	339
28	876	346	6214	615
34	2267	824	19067	1732
41	6116	2143	24826	1400
50	201(192)	45(41)	1667 (1883)	9 (9)
56	111(74)	8(11)	749 (852)	59 (37)

Cer. (cerebroside); MGD (monogalactosyl diacylglycerol); SG-Ceram. (sulfogalactosyl ceramide); and SGG-lipid (sulfogalactosyl diacyl- and monacylmonoalkylglycerol). Data in parentheses are duplicate values. The cells were grown in the presence of the [³H]-galactose for 16 hrs. (21) Brain Research, Table I.

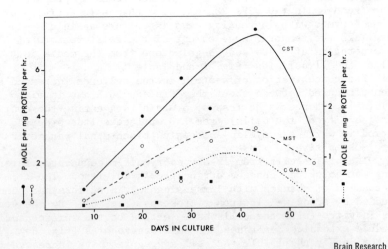

Brain Research

Figure 4. Activities of cerebroside; PAPS sulfotransferase (CST); monogalactosyl diacylglycerol: PAPS sulfotransferase (MST); and hydroxy fatty acyl ceramide: UDP-galactose galactosyltransferase (C Gal. T) in reconstituted homogenates of dissociated brain cells from 15-day embryonic mice grown for varying days in culture. Note that the activity of the galactosyltransferase is expressed on a different scale from that of the sulfotransferase (21).

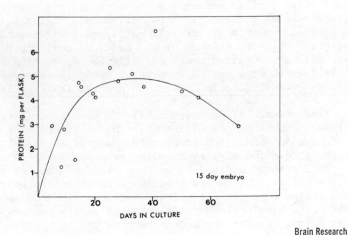

Brain Research

Figure 5. Concentration of protein per flask of dissociated cells from 15-day embryonic mice at different days in culture (21)

a peak at about 43 days in culture (fig. 4). The peak activities
appeal to be the same order of magnitude found in preparations
from fresh brain (30, 29, 28, 5). The change in activity of
these enzymes with increasing days in culture again closely
parallels the temporal development of the relative rate of
incorporation of $^{35}SO_4^=$ and ^{3}H-galactose from the media into
sulfo- and glycolipids (fig. 1 and Table I).

The concentration of protein in the cultures increased and
then plateaued between about the 20th and 56th day in vitro
(fig. 5). There was no loss of protein (suggesting the continued viability of the cells) from the surface cultures during the
period in which the myelin-associated biochemical parameters
were most active.

The brain cells from 15 day embryonic mice undergo interesting morphological changes during growth in vitro. The
succession of changes with increasing age in culture had been
studied by scanning electron microscopy (fig. 6). By the 4th
day in vitro (DIV) the cells have settled on the surface and
produce extensive membranes and small aggregates (fig. 6a). By
approximately the 15 DIV the cell aggregates have increased in
size and have coalesced to form nests of cells (fig. 6b). During
this stage the surface is initially filamentous but by the 29 DIV
appears smoother as if covered with a martrix (fig. 6c). By the
43 DIV the nests of cells and the fibrous character of the surface have disappeared, apparently covered by a membrane- like
substance, leaving a relatively smooth surface (fig. 6d). Higher
magnifications of each of the panels of fig. 6 are shown in
fig. 7 which further emphasizes the striking temporal morphological changes observed in these cultures. Figure 8 illustrates
the general features of the cell nests and their morophological
diversity. The key feature of all nests of cell groups concerns
their ability to extend membranes (fig 8b, d, e, f). However,
the cell morphology varies from being round to flattened. Occas-
ionally a group of cells is seen to produce thick membrane (fig.
8c).

The sequence of development of the membranous material
observed in culture correlates with the progression of activities
of the myelin-associated parameters measured above. Support
for the production of a myelin-like membrane in these surface
adhering, primary cultures have come from the studies of Yavin
and Yavin (25). Using transmission electron microscopy they
demonstrated that similar primary cultures produced myelinated
axons. The bimolecular myelin membrane of the culture derived
preparations appeared to be the same as in vivo derived preparations.

The culture system used in this study has proven suitable
for studying the regulation, especially by hormones, of myelination in vitro. Initial studies (35-36), however, showed no
effect of 3, 5, 3'-triiodothyronine (T_3) on sulfolipid synthesis
by dissociated brain cells grown on medium containing 20% calf

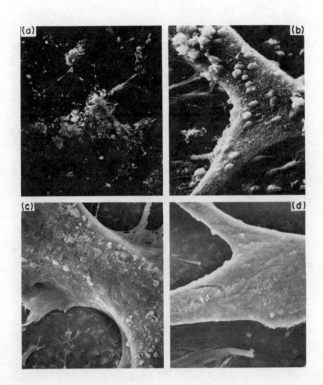

Figure 6.   Microphotographs illustrating the morphological changes associated with different stages of the in vitro growth patterns for cell populations isolated from brains of 15-day old mouse embryos (21)

(a): After four DIV a combination of membranous structure and cell aggregates form (×140). (b): After 15 DIV super aggregates develop in addition to nests of small cells (×420). (c): After 29 DIV, aggregates have become very large, and their surfaces are covered with matrix and single cells (×210). (d): By 43 DIV the surface of the aggregates is smooth and characterized by the absence of single cells (×210).

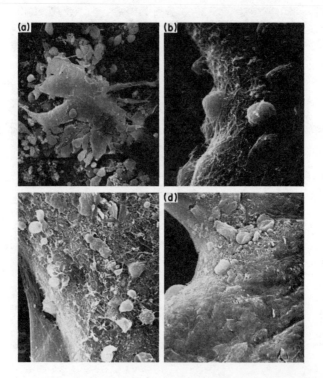

Figure 7. Microphotographs illustrating the morphological features associated with changes in Figure 6; at higher magnification topographical detail is clearer.

(a): Four DIV, note the membrane covering this aggregate (×700). (b): 15 DIV, note the filar nature of the surface (×1400). (c): 29 DIV, surface topography is less filamentous, cells are still readily visible (×700). (d): 43 DIV, surface topography is quite smooth, absence of filament and individual cells are not readily visible (×700).

serum. It should be noted that the endogenous concentrations of
hormones and growth factors in the serum may be sufficiently
high (36) to preclude observing an effect by exogenous hormones.
This condition could not be differentiated from that in which the
added hormone is completely nonassociated with the process being
measured. The role of thyroid hormone may be tested best by
using calf serum obtained from a thyroidectomized animal.
Depressed activities observed in the hypothyroid state should be
restored by exogenous hormones. This experimental design has
been used successfully by Samuels, et al. (37) to study the
effect of thyroid hormone on the metabolism of pituitary tumor
cell line in culture.

The hormone concentrations in the normal and hypothyroid
calf sera as determined by radioimmunoassay are given in Table
II. Both thyroxine (T_4) and T_3 values are far below normal in
hypothyroid calf serum. In the experiments described, the
growing brain cells in culture were challenged with hypothyroid
calf serum and hormone supplementation. The accumulation of
sulfatides (cerebroside sulfate and monogalactosyl diacylglycerol)
was studied by using $H_2[^{35}S]O_4$ and $[^3H]$-galactose as the labeled precursors. The synthesis of these lipids has been used as
an index for following myelination (11, 12, 2, 3, 4).

Table III gives the results of an experiment in which the
effect of serum manipulation on sulfolipid synthesis by dissociated brain cells was examined. The cells isolated from the
embryonic mouse brain were grown on the medium containing calf
serum (20%) for 3 days by which time most of the cells would
have attached to the substratum. On the 4th day, the medium
was replaced by fresh medium containing calf serum, calf serum +
T_3 (2 x 10^{-8}M), hypothyroid calf serum, or hypothyroid calf
serum + T_3 (13 ng/ml). Cultures were grown for another week and
then labeled with 400 µCi $H_2[^{35}S]O_4$, for 16 h. The lipids
isolated from the cultures were analyzed by thin layer chromatography and the radioactivity was determined. As is clear from
the results, when T_3 was added to the cultures grown on media
containing normal calf serum, hardly any effect was discernible.
On the other hand, presence of hypothyroid calf serum caused a
reduction in the synthesis of sulfolipids. This inhibition could
be reversed by including T_3 in the deficient medium.

In Table IV is described the effect of hormone manipulations
on myelin lipid synthesis at a later stage, namely, 11th day in
culture. The cultures were exposed to the effectors for 3 days
and then the synthesis of glycolipids was followed by labeling
the cells with $H_2[^{35}S]O_4$ and $[^3H]$-galactose. The total lipid
extract was analyzed for cerebrosides, monogalactosyl diacylglycerol, cerebroside sulfate, and monogalactosyl diacylglycerol
sulfate. As expected, there was about a 3- to 4-fold increase
in the rate of synthesis of sulfolipids on the 15th day as compared to 10 days in culture. The synthesis of all the four lipid
classes studied appears to be affected by thyroid hormone level

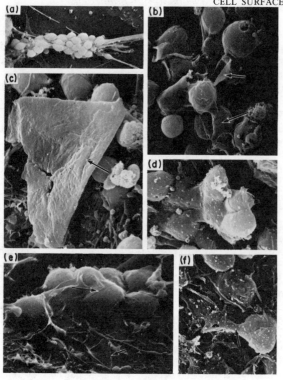

*Figure 8.   Microphotographs illustrating the morphological features of cell groups/
nests representative of activities after nine DIV.   Arrows point to the different
membranous structures characteristic of cell nests.   (a): $\times 700$; (b)–(f): $\times 2100$.*

Table II

Values of Thyroxine (T_4) and Triiodothyronine (T_3) in normal
and hypothyroid calf-sera (22).

	Hormone Concentration	
	T_4, μg/ml	T_3, ng/100 ml
Calf serum	5.8	110
hypothyroid calf serum	1.2	<25

J. Biol. Chem. Table I

Table III

Effect of serum manipulation and T_3 addition on sulfolipid synthesis by the dissociated brain cells treated on 4th day culture.

Cultures were grown for 3 days on medium containing normal
calf serum. On the 4th day, the cells were fed medium containing
normal or hypothyroid calf serum or without T_3 (2 x 10^{-8}M) supplementation. On the 10th day in culture the cells were labeled
with $H_2{}^{35}SO_4$ (400 µCi) (50 µCi/umol) for 16 hrs. The labeled
lipids were extracted from the cells and the radioactivity
determined. Values are mean ± s.d. of 4 experiments (22).

	Total cell protein mg/flask	$H_2{}^{35}SO_4$ incorporated into the lipids cpm/mg protein
calf serum (cs)	2.09 ± 0.20	4154 ± 642
CS T_3	2.10 ± 0.22	4122 ± 610
Hypothyroid calf serum (HCS)	1.64 ± 1.31	2231 ± 300
HCS T_3	1.88 ± 0.25	4142 ± 416

J. Biol Chem. Table II

TABLE IV

Effect of serum manipulation and T_3 addition on the incorporation of [^{35}S] H_2SO_4 and [^{3}H]galactose into glycolipids by dissociated brain cells treated on 11th day in culture.

After growing for 11 days on the medium containing sera, the cells were treated with hypothyroid calf serum and T_3 as described in Table 3. Three days later the cells were exposed to $H_2^{35}SO_4$ (400 μCi) (50 μCi/μmol) and [^{3}H] galactose (10 μCi) to (2.1 Ci/nmol) for 16 hrs. The cells were then extracted and analyzed for lipid content as described in Materials and Methods. (22).

Exp.		Total protein mg/flask	Incorporation of [^{3}H]galactose and [^{35}S] into the lipids, cpm/mg protein					
			[^{3}H]gal.				[^{35}S]H_2SO_4	
			cerobroside	MGD	CerSO$_4$	MGD-SO$_4$	CerSO$_4$	MGD-SO$_4$
CS	1	2.36	4670	902	22265	3644	14207	2928
	2	2.36	3710	706	24142	3992	11845	2684
CS+ T$_3$	1	2.85	4187	1143	23428	4354	15120	3146
	2	2.50	4955	1096	23321	4719	13196	2951
HCS	1	2.14	3066	313	9660	2256	7061	1735
	2	2.29	2751	266	7886	1584	6707	1782
HCS + T$_3$	1	2.50	4171	616	24735	3042	22142	3876
	2	2.50	3559	500	23963	3510	19925	4538

MDG (monogalactosyl diacyl- and monoacylmonoalkylglycerol),Cer-SO$_4$ (cerebroside sulfate), and MGD-SO$_4$ (monogalactosyl diacyl- and monoacylmonoalkylglycerol sulfate).

J. Biol. Chem. Table III

in the medium. The concentration of the hormone capable of eliciting a response in vitro was similar to the concentration found in vivo, indicating that the responsiveness of the culture system is quite sensitive to thyroid hormone.

Discussion

The temporal appearance of myelin-related lipids and enzymes in the cultures of dissociated fetal mouse brain cells mimics the temporal development of these parameters in normal mouse or rat brain (12, 5, 29, 28). The types of myelin-related lipids and the order of magnitude of the activities of the enzymes producing some of these lipids are the same as those found in brain in vivo (30, 29, 28).

The biochemical parameters of myelination appear to be controlled in a coordinated manner. That is, the parameters we measured responded in the same direction and degree to a certain condition of growth. For example, the cyclic nucleotide phosphohydrolase and incorporation of ^{35}S into lipids both were less active in cultures from the newborn mouse than from the 15 day fetus. In addition the developmental pattern of all of the activities measured occurred in parallel and were all highest at about the same growth period. This coordinated control suggests that they are derived from a central source, such as a single cell type (for example, oligodendroglia) of brain.

There is a well defined progression of morphological changes during the in vitro growth of brain cell population isolated from fifteen day old mouse embryos. The morphological changes correlated with the biochemical findings. During the first seven to eight days, there is relatively little synthesis of myelin associated parameters. This period is characterized by cells which adhere to the polylysine-coated plastic surface and synthesize extensive but thin membranes, and other membrane synthesizing cells adhere to preexisting membrane and begin to form aggregates. During the next fifteen to twenty days there is an increase in all parameters measured. The morphological features during this time show active growth and coalescence of aggregates and the synthesis of membrane is ubiquitous. Between 35-40 DIV the myelin-associated synthetic activities reach a maximum and the aggregates show distinct surface changes, the disappearance of cell nests, the loss of filamentous structures and the development of a rather smooth topography. The relative significance of our findings is the visualization of membrane synthesis and profusion which corresponds to the biochemical analysis.

The results on the effect of thyroid hormone on myelin lipid synthesis correlate well with the in vivo changes induced under altered thyroid functions, proving, thereby, the direct influence of T_3 on brain maturation. Another in vitro system used to demonstrate such a direct effect of thyroid hormone on myelination was an explant culture of cerebella obtained from newborn rats.

Using this system, Hamburg (13) could show an acceleration of
myelinogenesis by T_4 (1.5 to 3 µg/ml) addition. The mechanism
whereby T_3 influences myelination includes (a) differentiation of
the neuroglial cell population responsible for myelin synthesis
(38), (b) induction of such differentiated cells to synthesize
myelin components, and (c) assembling of the various components
to form the complex myelin membrane. Studies are underway to
examine the latter possibilities.

Acknowledgment

    This work was supported by the Kroc Foundation and in part,
by Research Grants NS-10221, AI-05730, and RR05417 from the
United States Public Health Service.

Literature Cited

1. McIlwain, H., and Bachelard, H. S., Biochemistry and the
Central Nervous System, Churchill Livingstone, London, 1971.
2. Wells, M. A., and Dittmer, J. C., Biochemistry, 1967, 6, 3169.
3. Inoue, T., Deshmukh, D. S., and Pieringer, R. A., J. Biol
Chem. 1971, 246, 5688.
4. Pieringer, J. A., Subba Rao, G., Mandel, P., and Pieringer,
R. A., Biochem, J., 1977, 166, 421.
5. Sarlieve, L. L., Neskovic, N. M., Rebel, G., and Mandel, P.,
Neurobiology, 1972, 2, 70.
6. Neskovic, N. M., Nussbaum, J. L., and Mandel, P., Brain
Research, 1970, 21, 39.
7. Braun, P. E., and Brostoff, S. W., Myelin, P. Morell (Ed.).
Plenum Press, New York, 1977, p. 201.
8. Kurihara, T., and Tsakada, J. Neurochem., 1968, 15, 827.
9. Balazs, R., Brooksbank, B. W. L., Davison, A. N., Eayrs, J.
T., and Wilson, D. A., Brain Res., 1969, 15, 219.
10. Balazs, R., Brooksbank, B. W. L., Patel, A. J., Johnson, A.
L., and Wilson, D. A., Brain Res., 1971, 30, 273.
11. Walravens, P., and Chase, H. P., J. Neurochem, 1969, 16, 1477.
12. Flynn, T. J., Deshmukh, D. S., and Pieringer, R. A., J. Biol.
Chem., 1977, 252, 5864.
13. Hamburgh, M., Dev. Biol. 1966, 13, 15.
14. Sensenbrenner, M., Cell, Tissue and Organ Cultures in Neurobiology, In S. Fedoroff and L. Hertz (eds.), Academic Press,
New York, 1977, p. 191.
15. Yavin, E. and Yavin, Z., J. Cell Biol., 1974, 62, 540.
16. Bornstein, M. D., and Model, P. G., Brain Res., 1972, 37, 287.
17. Seeds, N. W., and Vatter, A. E., Proc. Natl. Acad. Sci. 1971.
68, 3219.
18. Sheppard, J. R., Brus, D., and Wehner, J. M., J. Neurobiol,
1978, 9, 309.
19. Matthieu, J. M., Honeggar, P., Trapp, B. D., and Cohen, S. R.,
Neuroscience, 1978, 3, 565.

20. Matthieu, J.M., Honeggar, P., Favrod, P., Gautier, E., and Dolivo, M., J. Neurochem., 1979, 32, 869.
21. Sarlieve, L.L., Subba Rao, G., Campbell, G.LeM., and Pieringer, R.A., Brain. Res. 1979 in press.
22. Bhat, N.R., Sarlieve, L.L., Subba Rao, G., and Pieringer, R.A., J. Biol. Chèm., 1979, 254, 9342.
23. Paul, J., Cell and Tissue Culture, Churchill Livingstone, New York, 1975., p. 191.
24. Sensenbrenner, M., Booher, J. and Mandel, P., Z. Zellforsch, 1971, 117, 559.
25. Yavin, Z. and Yavin, E., Exp. Brain Res. 1977, 29, 137.
26. Bligh, E.G. and Dyer, W.J., Can. J. Biochem. Physiol, 1959, 37, 911.
27. Neskovic, N., Sarlieve, L.L., Nussbaum, J.L., Kostic, D., and Mandel, P., Clin. Chim. Acta, 1972, 38, 147.
28. Subba Rao, G., Norcia, L.N., Pieringer, J. and Pieringer, R. A., Biochem. J., 1977, 166, 429.
29. Sarlieve, L.L., Neskovic, N.M., Rebel, G. and Mandel, P., Exp. Brain Res., 1974, 19, 158.
30. Neskovic, N.M., Sarlieve, L.L., and Mandel, P., Biochim. Biophys. Acta, 1974, 334, 309.
31. Prohaska, J.R., Clark, D.A., and Wells, W.W., Anal. Biochem., 1973, 56, 275.
32. Lowry, O.H., Rosebrough, A.L., Farr, A.L., Randall, R.J., J. Biol. Chem., 1951,193, 265.
33. Campbell G. LeM and Williams, M.P., Brain Res., 1978, 127, 69.
34. Silberberg, D.H., J. Neuropath. Exp. Neurol., 1975, 34, 189.
35. Sarlieve, L.L., Subba Rao, G., and Pieringer, R.A., Proceedings of the European Society for Neurochemistry, 1978, Vol. 1, p. 140 Abstract.
36. Subba Rao, G., Sarlieve, L.L., and Pieringer, R.A., Fed. Proc., 1978, 37, 1627, Abstract
37. Samuels, H.H., Tsai, J.R., and Cintron, R., Science, 1973, 181, 1253.
38. Hamburgh, M., Curr. Top. Dev. Biol., 1969, 4, 109.

RECEIVED January 2, 1980.

Gangliosides and Associated Enzymes at the Nerve-Ending Membranes

G. TETTAMANTI, A. PRETI, B. CESTARO, M. MASSERINI, S. SONNINO, and R. GHIDONI

Department of Biological Chemistry, The Medical School, University of Milan, Milan, Italy

Gangliosides are characteristic glycolipid components of the plasma membranes of mammalian cells. They are particularly abundant in the nervous tissue, specially the grey matter, where their concentration is about one tenth that of total phospholipids. The evidence concerning the high content of gangliosides in the neuronal membranes, and of their peculiar location in the outer membrane surface, stimulated research and speculation on the possible involvement of gangliosides in brain specific functions. However, in order to provide a plausible working hypothesis for such involvement a more precise knowledge on the contribution given by gangliosides to the local environment of the neuronal membrane is required.

Chemical and physico-chemical properties of gangliosides: a molecular introduction to ganglioside behavior in cell plasma membranes.

Gangliosides are a family of glycosphingolipids which contain at least one residue of sialic acid. The number of sialic acid residues per ganglioside molecule varies from 1 to 7, with an average content of 2-2.5 in the brain gangliosides of most verte-brates (1). The sialic acid residue(s) is(are) attached to the neutral oligosaccharide core which may contain glucose, galactose, N-acetylhexosamine (generally N-acetylgalactosamine) and fucose. The most abundant oligosaccharide core occurring in brain ganglio-sides is ganglio-N-tetraose, gal(β, 1→3)galNAc(β, 1→4)gal (β, 1→4)glc. The acidic oligosaccharide is β-glycosidically linked to ceramide, formed by a long chain fatty acid (primarily C 18:0) and a long chain, mainly unsaturated, base (C 18 and C 20) linked together by an amide bond.

0-8412-0556-6/80/47-128-321$5.75/0
© 1980 American Chemical Society

The oligosaccharide portion of gangliosides, responsible for the high hydrophilicity of the molecule, displays a double potentiality of interactions: hydrogen bonds and ion bonds (cation binding sites). These bonds are more likely to occur between the saccharide chains of adjacent gangliosides (or other glycoconjugates) molecules than within the same ganglioside molecule. Thus they play an important role in any process of ganglioside association. The ^{13}C NMR studies performed by Sillerud et al. (2) on ganglioside G_{M1} -in micellar form- lead to visualize the cation binding sites of this ganglioside as oxygen rich surfaces involving not only the sialic acid (with its carboxylic group), but also the N-acetyl-galactosamine and the terminal galactose residue present in the molecule. The occurrence of these additional oxygen ligands may explain why the affinity of G_{M1} for cations is much larger than that exhibited by free α-, and β-methyl glycoside of sialic acid. The oxygen rich surfaces described in ganglioside G_{M1} are expected to be present in all gangliosides and to constitute a general feature of ganglioside chemistry.

The apolar chains of the ceramide portion of gangliosides are responsible for the hydrophobic properties of gangliosides and for their availability to hydrophobic interactions. The formation of an hydrogen bond between the 3-hydroxyl group of the sphingosine and the carbonyl oxygen of the fatty acid, would tend to spread the two hydrocarbon chains reducing their tendency for mutual association, and thus to promote association of each chain with other molecules (3).

The presence in the ganglioside molecules of an hydrophilic and an hydrophobic portion of approximately equal volume gives them strong amphiphilic properties and leads them to associate in water. In the presence of small quantities of water (hydration range : 18-50 %) hexagonally packed cylinder structures of gangliosides were described (4) in which the apolar chains radiate from the center of the rods, with the sugar groups on the cylinder surface in contact with water. In these structures the radius of the all ganglioside molecule is, at 37°C, about 30 Å, that of the lipid core 20 Å and the annulus formed by the sugar head groups 10 Å (see Figure 1). In dilute aqueous solutions micelles of large molecular weight are formed. The literature values for the critical micellar concentration (cmc) of gangliosides are in the range 10^{-4} - 10^{-5} M (4, 5, 6, 7, 8), except for the recent works of Schwarzmann et al. (9) and of Formisano et al. (10), reporting a cmc of 10^{-8} - 10^{-9} M. Laser light scattering investigations on the micellar properties of gangliosides, recently performed in our laboratory, showed that ganglioside G_{M1} and G_{D1a} in the concentration range

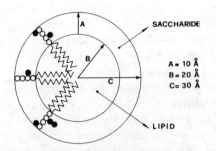

Figure 1.   Dimensions of the saccharide and lipid portion of a ganglioside in a cylinder structure (adapted from Curatolo et al. (4))

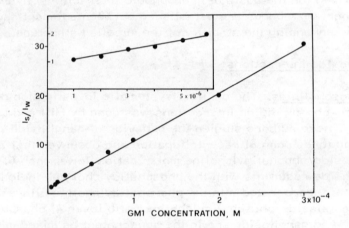

GM1 CONCENTRATION, M

Figure 2.   Laser light scattering of ganglioside G_{M1} in aqueous solution

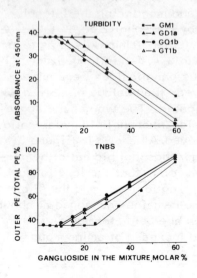

GANGLIOSIDE IN THE MIXTURE, MOLAR %

Figure 3.   Physicochemical features of mixed aggregates of phosphatidylcholine, phosphatidylethanolamine (PE, used as surface marker), and gangliosides (G_{M1}, G_{D1a}, G_{T1b}, G_{Q1b}) at increasing proportions of ganglioside. Highest value of the outer PE/total PE ratio corresponds to liposomes. Lowering of turbidity and concurrent enhancement of ratio indicate presence of micelles. "Break" point is indicated as the "transition ganglioside/phospholipid molar ratio."

10^{-6}-10^{-4} M, which was examined, are present as micelles (see
Figure 2). Thus the cmc should be lower than 10^{-6} M. According
to our measurements the hydrodynamic radius of the G_{M1} and
G_{D1a} micelles, at 10^{-5}-10^{-4} molarity, is 60 Å , a value which
doubles that exhibited by cylinder structures. This may support
the idea that in dilute solutions ganglioside micelles are not
spherical but prolate ellipsoids (rodlike micelles) or, better, disk-
like micelles. While ganglioside monomers rapidly associate to
form micelles, the dissociation of micelles to monomers is very
low (10). This can be interpreted assuming that once the micelles
are formed (on the basis of hydrophobic interactions) hydrogen
and other weak bonds are established between adjacent saccharide
chains, enhancing the stability of the micellar structure.

Ganglioside interactions

  Phospholipids. The earliest systematic investigation on gan-
glioside-phospholipids interactions was done by Hill and Lester
(11). These authors studied the behavior of ganglioside-phospholipid mixtures upon ultracentrifugation and observed that at low
ganglioside/phophatidylcholine molar ratios(lower than 0. 05) the
ganglioside sediments with the phosphatidylcholine, indicating
that it is incorporated into the phosphatidylcholine bilayer. At
high ganglioside/phosphatidylcholine ratio (over 4) phosphatidylcholine and ganglioside are in the supernatant as mixed micelles.
At intermediate ratios less defined phases or mixtures of phases
are present. A more detailed study on ganglioside-phospholipid
interactions was recently undertaken in our laboratory. For this
we used the technique of preparing and monitoring the integrity
of unilamellar liposomes described by Barenholtz et al. (12).
Phosphatidylcholine and phosphatidylethanolamine (used as a surface sided marker to be revealed with 2, 4, 6-trinitrobenzene
sulfonic acid, TNBS) and the various gangliosides were mixed
together, the organic solvent removed, the residue dissolved with
a proper buffer at pH 7. 0, and sonicated. The type pf the aggregate formed depends on the ganglioside/phospholipid molar ratio
and on the ganglioside species employed.As shown in Figure 3
till a certain value of the ganglioside/phospholipid ratio the process of aggregation leads to liposomes. In fact the level of turbidity is the same as in the absence of gangliosides, and the % of
outer sided phosphatidylethanolamine remains unchanged at about
60%. Over that value mixed micelles are being formed as shown
by gradual decrease of turbidity and increase of aminogroups

available for TNBS. The point at which micelles start being formed appears to be quite sharp. The critical value of the ganglio-
side/phospholipid molar ratio for transition, or "transition
ratio", varies with the different gangliosides: from 0.25 for G_{M1}
to 0.10 for G_{T1b}. In other words it rises by decreasing the sialic
acid content of gangliosides. In addition, while monovalent cations
do not influence the "transition ratio", divalent cations, particularly Ca^{2+}, at concentrations within the physiological range,
almost double it: from 0.25 to 0.45 for G_{M1} ; from 0.1 to 0.2 for
G_{T1b}. This means that Ca^{2+} ions determine a liposomal supramolecular organization in the presence of amounts of gangliosides
which, in the absence of Ca^{2+}, would give a micellar organization.
An attempt to verify the dynamics of gangliosides in phospholipid
layers was made by using spin-labeled gangliosides and recording
the EPR signals (13). The results of this investigation indicate
that in artificial phosphatidylcholine bilayers (at physiological
pHs) gangliosides, even at low concentrations (1.5 % of the
phospholipids, in molar terms) show a measurable tendency toward
cooperative interaction among themselves, likely through the formation of hydrogen bonds between adjacent saccharide chains. As
a consequence of this interaction the mobility of gangliosides in
layers tends to diminuish. The presence of physiological levels
of Ca^{2+}, or Mg^{2+}, ions increases the tendency to interactions, by
crosslinking the sugar head groups of gangliosides, and decreases further ganglioside mobility or causes ganglioside immobilization at lower ganglioside concentrations. The addition of glycophorin, a sialic acid rich erythrocyte membrane glycoprotein,
enhances the magnitude of the divalent cations effect. Probably
glycophorin takes part in the crosslinking process leading to a
more packed assembly of gangliosides. All these evidences are
consistent with the hypothesis (see Figure 4) that gangliosides
tend to form, on fluid lipid layers, "clusters", the magnitude and/
or stability of them being enhanced by crosslinking agents.

   Proteins. Gangliosides easily bind proteins. All published
studies on the interactions of gangliosides, either pure or mixed,
with microbial toxins, hormones, interferon, wheat germ agglutinin, deal with ganglioside-protein interactions. Considering the
ganglioside concentrations used in these studies, which abundantly exceeded 10^{-6} molarity, the described interactions pertain to
micellar rather than to monomeric gangliosides. Of course this
does not mean that monomeric gangliosides cannot interact with
proteins. A general model for protein-ganglioside (micellar and
monomeric) binding has not yet been worked out. In a recent our

investigation on the binding of G_{M1} ganglioside with bovine serum albumin, differential UV absorption, fluorescence and ultracentrifugation studies, associated with chromatographic evidences, showed that at least three G_{M1} -albumin complexes are formed, differing markedly in their molecular weight and molecular conformation. One form is the result of the interaction between albumin and G_{M1} monomers. The other two complexes are characterized by a G_{M1}/albumin ratio of one ganglioside micelle per albumin polypeptide chain: one complex polymerizes slowly and irreversibly to the other one, which is a dimer. These two complexes, which result from hydrophobic interactions are actually mixed ganglioside-albumin micelles, in which the original protein conformation has been extensively rearranged.

Monomeric gangliosides. The availability of labelled gangliosides with a very high specific radioactivity enabled to inspect at the behavior of gangliosides in the monomeric form. Ganglioside monomers appear to be capable to adhere potentially to all surfaces, glass and plastic walls included (14). In all these processes the sugar head groups appear to be exposed on surface, indicating the apolar portion of the ganglioside molecule to be responsible for the interaction.

Gangliosides, sialidase and sialyltransferase in the membranes surrounding nerve endings (synaptosomal membranes)

Gangliosides are present in the plasma membranes of all vertebrate cells. The highest ganglioside content is displayed by nervous tissue cells, likely the neurons. The all neuronal plasma membrane contains gangliosides (15). However the portion of it surrounding nerve endings (synaptosomal membrane) was shown not only to carry large amounts of gangliosides (25-30 nmoles of bound N-acetylneuraminic acid / mg protein) (16), but to have a ganglioside content much higher than elsewhere in the nervous tissue (17). An evaluation, based on the yield of synaptosomal membranes obtained in a conventional separation procedure, makes acceptable a 5-fold enrichment of gangliosides on these membranes (18). The gangliosides of the synaptosomal membranes, similarly to all cell surface glycolipids (19, 20), appear to expose their oligosaccharide chains to the outer membrane side (17, 21). Synaptosomal membranes carry sialidase (neuraminidase) activity, able to remove sialic acid from sialylglycoconjugates -gangliosides included- either intrinsic to the membrane or added (22, 23, 24) .

In parallel with gangliosides this sialidase activity, which is present along the all neuron surface (25) , appears to be enriched
in the synaptosomal membranes (22, 23). The assessment of the
occurrence of a sialyltransferase activity in the synaptosomal
membranes encountered much more technical difficulties than that
of sialidase. In fact the Golgi apparatus is known to be the main
location site for glycosyltransferases. Therefore the plasma membranes to be used should be devoid of fragments of the Golgi apparatus : which is not an easy task. We recently (26) approached
the problem with calf brain cortex. In setting up the strategy of
our experimental approach we considered that the Golgi complex
has no morphological and cytoenzymatic evidences to occur inside
the nerve endings. Therefore the only possibility for nerve ending
preparations to contain Golgi complex material is to carry light
membranes of intracellular origin, formed during homogenization.
Relying on this we focused our attention and efforts on removing
all possible contamination of light membranes when preparing nerve endings. The preparation procedure which came out, perfecting
previous methods, has been described alsewhere (26). As shown
on Figure 5 the nerve ending preparation we obtained appeared,
morphologically, fairly homogeneous. Then, by submitting the
nerve ending fraction to hypoosmotic shock (27, 28) , and the
hypoosmotically treated material to a series of density gradient
and diffrential centrifugations, a highly homogeneous preparation
of synaptosomal membranes was obtained (see Figure 5). This
preparation , when submitted to biochemical analyses and compared to the starting nerve ending fraction, showed (see Table I):
(a) no "trapped" lactate dehydrogenase (LDH) activity; this ex-
cludes the presence of unruptured nerve endings ; (b) markedly
enhanced specific activity of authentic plasma membrane markers
(ATP-ase, acetylcholine –Ach- esterase, 5'-nucleotidase), this
qualifying them as plasma membranes ; (c) very low absolute content , and lowered specific activities of intracellular membrane
markers (NADH–Cyt. C reductase, NADPH–Cyt. C reductase),
this proving a low contamination of membranes of other origin ;
(d) enhancement of the specific concentration of gangliosides and
of the specific activity of sialidase parallel to that of authentic
membrane markers. The same preparation of synaptosomal membranes contains substantial sialyltransferase activity, displaying
the same enrichment as gangliosides and sialidase, and bearing
the properties exposed in Figure 6. This evidence, which should
of course be confirmed and corroborated by other proofs, strongly
consists with the hypothesis that in the nervous tissue sialyltrans-

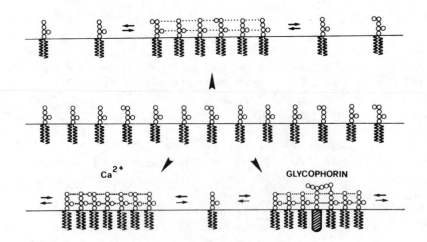

OUTER VESICLE LAYER

Figure 4.   Formation of ganglioside clusters

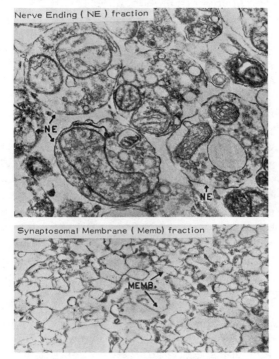

Figure 5.   Electron microscopic examination of the "nerve ending fraction" (×13,000) and of the "synaptosomal membrane fraction" (×7150)

TABLE I

Biochemical characteristics of the "Nerve ending fraction" and of the "Synaptosomal fraction", obtained from calf brain. Gangliosides are expressed as nmoles bound N-acetylneuraminic acid ; enzyme activities in milli International Units (1 nmole transformed substrate min^{-1} at 37°C - 30° for NADH-, and NADPH-Cyt C reductase and LDH). The data shown, referred to 1 g starting fresh tissue, are the mean values of 6 experiments; the S.E. was in all cases lower than $\pm$ 10 % of the mean values.

Parameter	Nerve ending fraction Activity (or concentration)		Synaptosomal membrane fraction Activity (or concentration)		
	total	specific	total	specific	Enrichment
"Occluded" LDH	4.4	0.67	0.003	0.016	0.024
ATP-ase	1491.0	226.0	161.0	865.6	3.83
Ach-esterase	34.2	5.4	3.4	18.4	3.41
5'-nucleotidase	101.2	15.3	13.2	71.1	4.65
Gangliosides	171.6	26.0	16.83	90.5	3.48
Neuraminidase	3.17	0.48	0.43	2.3	4.79
NADH-Cyt. C reductase	165.0	26.0	3.50	19.1	0.73
NADPH-Cyt. C reductase	15.18	2.3	0.316	1.7	0.74
Sialyltransferase[+]	119.9	19.0	14.1	76.3	4.02

[+] Sialyltransferase activity expressed as c.p.m. min^{-1} of incubation using ^{14}C-NeuAc-CMP and lactosylceramide as substrates

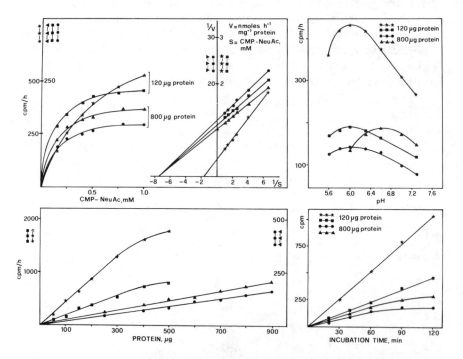

Figure 6. *Effect of CMP-NeuAc concentration (V/S), of pH (V/pH), of enzymatic protein concentration (V/protein), and of incubation time (V/t) on the activity of synaptosomal membrane-bound sialyltransferase. Calf brain cortex. Acceptor substrates for sialyltransferase: (★) lactosylceramide; (■) desialylated fetuin; (●) endogenous glycoprotein; (▲) endogenous glycolipids.*

ferase has, as one of its sites of subcellular location, the synaptosomal membranes . Thus the synaptosomal membranes feature the biochemical potentiality for a "sialylation–desialylation cycle" of gangliosides (as well as of other sialylglycoconjugates). This cycle is schematically depicted in Figure 7.

In conclusion, the synaptosomal membranes differentiate from the other plasma membranes of brain tissue for enabling neurotransmission. A biochemical correlation to this functional specialization is the striking enrichment in gangliosides and in the enzymes capable to modify the sialic acid / ganglioside ratio; hence the ion complexing and crosslinking capacity of the saccharide chains. The fact that sialidase works best at acidic pHs (4.0), and sialyltransferase at neutral pHs, may correlate the optimal functionality of these enzymes, hence the sialylation–desialylation cycle, to local fluctuations of pH value. A schematic picture of the location of ganglioside, sialidase, and sialyltransferase in the nerve ending membranes is shown in Figure 8. The location of gangliosides and sialidase in the outer plasma membrane surface has a consistent support. The sidedness of sialyltransferase has been not yet ascertained. However, the reported occurrence of sialyltransferase at the external surface of a number of cells (29, 30, 31) makes this assignement very probable.

Ganglioside contribution to the supramolecular organization of synaptosoaml membranes

Highly purified synaptosomal membranes, prepared from rat brain cortex, were shown (16) to contain 0.73–0.93 mg of total phospholipids and 0.073–0.125 mg of gangliosides per mg protein. The ganglioside/phospholipid molar ratio, established in this material – 1/8– is by far (50–100 fold) the highest observed in plasma membranes obtained from non neural cells of vertebrates. This figure might be considered of general value for synaptosomal membranes, since the specific concentration of both gangliosides and phospholipids in the brain of different animals varies within a reasonably low range. Due to the ganglioside asymmetrical location the ganglioside/ phospholipid ratio in the outer layer of the membrane should be twice as much greater, i.e. 1/4. This molar ratio is in the range of the "transition ratio" discussed above. Likely the chemical nature of the individual phospholipids occurring in synaptosoaml membranes, the presence of cholesterol and of hydrophobic proteins, can be accounted for enhancing the relative quantities of gangliosides required for causing a bilayer-

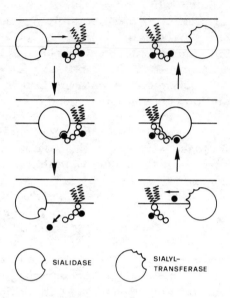

Figure 7.   Sialylation–desialylation cycle of gangliosides

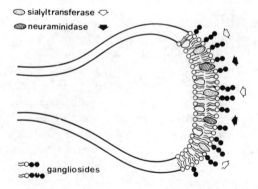

Figure 8.   Location of gangliosides, sialidase, and sialyltransferase at the nerve-ending membrane

micellar transition. However a consideration should be made. Gangliosides tend to form clusters : thus their distribution on the membrane surface may not be even. The formation of such clusters is mainly dependent on the ganglioside concentration and on the presence of crosslinking agents, like divalent cations. In analogy with what has been suggested for capping of membrane receptors (32), below a certain "critical" concentration of gangliosides, or in the absence of crosslinking agents , the process of ganglio- side clustering leads to a patch which is less than a certain size; it cannot stay in equilibrium with the surrounding solution of mob- ile, laterally diffusing, ganglioside molecules, and should dissol- ve. Over the critical concentration, or in the presence of ions or other crosslinking agents which facilitate mutual interactions, clusters greater than the critical size are formed: these, thermo- dynamically favoured, would be sufficiently stable to survive. Now, as reported by Sharom and Grant, in their already quoted investigation (13), signals of ganglioside immobilization started being recorded at a ganglioside concentration in the layer of 1.5 % with respect to phospholipids. This concentration is abundantly exceeded in the synaptosomal membranes. Moreover Ca^{2+} ions are present on the membrane surface, which could serve as cross- linking agents. Therefore the synaptosomal membranes appear as ideal candidates for the formation of stable ganglioside clusters.

One problem, essential from a physiological point of view, is whether the formation of stable ganglioside clusters would occur in any sites, or in preferential sites of the membrane. In this respect we should remind that gangliosides can interact either with membrane embedded proteins (hydrophobically), or with membrane glycoproteins (by mutual carbohydrate–carbohydrate interactions). Thus proteins (carrying high ganglioside binding affinity) and glycoproteins, which have a defined location on the membrane, may easily serve as focal points which direct the packing of gangliosides in given sites. In conclusion proteins would be the molecules governing and giving a possible functional signi- ficance to the process of ganglioside clustering. An important point to be kept in mind, in addition, is that the forces involved in ganglioside clustering are non-covalent and could break easily under appropriate stress and reform when the stress is removed, or viceversa. In other words this kind of supramolecular organiz- ation displays the characteristics of great flexibility and of rever- sible phase transitions. (See Figure 9).

Several are the consequences which can be expected from the cluster organization of carbohydrate carrying molecules in the

membrane surface. First, both gangliosides and glycoproteins
have been indicated as receptor sites at cell surfaces, their
carbohydrate portions being the instruments for determining specificity. The aggregation of carbohydrate chains on surface might
not only facilitate receptors binding but, also, by modulating the
extent of mutual interactions, give to the binding kinetics a cooperative nature. Second, the patch organization of the glycocalyx
would cause the formation of oligosaccharide-free areas, enabling
easier collision with apolar ligands. Finally let us look at the lipid composition in correspondence to a cluster. Here the ganglio-
side/phospholipid ratio would greatly increase till reaching, or
exceeding, the transition value. The lipids would be forced to
rearrange toward a micellar kind of aggregation. This organization, likely involving proteins, may be the molecular basis for
the formation of polar channels (see Figure 10). An indication
in this sense can be seen in the recent report by Tosteson and
Tosteson (33) , describing the development of channels when
bilayers of glycerolmonooleate, containing ganglioside G_{M1}, were
exposed to cholerae toxin. According to the above view cholerae
toxin, functioning as a crosslinking agent for G_{M1}, leads to the
formation of G_{M1} clusters. The following rearrangement of the
lipid matrix would result in the formation of channels.

An experimental model for the study of ganglioside behavior in
synaptosomal membranes

      The cluster hypothesis of ganglioside distribution on the sur-
face of synaptosomal membranes needs, of course, precise experimental supports. For this the availability of a study model mimic-
ing the cell membrane would be of great help. The model which
provided the evidence of ganglioside immobilization on lipid layers
is the liposome. This model, as used by the quoted authors (13) ,
suffers from some important limitations. In fact the liposomes
were prepared from dispersions of phospholipids and gangliosides
and yielded as multilamellar vesicles, carrying gangliosides on
both sides. As determined by Cestaro et al. (34) on a substantial-
ly similar system, about 60% of gangliosides are outer sided, the
remainder being located on the inner side. Of course an asymmetrical location of gangliosides on the outer liposome layer would ren-
der this model much more suitable for the above purposes. We worked on this direction, using monolamellar phospholipid vesicles,
of small and homogeneous size, prepared according to Barenholtz
(12). These vesicles, containing phosphatidylcholine (carrying

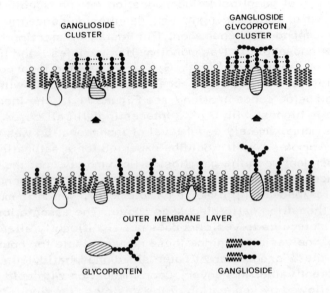

GANGLIOSIDE
CLUSTER

GANGLIOSIDE
GLYCOPROTEIN
CLUSTER

OUTER MEMBRANE LAYER

GLYCOPROTEIN

GANGLIOSIDE

Figure 9. Formation of stable ganglioside clusters: role of proteins (hatched irregular circles) and of glycoproteins as focal points of clustering

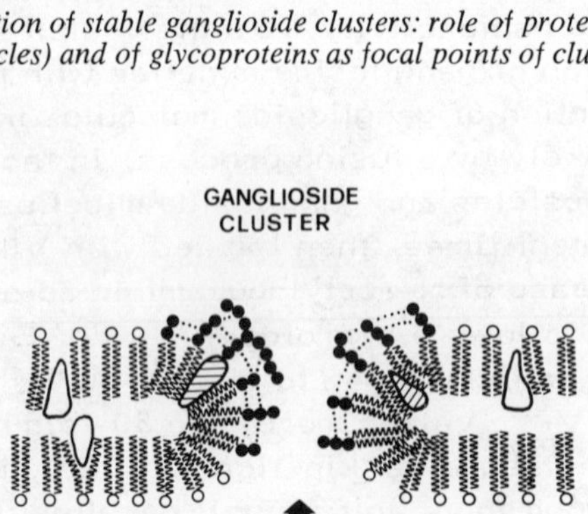

GANGLIOSIDE
CLUSTER

CHANNEL

Figure 10. Formation of a polar channel in correspondence of a ganglioside cluster. Note the presence of proteins and glycoproteins (hatched irregular circles).

^{14}C choline) and phosphatidylethanolamine (95/5, by mol) upon
incubation in the presence of ganglioside micelles (containing
tritium labelled gangliosides) do incorporate gangliosides (35).
The process of ganglioside incorporation into phospholipid vesicles is a time, ionic strength, pH, ganglioside concentration, temperature dependent phenomenon. For instance, starting from 0.9
μmoles of phospholipid (as monolamellar vesicles) and from 0.5-
2 μmoles of ganglioside (G_{M1}, G_{D1a}, G_{T1b}), the incorporation
of ganglioside into vesicles proceeds proportionately with time
and ganglioside concentration (see Figure 11). The incorporation rate is highest with G_{T1b}. Interestingly, in all cases, a maximum and approximately equal level of incorporation was reached
(0.6-0.7 μmoles) as it would be expected for a saturation process dependent upon the ganglioside lipid moiety, not the quality
of the saccharide chain. This saturation level corresponds to a
ganglioside/phospholipid molar ratio of 0.07, which is much lower
than the transition ratio discussed above. The association of ganglioside molecules to vesicles does not significantly alter the integrity of the vesicles, since these remain stable for hours, and
maintain the % proportion of outer sided phosphatidylethanolamine
at a constant value. Moreover, upon incubation with cold gangliosides, followed by separation of the vesicles, no significant loss
of the ganglioside radioactivity from the vesicles was observed.

The interaction of ganglioside micelles with phospholipid vesicles lead to insertion of ganglioside molecules into the lipid matrix
of the vesicle, likely by a fusion process. In fact, when mixtures
of phospholipid vesicles and ganglioside micelles are incubated
at 37°C for different times, then treated with Vibrio cholerae
sialidase , a release of N-acetylneuraminic acid (NeuAc) is
recorded, which follows a sigmoidal kinetics. Since Vibrio Cholerae sialidase was found (34) to display on G_{D1a}-phospholipid
mixed liposomes V_{max} values more than 50-fold higher than on
G_{D1a} micelles, the sigmoidal kinetics are likely the expression of
the following phenomenon. Initially all ganglioside is present in
micellar form, this yielding the lowest record of neuraminidase
activity. By allowing interaction with phospholipid vesicles gangliosides become inserted into vesicles, this yielding a mixed vesicle
which is a better substrate for the enzyme; therefore the rate of
NeuAc release increases. As a further proof, when phospholipid
vesicles, which incorporated after incubation a certain amount of
G_{D1a},were isolated and submitted to the action of Vibrio Cholerae
sialidase the kinetics are hyperbolic, the same exhibited by mixed
G_{D1a}-phospholipid vesicles prepared by sonication. This means

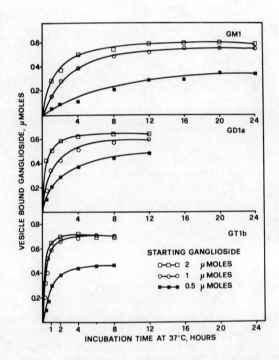

Figure 11. Effect of incubation time (at 37°C) and of ganglioside concentration on the incorporation of gangliosides (G_{MI}, G_{D1a}, G_{T1b}) into phosphatidylcholine monolamellar vesicles. Phosphatidylcholine (as vesicles): 9 μmol. Ganglioside: from 0.5 to 2 μmol. After incubation the mixtures were passed through a 1 × 20 cm Sepharose 4B column to separate vesicles from ganglioside micelles.

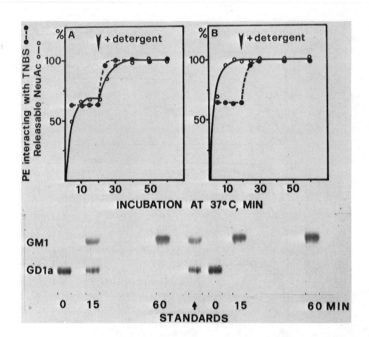

Figure 12. Time course of NeuAc release from liposome-associated ganglioside
G_{D1a} by the action of Vibrio cholerae sialidase

Incubations done at 37°C in 0.05M Tris-HCl buffer, pH 6.8, with 1 IU of enzyme
(Behringwerke). Released NeuAc determined by method of Warren (37); available
amino groups (carried by phosphatidylethanolamine) by TNBS method (12). Arrow
indicates addition of detergent (Triton X-100, 0.5%). Ganglioside pattern during enzyme
hydrolysis was monitored by TLC (silica gel plates; solvent: chloroform/methanol/0.3%
aqueous $CaCl_2$, 60/35/8, by vol, 2-hr run; spots detected by spraying with Ehrlich's
reagent and heating at 110°C for 10 min).

(A): liposomes containing phosphatidylcholine, phosphatidylethanolamine, and G_{D1a}
(90/3/7, by mol) and prepared by the sonication method (12)

(B): liposomes containing phosphatidylcholine and phosphatidlyethanolamine (90/5, by
mol), prepared by sonication, were incubated in 0.05M Tris-HCl buffer (pH 6.8) with
G_{D1a} micelles for 1 hr, then separated by Sepharose 4B column chromatography. These
liposomes contained about 5% (by mol) of incorporated ganglioside.

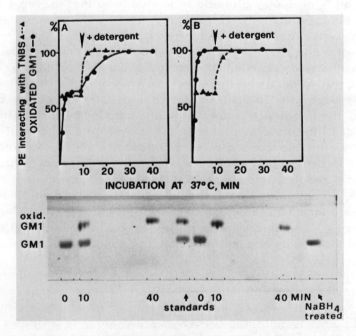

Figure 13. Time course of oxidation of the terminal galactose residue of liposome-associated ganglioside G_{M1} by the action of galactose oxidase

Incubations done in 0.05M Tris-HCl buffer (pH 6.8) at 37°C, with 1 IU of enzyme (Kabi). Oxidation was followed by the coupled o-anisidine peroxidase procedure. Formation of oxidized G_{M1} was also monitored by thin-layer chromatography, under the conditions described in Figure 12. Note that oxidized G_{M1} could be reduced to the starting G_{M1} by $NaBH_4$ treatment. All other conditions as described in Figure 12.

that in both vesicle species (prepared by sonication or by absorption) the insertion of gangliosides in the lipid layer leads the carbohydrate chains to protrude on the layer surface.

As shown in Figure 12 when liposomes in which G_{D1a} was introduced by sonication, or by absorption, are submitted to the action of Vibrio Cholerae sialidase the following differential behavior is observed. In the first case only about 60 % of releasable sialic acid is split by sialidase, meaning, as expected, that only about 60 % of the ganglioside is available to the enzyme (the outer sided ganglioside). Noteworthy, during sialidase treatment liposomes maintain their integrity as indicated by TNES measurements of outer sided phosphatidylethanolamine aminogroups, which give unchanged records. Of course, on addition of 0.5 % Triton X-100, which destroys liposome (as shown by the behavior of aminogroups with TNBS) and forms micellar dispersions of the lipids, the remainder NeuAc (the one carried by inner sided gangliosides) becomes available to the enzyme and is split off. Conversely, in the second species of liposomes 100 % of releasable NeuAc is released by sialidase before the addition of Triton X-100. Identical results, showing this differential bevavior, were obtained with G_{M1} containing liposomes, in which G_{M1} was monitored as available terminal galactose to galactose-oxidase oxidation (see Figure 13). All this means that liposomes in which gangliosides are introduced by absorption carry the ganglioside units only in the outer layer.

The formation of phospholipid monolayer vesicles, carrying gangliosides asymmetrically located on the outer side, is suggested to occur as follows. An initial contact, or adhesion, mainly determined by the ganglioside carbohydrate chains, is followed by stable insertion of the ganglioside lipid moieties into the lipid matrix of the vesicle. This latter process leads to ganglioside diffusion on the outer lipid layer, the entry of ganglioside units into the inside layer being prevented by the high energetic requirements for the movement of the large polar head groups across the lipid layer. After a certain amount of ganglioside has been incorporated into the vesicle, hydrogen and other weak bonds are being formed within the oligosaccharide chains, this resulting in a stabilization of the vesicle structure. On the other hand the acquired surface charge prevent further adhesion (and incorporation) of ganglioside micelles, this mimicking a saturation process.

Conclusion

The studies on the physico-chemical properties and behavior of gangliosides in artificial membranes became more frequent in these recent years, surely stimulated by the increasing evidences on the involvement of gangliosides in a number of biologically important phenomena. Moreover more sophisticated and adequate experimental models have been developed. Thus it is reasonable to expect that, in the near future, integrated progress of research in different fields – physico-chemistry of gangliosides in artificial membranes; enzyme events occurring at the membrane surface; ligands –glycocalyx interactions – will provide enough information to figure out the role played by gangliosides in synaptosomal membranes.

Symbols used

The ganglioside nomenclature proposed by Svennerholm (36) was followed.

Acknowledgements

The experimental data reported in this paper pertain to work supported by grants from the Consiglio Nazionale delle Ricerche (C.N.R.), Italy.

Literature cited

1 . Wiegandt, H. Advances Lip.Res. , 1971, 9, 249.
2 . Sillerud, L.O.;Prestegard, J.H.;Yu, R.K.;Schafer, D.E.; and Konigsberg, W.H. Biochemistry, 1978, 17, 2619.
3. Howard, R.E.;and Burton R.M. Biochim. biophys. Acta, 1964, 84, 435.
4 . Curatolo W.;Small D.W.;Shipley G.G. Biochim. Biophys. Acta 1977, 468, 11.
5 . Gammack, D.B. Biochem. J. , 1963, 88, 373.
6 . Rauvala, H. FEBS Lett. , 1976, 65, 229.
7 . Yohe, H.C.; and Rosenberg, A. Chem. Phys. Lipids, 1972, 9, 279.
8 . Yohe H.C.; Roark, D.E.; and Rosenberg, A. J. Biol. Chem. 1976, 251, 7083.
9 . Schwarzmann, G.; Mraz, W.; Sattler, J.; Schindler, R.; and Wiegandt, H. Hoppe-Seyler's Z.Physiol.Chem.1978, 359, 1277

10. Formisano,S.; Johnson,M. L.;Lee, G.;Aloj, S.M.;and Edel-
     hoch, H. Biochemistry, 1979, 18, 1119.
11 . Hill, M. W.;and Lester, R. Biochim. Biophys. Acta, 1972, 282,
     18.
12 . Barenholtz, Y.;Gibbs, D.;Sitman, B. J.;Goll, J.;Thompson, T. E.;
     and Carlson, F. D. Biochemistry, 1977, 16, 2806.
13 . Sharom F. J.;and Grant, C. W. M. Biochim. Biophys. Acta, 1978,
     507, 280.
14 . Holmgren, J. Proceedings CNRS International Symposium
     "Structure and function of gangliosides", Le Bischenberg,
     France, April 1979.
15 . Ledeen, R. W. J. Supram. Struct., 1978, 8, 1.
16 . Breckenridge, W. C.;Gombos, G.;and Morgan, I. G. Biochim.
     Biophys. Acta, 1972, 266, 695.
17 . Hansson, H. A.;Holmgren, J.;and Svennerholm, L. Proc. Natl.
     Acad. Sc. USA., 1977, 9, 3782.
18 . Tettamanti , G.;Preti, A.;Cestaro, B.;Venerando, B.;Lombardo,
     A.;Ghidoni, R.;, and Sonnino, S. Proceedings CNRS Interna-
     tional Symposium "Structure and function of gangliosides",
     Le Bischenberg, France, April 1979.
19 . Bretscher, M. S. Science, 1973, 181, 622.
20 . Yamakawa, T.;and Nagai Y. TIBS, 1978, 3, 128.
21 . Rosenberg, A. Adv. Exptl. Med. Biol. 1978, 101, 439.
22 . Schengrund C. L.;and Rosenberg, A. J. Biol. Chem. 1970,
     254, 6196.
23 . Tettamanti , G.;Morgan, I. G.;Gombos, G.;Vincendon, G.;and
     Mandel, P. Brain Res. 1972, 47, 515.
24 . Tettamanti, G.;Preti, A.;Lombardo, A.;Suman, T.;and Zambot-
     ti, V. J. Neurochem., 1975, 25, 451.
25 . Tettamanti , G.;, Preti, A., Lombardo, A.; Bonali, F.;and Zam-
     botti, V. Biochim. Biophys. Acta., 1973, 306, 466.
26 . Preti, A.;Fiorilli, A.;Lombardo, A.;Caimi, L.;and Tettamanti
     G. submitted for publication.
27 . Ledeen, R. W.;Scrivanek, L. J.;Tirri, R. K.;Margolis, R. K.;
     and Margolis, R. U. Adv. Exptl. Med. Biol., 1976, 71, 83.
28 . Venerando, B.;Preti, A.;Lombardo, A.;Cestaro, B.;and Tetta-
     manti, G. Biochim. Biophys. Acta, 1978, 527, 17.
29 . Shur, B. D.;and Roth, S. Biochim. Biophys. Acta, 1975, 415, 473.
30 . Porter, C. W.;and Bernacki, R. J. Nature, 1975, 256, 648.
31 . Colombino, L. F.;Bosmann, H. B.;and Mc Lean, R. J. Exptl. Cell
     Res., 1978, 112, 25.
32 . Gershon, N. D. Proc. Natl. Acad. Sc. USA, 1978, 75, 1357.

33. Tosteson, M. Y.; and Tosteson, D. C. Nature, 1978, 275, 142.
34. Cestaro, B.; Barenholtz, Y.; and Gatt, S., submitted for publication.
35. Cestaro, B.; Ippolito, G.; Ghidoni, R.; Orlando, P.; and Tettamanti, G. Bull. Mol. Biol. Med, 1979, in press.
36. Svennerholm, L. J. Lipid Res., 1964, 5, 145.
37. Warren, L. J. Biol. Chem., 1959, 234, 1971.

RECEIVED December 10, 1979.

Specificity and Membrane Properties of Young Rat Brain Sialyltransferases

SAI-SUN NG[1] and JOEL A. DAIN

Department of Biochemistry and Biophysics, University of Rhode Island, Kingston, RI 02881

Sialyltransferases are a group of soluble or membrane bound enzymes that transfer sialic acid from CMP-sialic acid to acceptor molecules. The acceptor molecules may be low molecular weight oligosaccharides or higher molecular weight glycolipids and glycoproteins. Twelve activities typical of sialyltransferases have been described and these activities probably represent eight separate and distinct enzymes (1). Our interest in rat brain sialyltransferases stemmed from our work on the ganglioside biosynthetic pathways. These studies by us (2,3) and others (4,5) suggested the existence of more than one pathway for the synthesis of polysialogangliosides depending on when a sialyltransferase is brought into action after other sugar residues have been added to a precursor glycolipid. These early studies had also documented that the rat brain sialyltransferases are mainly membrane-bound. This is of interest because the neuraminidases in brain tissues are also membrane bound.

The neuraminidases together with gangliosides have been localized in the nerve ending structures (6,7). Theoretically the sialylation and desialylation cycle may mediate a cyclic reaction at a very important locale in a nerve synaptic structure. This hypothetical involvement of sialic acid metabolism in synaptic transmission has gained support from several studies which have suggested a synaptic localization of the glycosyltransferases (8, 9,10,11) and from proposed theoretical models in which the sialoglycolipids are considered an important constituent in the functional units of neuronal membranes (12,13,14).

It was apparent, however, that the rat brain sialyltransferases have not been sufficiently characterized for the postulation of a biological role for the sialylation-desialylation cycle. Consequently, we concentrated our efforts to characterizing the general behaviors of the sialyltransferases in their membrane environments in the rat brain. What follows is a summary of our

[1] Department of Biochemistry, McGill University, Montreal, Quebec H3C 3G1, Canada.

results on the properties of rat brain membrane-bound sialyltransferases, their sub-cellular localization and our initial attempts
to solubilize and purify these enzymes. Part of the results have
been published previously (15,16).

Methods

In all experiments to be described, brains of 11-15 day old
albino rats (Sprague-Dawley) were used. Young rats of this age
were chosen because rapid accumulation of gangliosides and sialo-
proteins have been reported to occur around this period (17).
For most studies, total brain homogenates were centrifuged at
105,000 g for 60 min and the pellet used as the enzyme source.
Details of the conditions for the enzymic assays have been reported (15,16). The four reactions below were investigated. The
abbreviations for gangliosides are those proposed by Svennerholm
(18).

(A) Endogenous glycolipids
$$\text{Cer-Glc-Gal} \longrightarrow \underset{\underset{\text{NeuNAc}}{|}}{\text{Cer-Glc-Gal}}(GM_3)$$

$$\underset{\underset{\text{NeuNAc}}{|}}{\text{Cer-Glc-Gal-GalNAc-Gal}}(GM_1) \longrightarrow \underset{\underset{\text{NeuNAc}}{|}\quad\underset{\text{NeuNAc}}{|}}{\text{Cer-Glc-Gal-GalNAc-Gal}}(GD_{1a})$$

(B) Exogenous glycolipid
$$\underset{\underset{\text{NeuNAc}}{|}}{\text{Cer-Glc-Gal-GalNAc-Gal}}(GM_1) \longrightarrow \underset{\underset{\text{NeuNAc}}{|}\quad\underset{\text{NeuNAc}}{|}}{\text{Cer-Glc-Gal-GalNAc-Gal}}(GD_{1a})$$

(C) Endogenous glycoproteins
$$\text{Glycoproteins} \longrightarrow \text{Glycoproteins-NeuNAc}$$

(D) Exogenous glycoprotein
$$\text{Desialated (DS) fetuin} \longrightarrow \text{DS-fetuin-NeuNAc}$$

Results and Discussion

Kinetic Properties of Sialyltransferases. The sialyltransferase activities with the endogenous glycoprotein and glycolipid acceptors in the standard assays (15) were linear with
time for at least 60 min, while those with the exogenously added
GM_1 and DS-fetuin were linear with time only for about 30 min
(Figure 1). Activities were directly proportional to the amount
of enzyme added up to 0.75 mg protein/assay (Figure 2).
The enzyme activities, expressed as nmol of NeuNAc incorporated per 0.5 mg protein per 30 min at 37C and pH 6.3, were 0.095,
0.039, 0.17 and 0.64 with the endogenous glycolipids, the endogenous glycoproteins, the exogenous GM_1 and exogenous DS-fetuin,
respectively. Incorporation into the endogenous glycolipids was
always higher than the incorporation into the endogenous glyco-

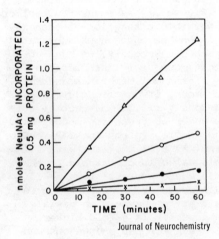

Journal of Neurochemistry

Figure 1. Effect of incubation time on sialyltransferase activities. Incorporation of NeuNAc into the following substrates was determined: ✕, *endogenous glycoproteins;* △, *endogenous glycoproteins plus exogenous DS-fetuin;* ◯, *endogenous glycolipids;* ◯, *endogenous glycolipids plus exogenous GM₁. (15).*

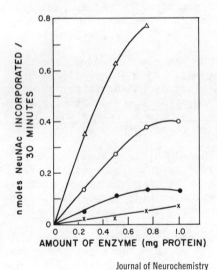

Journal of Neurochemistry

Figure 2. Effect of enzyme concentration on sialyltransferase activities. Incorporation of NeuNAc into glycoproteins and glycolipids was analyzed: ✕, *endogenous glycoproteins;* △, *endogenous glycoproteins plus exogenous DS-fetuin;* ◯, *endogenous glycolipids;* ◯, *endogenous glycolipids plus exogenous GM₁. (15).*

proteins. This is different from the case in neuronal and glial
cell preparations from the rat brain (19). The discrepancy does
not appear to be a result of an extensive loss of synaptic membranes causing a decrease in the endogenous glycolipids in the
neuronal and glial preparations. There are two reasons for this
inference. Firstly, the linearity of sialyltransferase activities versus enzyme concentration (Figure 2), as suggested earlier
by Arce et al., (20) indicates a possible cis action of the enzyme systems: the enzymes act only on the endogenous substrates
located on the same membrane fragments. Secondly, we have obtained evidence which indicates that most sialyltransferases in
rat brain are associated with the microsomal membranes and not
with the synaptosomes. The large amount of gangliosides in synaptic membranes as described previously (12,21) may not be uti-
lizable by the sialyltransferases under our reaction conditions.

The apparent Km values of the enzyme systems for CMP-NeuNAc
assayed with 0.5 mg enzyme protein, was 0.13 mM (same with all
four types of acceptors (15)). This value is comparable to that
(0.15 mM) obtained (22) in cultured mouse neuroblastoma cells
with lactosylceramide as the NeuNAc acceptor. The Km value reported previously (23) for the calf brain enzyme with desialylated α_1-acid glycoprotein as the acceptor is 4-fold higher.
Other values obtained (20) in rat brain for the endogenous and
exogenous glycolipid acceptors were 3-40 times smaller than the
value reported here. The reasons for such marked discrepancies
between the reported Km values are not clear. Conceivably, tissue differences and assay conditions in these other systems can
contribute to such differences. Detergents, for example, were
not used by Arce et al., (20).

It is interesting that all four reactions studied gave the
same apparent Km values for CMP-NeuNAc. This observation suggests that the CMP-NeuNAc binding sites of the different species
of sialyltransferases are similar in structure and affinity properties, although the binding sites for the acceptors and the
catalytic sites may be very different, providing a basis for substrate specificities.

The Km value for exogenously added GM1 is 0.2 mM, which is
identical to the value obtained with the chick embryonic brain
(24) but is lower than the value reported for the rabbit neuro-
hypophysis (25). Substrate inhibition at high glycolipid concentration as observed with other glycosyltransferases (22,26,27)
was also noted in this study when the concentration of GM1 was
above the Km (15). The Km value of the exogenously added DS-
fetuin is 0.15 mM or 1.2 mM in terms of acceptor sites (15). This
value is only one third of that obtained for desialylated α_1-acid
glycoprotein which is a much less efficient acceptor than DS-
fetuin (23).

                    Analyses of the Reaction Products. About 90% of the
label in the endogenous glycolipid products was releasable by

treatment with <u>Vibrio cholera</u> neuraminidase. The label, however, was completely labile to dilute acid hydrolysis with 0.05M H_2SO_4 at 80C for 60 min. Acid hydrolysis of the endogenous glycoprotein products solubilized in 4% SDS also resulted in complete removal of the label.

Radioactivity scanning of a thin layer chromatogram indicated that GM_3 and GD_{1a} were labeled (15). This suggested that the endogenous substrates were lactosylceramide and GM_1. Under the standard assay conditions, incorporation into GM_3 and GD_{1a} was about 60% and 40%, respectively. Incorporation into other gangliosides was not significant. Exogenously added GM_1 was converted to GD_{1a} without any apparent effect on the magnitude of the endogenous GM_3 peak, an indication that sialyltransferases acting on lactosyl-ceramide and GM_1 are different enzymes.

We have noted that Arce et al. (20) demonstrated a similar level of incorporation of NeuNAc into the endogenous glycolipids and, in addition, effected high levels of incorporation into GD_{1b} and GT_{1b} which were not observed in our assays. We believe that the use of detergents in our assay procedures, although not completely abolishing the membrane nature of the enzyme systems, was responsible for certain subtle changes such that GD_{1b} and GT_{1b} were not formed in significant amounts. Under these conditions, tight physical association between the glycosyltransferases and their substrates as may be the situation <u>in vivo</u> (20,28) appears not to be essential for the glycosyltransferase actions <u>in vitro</u> and thus exogenous substrates can also be efficiently utilized. It is also possible that our failure to detect GD_{1b} and other gangliosides was due to the activities of membrane associated neuraminidases which, when activated by detergents (29,30) degraded these gangliosides once they were formed under the assay conditions.

Detailed studies on the nature of the endogenous glycoprotein sialic acid acceptors have not been reported in literature. We have analyzed the endogenous glycoprotein reaction products by SDS-polyacrylamide gel electrophoresis and detected about 20 species of labeled polypeptides of molecular weights ranging from 20,000 to over 120,000 (15). Neuraminidase pretreatment of the enzyme preparation decreased the incorporation of NeuNAc into the high molecular weight polypeptides but increased the incorporation into polypeptides of lower molecular weights, as well as two polypeptides of intermediate molecular size. The action of neuraminidase may be nondiscriminative to all susceptible glycoproteins. The increase in incorporation into polypeptides of lower molecular weights may be due to their faster diffusion rate in membrane, thus allowing them to undergo faster sialylation than the higher molecular weight species. This proposition is based on the assumption that there is only one species of glycoprotein sialyltransferase, an assumption, which is evident later, may not be true.

The number of proteins in the total membrane preparation of

15-day old rat cerebra is about 25 or higher (31). Apparently
half of these are glycoproteins as indicated by periodic-acid-
Schiff stain reaction (32). If consideration is given for the
possible existence of oligomeric proteins and microheterogeneities
of these species, the number of sialylated polypeptides observed
by us may not be as great as it appears.

Multiplicity of Sialyltransferases. The multiplicity of
the glycosyltransferases and their substrate specificities are
important in the biosynthesis of the multitudinous and diversi-
fied species of complex carbohydrates. We (15) were able to distinguish at least two glycolipid sialyltransferases and one glycoprotein sialyltransferase. The sialyltransferases acting on
lactosylceramide and GM_1 can be differentiated because the addition of exogenous GM_1 did not decrease the incorporation of
NeuNAc into lactosylceramide (15). Furthermore, the heat inactivation experiment indicates that the exogenous activity with GM_1
acceptor was more heat labile than the endogenous acceptor activity in which lactosylceramide is the major acceptor (15). The
differentiation of the two glycolipid sialyltransferases is in
agreement with a previous report using exogenous substrates (33,
34). A third sialyltransferase which acts specifically on GM_3
has also been reported (34).

The glycoprotein sialyltransferase appears to be different
from the glycolipid sialyltransferases since varying amounts of
exogenously added DS-fetuin do not compete with both the exogenous and the endogenous glycolipid acceptors for CMP-NeuNAc (15).
Studies with extraneural tissues have favored the existence of at
least three glycoprotein sialyltransferases (35,36,37). Our own
studies on rat brain sialyltransferases, as will be described below, have suggested the existence of multiple species of brain
glycoprotein sialyltransferases.

The clear distinction between the glycoprotein and glycolipid sialyltransferases suggests that the extent of sialylation
of the two classes of complex carbohydrates in vivo may be under
different metabolic controls. This notion is supported by the
observation that the developmental profiles of the glycoprotein
and glycolipid sialyltransferases are distinctly different.
Whereas the activities of CMP-NeuNAc: lactosylceramide sialyltransferase increase postnatally (38), the activities of the CMP-
NeuNAc: glycoprotein sialyltransferase decline dramatically during early development (17). Separate regulatory controls on the
glycoprotein and glycolipid metabolism have also been implicated
in a study on the brain tissues of Tay-Sachs disease patients.
The carbohydrate content and composition, including the sialic
acids of the glycopeptides derived from the Tay-Sachs brain tissues, were found to be the same as those in the normal control although the patterns of gangliosides were markedly different due to
excessive accumulation of GM_2 in the Tay-Sachs brain (39).

Subcellular Localization of Sialyltransferases. The
subcellular localization of the glycosyltransferases in brain
tissues has been a subject of much controversy. Some of the enzymes have been reported to be present in all major subcellular
structures (20,38,40,41). Several studies have indicated that
these enzymes are concentrated in the membranes of the synaptic
complexes or synaptic vesicles (8,9,10,11). In other studies,
however, the glycosyltransferase activities were found to be relatively low in the synaptic membranes (42). Both galactosyl and
N-acetylgalactosaminyl transferases have since been shown to be
localized in microsomal membranes (43,44). With regard to sialyltransferases, the non-synaptic localization is supported by the
following observations: (1) both neuronal and glial cell bodies
were shown to possess sialyltransferase activities (19); (2) different regions of the brain with different degrees of synaptic
densities were found to have comparable specific activities of the
enzymes (23); and (3) high glycoprotein sialyltransferase activities were observed in the newborn rat brain even though the synaptic structures were not developed. In fact, the sialyltransferase
level was found to decrease during postnatal development when
active synaptic formation took place (17).

The controversy over this issue appears to arise from the
fractionation techniques used. In all the reported studies on
cellular fractionation, differential centrifugations were made
followed by discontinuous sucrose density gradients. The pellet-
ing and resuspension procedures may cause unnecessary artifacts
in assessing the subcellular localization of the glycosyltransferases. With this understanding, we have approached the problem
using a continuous sucrose gradient centrifugation. Rat cerebral
homogenate with nuclei and large debris removed, was applied directly onto a continuous sucrose density gradient. Sialyltransferase activities were assayed in the fractions collected.

As indicated in Figure 3a, the distribution of all four sialyltransferase activities did not correspond to protein peaks III
and IV which, according to marker enzyme activities (Figure 3b;
45,46,47) and electron microscopic examinations (16) were enriched
in synaptosomes and mitochondria, respectively. All four sialyltransferase activities occupy the same range in the gradient,
overlapping with protein peak I which was enriched in myelin fragments. The exogenous sialyltransferase activities showed peak
activities at about 0.7M sucrose (Figure 3a). Electron microscopic examination (16) revealed that the exogenous sialyltransferase peak activity fraction (II) was associated with an enrichment in smooth membrane fragments and vesicles, some of which re-
sembled structures of Golgi complexes. There were very few synaptosomes and essentially no mitochondria and myelin in that fraction. From these observations, the conclusion can be drawn that
sialyltransferases are concentrated in membrane structures which
are not related to synaptosomes and mitochondria. These sialyl-
transferase-enriched structures are presumably derived from the

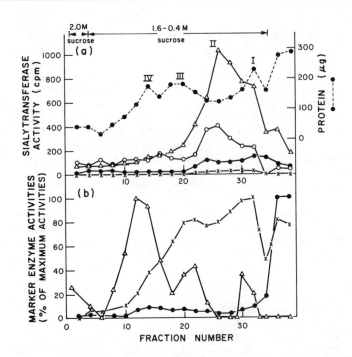

Journal of Neurochemistry

Figure 3.   (a): Distribution of endogenous sialyltransferase activities on continuous sucrose density gradient. Enzyme assays done under standard conditions using NeuNAc-4-¹⁴C-CMP-NeuNAc (specific radioactivity, 1.9 mCi/mmol). Volume of gradient, 80 μL except for endogenous glycoprotein sialyltransferase assay in which only 50 μL was used. Incubations at 37°C for 30 min. ✕, endogenous glycoproteins; △, exogenous DS-fetuin; ○, endogenous glycolipids; ◯, exogenous GM₁; - - -, protein.

(b): Marker enzyme activities. △, Succinate dehydrogenase; ✕, acetylcholinesterase; ●, lactate dehydrogenase (16).

endoplasmic reticulum, the Golgi complexes and the plasma membrane. The sialyltransferase activity associated with the myelin-enriched fractions (protein peak I) are believed to be due to the light microsomal membranes derived from the same structures which also gave rise to the enzyme peak II. In addition, the profile of the endogenous s̆ialyltransferase activities reflects a simultaneous occurrence of the sialyltransferases and the endogenous substrates.

The gradient pattern is highly reproducible. We have also demonstrated that UDP-galactose: GM_2 galactosyltransferase is similarly located in this gradient peak (48). The peak galactosyltransferase activities also corresponded to peaks of UMP-ase ('-ribonucleotide phosphohydrolase, EC 3.1.3.5) and UDP-phosphohydrolase (nucleosidediphosphate phosphohydrolase, EC 3.6.16) activities which are marker enzymes for the endoplasmic reticulum.

We have no information on the relative contribution of the various cell types in the observed sialyltransferase activities. Comparable enzyme activities have been reported in the neuronal and glial cell preparations (19). More definitive work will necessarily involve autoradiographic studies as initiated by some investigators (49).

The mechanisms which are responsible for the short- and long-range order (50) of the glycosyltransferases and their substrates and products within specific cellular structures are open to speculation. Results from our studies suggest that the sialyltransferases are integral proteins with their active sites deeply embedded in the lipid bilayer of cell membrane. Thus, factors located either external or internal to the lipid bilayer can bring about the short- and long- range order of the glycosyltransferase systems.

Membrane Properties of Sialyltransferases. The nonionic detergent mixture, Triton CF-54/Tween 80 (2/1, w/w), stimulated all four sialyltransferase activities, the effect being much more pronounced with the exogenous substrates. At 1 mg detergent mixture per mg enzyme protein, the percent increases in enzyme activities with the different substrates were: endogenous glycolipids, 100; endogenous glycoproteins, 50; GM_1, 700 and DS-fetuin, 230 (Figure 4).

The stimulatory effect of nonionic detergents on the sialyltransferase reactions may reflect an interaction of the hydrophobic environments of the active sites with the detergents, possibly by the insertion of the latter into the lipid bilayer surrounding the enzymes or by the formation of detergent-enzyme complexes, thus inducing more active enzyme conformations (51,52,53). The effect of nonionic detergents may be similar to the previously reported effects of phosphatidyl ethanolamine (34), CDP-choline and lysolecithin (54,55), phospho-diglycerol and cardiolipid (56).

Pretreatment of enzyme preparations with trypsin diminished all four sialyltransferase activities (16). The sensitivity to

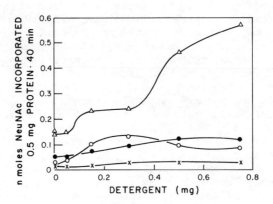

Journal of Neurochemistry

Figure 4. Effects of nonionic detergents on sialyltransferase activities. ✕, *Endogenous glycoproteins;* △, *exogenous DS-fetuin;* ●, *endogenous glycolipids;* ○, *exogenous GM₁ (16).*

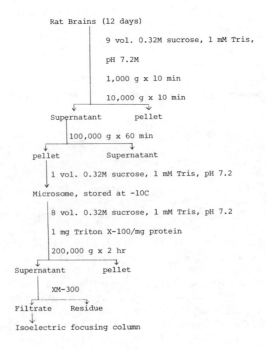

Figure 5. Solubilization and purification of rat brain sialyltransferases.

trypsinization was very apparent and the maximum decrease in enzyme activities was reached with the lowest concentration of trypsin used (0.025%, w/v). Residual activities were about 50% for
the endogenous glycoprotein reaction C and about 70% for the three
other reactions. This greater decrease in endogenous glycoprotein
activities was presumably due to a removal of endogenous substrates by trypsin pretreatment.

The residual sialyltransferase activities were not due to a
compartmentalization effect (16). Freezing and thawing twice and
six times did not change the enzyme activities. Furthermore,
trypsinization following freezing and thawing resulted in residual
activities similar to those with trypsinization alone. However,
when the enzyme preparation was pretreated with nonionic detergent
followed by trypsinization, nearly all enzyme activities were
abolished. The action of the detergent thus appeared to be on the
part of the enzyme molecules bearing the active sites. These regions of the enzyme molecules appeared to be masked in the native
cell membrane and were thus resistant to tryptic action. The action of the nonionic detergents exposed these sites, making them
susceptible to the action of trypsin.

The hypothesis that the active sites of sialyltransferases
are situated in a hydrophobic environment in native membranes was
tested with the use of organic solvents of various dielectric constants (16). With all the seven organic solvents tested, marked
inhibition towards the endogenous glycolipid and glycoprotein
sialyltransferase activities was observed. The inhibitory effect
of the alcohols clearly increased with the chain lengths of the
homologs. Chloroform, the solvent with the lowest dielectric constant tested, was the most potent inhibitor. Chloroethanol had an
effect intermediate between n-butanol and n-propanol. The effect
of acetone is similar to ethanol. In general, the inhibitory effects of the organic solvents were inversely related to their dielectric constants (16). It is conceivable that the more lipophilic solvents, with lower dielectric constants, are more powerful perturbing agents on the hydrophobic environments of the enzyme molecules.

Solubilization and Purification of Sialyltransferases.
Our next objective was to solubilize and purify the sialyltransferases. We followed the procedure established (11) for the purification of fucosyltransferase. The details are outlined in
Figure 5.

Rat brain microsome preparations were conveniently stored at
-10C with retention of enzyme activity. Solubilization with Triton X-100 appears to be effective and the solubilized enzyme preparation, after filtration once with Amicon XM-300 diaflo membrane, was introduced into an isoelectric focusing column (LKB)
with an ampholine pH range of 3.5-10.

During isoelectric focusing, a thick band of protein precipitation occurred close to the bottom anode. In order to eli-

minate interference of this precipitation band on the upper part
of the column, the column was eluted from the top by pumping a
dense sucrose solution into the bottom of the column.

Sialyltransferase activities were assayed with exogenous DS-
fetuin as well as mixed beef brain gangliosides. The results
were quite unexpected. Sialyltransferase activities using exogenous DS-fetuin and mixed gangliosides were found throughout the
whole column from pH 2-12. With either substrate about 10 activity peaks were found. Furthermore most of the glycoprotein
activity peaks do not overlap with the glycolipid activity peaks.
This is especially interesting because it is the first time that
we note a higher glycolipid activity than the glycoprotein activity in a particular enzyme preparation.

The basis for the multiplicity of the sialyltransferase
activities remains to be elucidated. We plan to purify these enzyme species to homogeneity, using isoelectric focusing columns of
smaller pH ranges in conjunction with affinity chromatography
which has been successfully used to purify the soluble sialyltransferases from bovine colostrum (57). Possibility exists that
the heterogeneity of sialyltransferase activities as observed is
due to differences in polypeptide sequences, carbohydrate content,
or non-covalent interactions with other membrane components, and
these possibilities can be clarified only with highly purified enzyme preparations.

Summary

    1. The endogenous glycolipid acceptor sites for sialic acid
in 11-15 day old rat brain were identified as lactosyl ceramide
and GM_1 ganglioside. These glycolipids comprise about 67% of the
available endogenous sialic acid acceptor sites. The remaining
acceptor sites are glycoproteins.
    2. The exogenous glycoproteins and glycolipid sialyltrans-
ferases can be solubilized and separated from each other on an
isoelectric focusing column. They can also be differentiated from
each other by competition experiments.
    3. A continuous sucrose gradient of the total homogenate
from young rat brain and electron microscopic examination of these
fractions found most of the sialyltransferase activities to be
localized in smooth microsomal membrane and Golgi complex derivatives and not associated with synaptosomes.

Acknowledgement

    This work was supported in part by NIH Grant NS-05104.

Literature Cited

1. McGuire, E.J., Ed. "Biological Roles of Sialic Acids";
    Plenum Press:New York, NY, 1976; p. 123.
2. Yip, M.C.M.; Dain, J.A. Lipids, 1969, 4, 270.
3. Dain, J.A.; DiCesare, J.L.; Yip, M.C.M.; Weicker, H. Adv.
    Exp. Biol. Med., 1972, 25, 151.
4. Yip, M.C.M. Biochim. Biophys. Acta, 1973, 306, 298.
5. Stoffyn, A.; Stoffyn, P.; Yip, M.C.M. Biochim. Biophys. Acta,
    1975, 409, 97.
6. Tettamanti, G.; Venerando, B.; Preti, A.; Lombardo, A.;
    Zambotti, V. Adv. Exp. Biol. Med., 1972, 25, 161.
7. Schengrund, C.; Rosenberg, A. J. Biol. Chem., 1970, 245,
    6196.
8. DiCesare, J.L.; Dain, J.A. J. Neurochem., 1972, 403, 410.
9. Bosmann, H.B. J. Neurochem., 1972, 19, 763.
10. Den, H.; Kaufman, B.; McGuire, E.J. J. Biol. Chem., 1975,
    250, 739.
11. Broquet, P.; Louisot, P. Biochimie, 1971, 53, 921.
12. Burton, R.M.; Howard, R.E.; Baer, S.; Balfour, Y.M. Biochim.
    Biophys. Acta, 1964, 84, 441.
13. Lehninger, A.L. Proc. Nat. Acad. Sci. U.S., 1968, 60, 1069.
14. Rahmann, H.; Rosner, H.; Breer, H. J. Theor. Biol., 1976, 57,
    231.
15. Ng, S.-S.; Dain, J.A. J. Neurochem., 1977, 29, 1075.
16. Ng, S.-S.; Dain, J.A. J. Neurochem., 1977, 29, 1085.
17. Van Den Eijnden, D.H.; Van Dijk, W.; Roukema, P.A.
    Neurobiology, 1975, 5, 221.
18. Svennerholm, L. J. Lipid Res., 1964, 5, 145.
19. Gielen, W.; Hinzen, D.H. Hoppe-Seyler's Z. Physiol. Chem.,
    1974, 355, 895.
20. Arce, A.; Maccioni, H.F.; Caputto, R. Biochem. J., 1966,
    116, 52.
21. Eichberg, J., Jr.; Whittaker, V.P.; Dawson, R.M.C. Biochem.
    J., 1964, 92, 91.
22. Kemp, S.F.; Stoolmiller, A.C. J. Neurochem., 1976, 27, 723.
23. Van Den Eijnden, D.W.; Van Dijk, W. Biochim. Biophys. Acta,
    1974, 362, 136.
24. Kaufman, B.; Basu, S.; Roseman, S., Eds. "Proceeding of the
    Third International Symposium on the Cerebral Sphingolipi-
    doses"; Pergamon Press:New York, 1966; p. 193.
25. Clarke, J.T.R.; Mulcahey, M.R. Biochim. Biophys. Acta, 1976,
    441, 146.
26. Chandrabose, K.A.; Graham, J.M.; Macpherson, I.A. Biochim.
    Biophys. Acta, 1976, 429, 112.
27. Fishman, P.H.; Bradley, R.M.; Henneberry, R.C. Arch. Biochem.
    Biophys., 1976, 172, 618.
28. Roseman, S. Chem. Phys. Lipids, 1970, 5, 270.
29. Ohman, R.; Rosenberg, A.; Svennerholm, L. Biochemistry, 1970,
    9, 3774.

30. Tettamanti, G.; Lombardo, A.; Preti, A.; Zambotti, V. Enzymologia, 1970, 39, 65.
31. Waehneldt, T.V.; Neuhoff, V. J. Neurochem., 1974, 23, 71.
32. Zanetta, J.P.; Morgan, I.G.; Gombos, G. Brain Res., 1975, 83, 337.
33. Den, H.; Kaufman, B.; Roseman, S. J. Biol. Chem., 1970, 245, 6607.
34. Kaufman, B.; Basu, S.; Roseman, S. J. Biol. Chem., 1968, 243, 5804.
35. Hudgin, R.L.; Schachter, H. Can. J. Biochem., 1972, 50, 1024.
36. Schachter, H.; Rosen, L., Eds. "Metabolic Conjugation and Metabolic Hydrolysis"; Academic Press:New York, 1973; p. 1.
37. Sader, J.E.; Paulson, J.C.; Hill, R.L. J. Biol. Chem. 1979, 254, 2112.
38. Duffard, R.O.; Caputto, R. Biochemistry, 1972, 11, 1396.
39. Brunngraber, E.G.; Witting, L.A.; Haberland, C.; Brown, B. Brain Res., 1972, 38, 151.
40. Arce, A.; Maccioni, H.F.; Caputto, R. Arch Biochem. Biophys., 1966, 116, 52.
41. Hilderbrand, J.; Stoffyn, P.; Hauser, G. J. Neurochem., 1970, 17, 403.
42. Morgan, I.G.; Reith, M.; Marinari, J.; Breckenridge, V.C.; Gambos, G. Adv. Exp. Med. Biol., 1972, 25, 209.
43. Ko, G.K.W.; Raghupathy, E. Biochim. Biophys. Acta, 1971, 244, 396.
44. Ko, G.K.W.; Raghupathy, E. Biochim. Biophys. Acta, 1972, 264, 129.
45. Whittaker, V.P. Biochem. J., 1959, 72, 694.
46. DeRobertis, E.; Deiraldi, P.; DeLores Arnaiz, G.R.; Salganicoff, L. J. Neurochem., 1962, 9, 23.
47. McIntosh, C.H.S.; Plummer, D.T. J. Neurochem., 1976, 27, 449.
48. Dain, J.A.; Hitchener, W. Trans. Am. Neurochem. Soc., 1977, 8, 237.
49. Porter, C.W.; Bernacki, R.J. Nature, 1975, 250, 648.
50. Singer, S.J.; Nicholson, G.L. Science, 1972, 175, 720.
51. Simons, K.; Helenius, A.; Garoff, H. J. Molec. Biol., 1973, 80, 119.
52. Utermann, G.; Simons, K. J. Molec. Biol., 1974, 85, 569.
53. Coleman, R. Biochem. Soc. Trans., 1974, 2, 813.
54. Mookerjea, S.; Yung, J.W.M. Can. J. Biochem., 1974, 52, 1053.
55. Mookerjea, S.; Cole, D.E.C.; Chow, A.; Letts, P.J. Can. J. Biochem., 1972, 50, 1094.
56. Keenan, T.W.; Morre, D.J.; Basu, S. J. Biol. Chem. 194, 249, 310.
57. Paulson, J.C.; Beranek, W.E.; Hill, R.L. J. Biol. Chem., 1977, 252, 2356.

RECEIVED December 10, 1979.

Modulation of Ganglioside Synthesis by Enkephalins, Opiates, and Prostaglandins

Role of Cyclic AMP in Glycosylation

GLYN DAWSON, RONALD W. McLAWHON, GWENDOLYN SCHOON, and
R. J. MILLER

Departments of Pediatrics, Biochemistry, and Pharmacology, University of Chicago,
Chicago, IL 60637

The ubiquitous occurrence of many different
species of glycosphingolipids on the outer surface of
eukaryotic cell membranes has suggested an important
role for glycolipids in a number of biological phenomena such as cell recognition, adhesion, ion transport
and receptors. However, despite many studies, the
role of G_{M1} ganglioside as the receptor for cholera
toxin remains the only unambiguous example of a
precise biological role (1,2). Much of the work in
attempting to understand the function of glycosphingolipids in the nervous system has been hindered by the
fact that very few molecules/cell are required to
mediate biological reactions. For example, less than
100,000 G_{M1} molecules/cell are required for a functional cholera toxin receptor, (3) an amount which is
chemically undetectable in less than 10^9 cells. This
problem can be overcome by working with tumor cell
lines which are highly metabolically active and incorporate sufficient amounts of isotopic precursors
(such as $[^3H]$ GlcN) to detect small changes in ganglioside composition. In addition, most tissues and
normal cells are extremely heterogeneous (e.g. very
few neurons respond to opiates). This latter problem
can be overcome by studying cloned mouse neuroblastoma
cell lines which are rich in gangliosides, express
many of the properties of mature neurons (4,5,6), and
can be reproducibly cultured in large quantities under
defined conditions. Many neuroblastoma cell lines
have been characterized pharmacologically with respect
to their ability to respond to neuro-active agents
such as catecholamines and opioid peptides (enkephalins), one such line being N4TG1 which has been
shown to possess enkephalin receptors (7).
An initial observation in this laboratory that
opiates such as morphine, levorphanol and fentanyl,

0-8412-0556-6/80/47-128-359$5.00/0
© 1980 American Chemical Society

and opioid peptides such as enkephalins
([D]Ala[2] - [D]Leu[5] enkephalin) and β-endorphin, specifically inhibited the glycosylation of glycolipids and
glycoproteins in opiate-receptor-positive cell lines
such as N4TGl (8) suggested an involvement of surface
carbohydrate moieties in the observed acute and
chronic action of opiates on cells. These observed
effects include changes in membrane conductance,
changes in ion fluxes, changes in sensitivity to other
neurotransmitters and an elevation in the membrane's
threshold for generating action spikes (9,10). All
these effects, as well as the inhibition of ganglioside and glycoprotein synthesis, are reversible by
naloxone.

Nirenberg, Klee, Hamprecht and co-workers (5,6,11)
reported that the initial effect of morphine on an
opiate receptor-positive neuroblastoma x glioma hybrid
cell line (NGl08-15) was the inhibition of prostaglandin E_1 (PGE_1) stimulation of adenylate cyclase
activity. We have confirmed that opiates and opioid
peptides inhibit both the basal and PGE_1-stimulated
synthesis of cyclic AMP in both NGl08-15 and another
opiate receptor-positive cell line, mouse neuroblastoma N4TGl. Further, we have shown that agents
(such as enkephalins) which depress cyclic AMP levels
in N4TGl cells depress the level of glycosylation at
the level of glycosyltransferase activity, whereas
agents which elevate cyclic AMP levels (such as prostaglandins, cholera toxin, phosphodiesterase inhibitors
and cyclic AMP analogs) stimulate glycosylation and
glycosyltransferase activity. We propose a theory
in which certain glycosyltransferase activities, especially G_{M3} : UDP GalNAc N-acetylgalactosaminyl-
transferase activity, may be activated by a cyclic AMP-
dependent protein kinase system.

Methods

Cell culture system.   Mouse neuroblastoma
(N4TGl, NB[2a]) and mouse neuroblastoma N18 x rat glioma
C-6 (NGl08-15) hybrid cell lines were grown as monolayer cultures on 100mm Falcon plastic dishes in modified Eagle's medium supplemented with 10% fetal calf
serum. Experiments were normally carried out 4 days
after initial plating and the cell density normally
increases from 2 to 10 x 10[6] cells/dish between day
4 and day 7. Drugs were added in 50% ethanol water
(taking care not to exceed ethanol concentrations of
10 μl/10ml of media) and isotopic precursors such as
[3H] GlcN, [14C] Gal, [3H] leucine or methyl-[14C]

thymidine were added as supplied by New England
Nuclear Inc. Cells were harvested mechanically,
sonicated briefly for 10 sec at power setting 4 with
a model W-185 sonifier (Heat Systems Inc.) and subjected to a variety of procedures.

Glycolipid analysis. Gangliosides were extracted with chloroform-methanol (2:1), purified on
Sephadex G-25 columns and separated into individual
species by thin-layer chromatography as described
previously (12).

Glycosyltransferase assays. Glycosyltransferase
assays were carried out on sonicated fresh or frozen
(-20°C) cell suspensions in 0.05$\underline{M}$MES.KCl buffer pH 6.5
containing 0.003$\underline{M}$ MnCl$_2$ as described previously (13).
For N-acetylgalactosyltransferase measurements the
donor was UDP-[^{14}C]GalNAc (45mCi/mmol) and the substrate G$_{M3}$ ganglioside (II^3 NeuAc-Lac Cer from dog
erythrocytes);for galactosyltransferase measurements
the donor was UDP-[^{14}C]Gal (200mCi/mmol) and the substrate G$_{M2}$ ganglioside (II^3-NeuAc-GgOse$_3$ Cer from Tay-
Sachs human brain). Substrates and membranes were
dispersed in a mixture of Cutscum (120 µg) and Triton
X-100 (60 µg) in a total reaction vol. of 0.055 ml (13)..
Labelled G$_{M2}$ and G$_{M1}$ were purified by Sephadex G-25 and
thin-layer chomatography as described previously
(12,13).

Cyclic AMP assay. Cyclic AMP levels were measured
on 6% cold trichloracetic acid extracts of sonicated
cell extracts as described previously (14). The
succinylated [^{125}I] tyrosyl derivative of cyclic AMP
was obtained from Schwartz-Mann Inc. and the anti-
cyclic AMP antibody was a generous gift from Dr. P.
Hofmann (Dept. Pharmacology, University of Chicago);
the radioimmunoassay was carried out as described
previously (14).

Results and Discussion

N4T$_{G1}$ cells contain G$_{D1a}$, G$_{M1}$ and G$_{M2}$ gangliosides in the ratio 10 : 2 : 6 as judged by quantitative analysis (12, 13) and the level of incorporation
of the isotopic precursors [^{3}H] GlcN, [^{3}H] Gal or [^{3}H]
GalN. Incubation of monolayer cultures with morphine,
levorphanol or [$_D$Ala^2, $_D$Leu^5] enkephalin (10µM) for
24 hrs resulted in a marked reduction in the amount of
G$_{D1a}$ G$_{M1}$ and G$_{M2}$ present in the cells. The addition
of further quantities (10µM) of morphine at daily

362

TABLE I. Effect of Opiate Agonists and Antagonists
on Ganglioside synthesis in N4TG1 Cells

Treatment (24 hr)	Gangliosides
	cpm x 10^{-3}/mg protein
Control	101
Morphine	59
Fentanyl	47
Etorphine	73
Enkephalin	54
β-Endorphin	44
Meperidine (Demerol)	93
Pentazocine	
50µM (antagonist)	94
0.5µM (agonist)	51
Nalorphine	
50µM (agonist	40
0.5µM (partial)	54

All drugs were used at 1 µM except where stated.

intervals resulted in the virtual disappearance of
GD_{1a} after 4 days (15). This is somewhat reminiscent
of the effects of viral transformation of cell lines
such as mouse 3T3 with polyoma or SV-40 virus (16).
The inhibitory effects of various opiates and opioid
peptides is shown in Table I and in general is in
agreement with their pharmacological potency both
in vivo and in in vitro preparations. Of particular
interest is pentazocine which is known to act as an
antagonist at high concentrations (no effect on ganglioside synthesis) and an agonist at low concentrations
(inhibits ganglioside synthesis). The inhibition of
ganglioside synthesis (as judged by the incorporation
of $[^3H]$ GlcN into gangliosides) was dose-dependent
(Fig. 1; A, B) and reversible by naloxone. Further
evidence for the stereospecificity of the phenomenon
was provided by the fact that the inactive isomer of
levorphanol, called dextrorphan, had no effect on
ganglioside synthesis (8,15).

Since Nirenberg and his collaborators (5,6,11)
have reported that in NG108-15 cells, opiates produce
an initial inhibition of cyclic AMP synthesis (0 to
30 min), followed by a period in which the cell ad-
justs and cyclic AMP levels return to normal (10 -
30 hrs), and finally a period where cyclic AMP levels
are elevated (when the drug is removed or metabolized;
supposed withdrawal) it was of interest to follow the
time-course of the inhibition.

Maximum inhibition of ganglioside synthesis by
either morphine or enkephalin occurred during the
first 30 hrs after which time an apparent increase in
the rate of synthesis occurred (Fig. 2) so that following 54 to 60 hrs of continuous exposure to the isotopic precursor it was not possible to distinguish
control from opiate-treated cells. This return to
normal levels of ganglioside synthesis could be accel-
erated by the removal of the drug, or the addition of
naloxone (50µM) but the inhibition could be prolonged
by the further addition of opiate (10µM). All these
detailed studies have been reported elsewhere (15).

Since opiates have been observed to cause the
inhibition of both cyclic AMP and ganglioside synthesis,
the next step was to see if the two phenomena were in
any way connected. N4TG1 neuroblastoma cells were
therefore incubated with a variety of agents known to
elevate cyclic AMP levels in cells possessing the
appropriate receptors (namely prostaglandin E, and
cholera toxin). It can be seen from Table II that
such agents caused an increase in the level of incorporation of $[^3H]$ GlcN into gangliosides and that

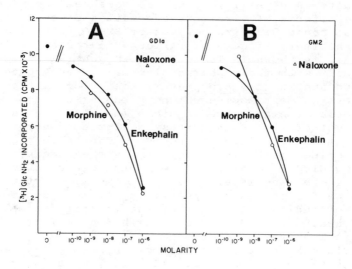

Figure 1.   Dose-dependent inhibition of ganglioside synthesis in N4TG1 cells by morphine and [D-Ala², D-Leu⁵] enkephalin under conditions described in text

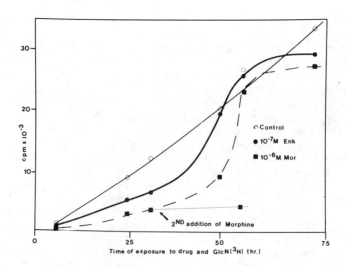

Figure 2.   Time-dependent inhibition of ganglioside synthesis in N4TG1 cells by morphine and [D-Ala², D-Leu⁵] exkephalin under conditions described in text

TABLE II. Ganglioside Synthesis is related to Cyclic AMP Synthesis

N4TGl cells were labelled with $[^3H]$ GLcN and cultured in the presence of different drugs for 24 hrs. Gangliosides were isolated as described in the text.

Treatment	Percent of normal ganglioside synthesis following 24 hr treatment of N4TGl cells.
Inhibitors of adenylate cyclase	
1μM Enkephalin	38
Stimulators of adenylate cyclase	
10μM Prostaglandin E_1 (+ IBMX)	190
Cholera toxin (+ IBMX)	152
Elevators of cellular cAMP	
100μM 8-Br cAMP + IBMX	135
100μM Bt_2 cAMP	200
Competition	
10μM Prostaglandin E_1 (+ IBMX)+ Enkephalin	100
100μM 8-Br cAMP+IBMX+Enkephalin	140

this increase could be blocked by the co-addition of
appropriate amounts of morphine or opiate peptide.
The addition of phosphodiesterase inhibitors such as
isobutylmethylxanthine (IBMX) and cyclic AMP analogs
such as the 8-bromo- and dibutyryl-derivatives resulted in a stimulation of synthesis which could not be
blocked by the co-addition of opiates. This implies
that opiates are exerting their effect at the level of
adenylate cyclase activation. The actual effect of
these drugs on cellular cyclic AMP levels was determined by radioimmunoassay as shown in Table III.
These data infer a direct correlation between cyclic
AMP levels and ganglioside synthesis. A precedent
does exist in that tumor cells generally have low
levels of cyclic AMP (17) and simple glycosphingolipid
composition as the result of the specific suppression
of transferase activities involved in the synthesis of
G_{M2} and G_{M1} (16). However, to our knowledge, no
direct correlation between reduced cAMP levels and
reduced glycosyltransferase activity has ever been
made. An inhibitory effect of opiates at the level of
ganglioside (and glycoprotein (8,15)) synthesis was
considered likely since most other possibilities had
been eliminated. Thus we had previously shown (8,15)
that the incorporation of isotopic precursors into
other cellular products such as proteins, nucleic
acids, proteoglycans and phospholipids was not suppressed, arguing against a transport defect, and that
no inhibition of lysosomal hydrolase activity could be
detected. We therefore attempted to verify this hypothesis by direct measurement of the two glycosyltransferases involved in the synthesis of G_{M2} and G_{M1}
gangliosides.
        Both N-Acetylgalactosaminyltransferase and gal-
actosyltransferase activities were suppressed when the
cells were pre-treated with opiates or opiate peptides
as shown in Table IV. The time course for such
studies is obviously crucial and Table V shows an example where the potent opiate fentanyl was used as the
inhibitor. It can be seen that the initial inhibition
is only observed for 24-48 hrs and that after this
time glycosyltransferase levels appear to be normal.

Conclusions

        If a direct correlation between ganglioside syn-
thesis and cyclic AMP levels does in fact exist then a
hypothesis must be developed to explain it. Based on
our current level of understanding, the most plausible
hypothesis involves linking the two phenomena by a

TABLE III. Regulation of c-AMP levels in mouse neuroblastoma N4TGl Cells.

N4TGl cells were exposed to drugs for 10 min and
the cyclic AMP level determined by radioimmunoassay
as described in the text.

Treatment	Percent of control after 10 min
Morphine (1μM)	41
Morphine + Naloxone	63
Morphine + Naloxone + IBMX	242
PGE_1 + IBMX	30,000
PGE_1 + IBMX + Morphine	335

TABLE IV. Modulation of Cyclic AMP levels in neuroblastoma cells is directly related to
ganglioside synthesis

N4TG1 cells were exposed to drugs for 8 hrs. and
the homogenates assayed for glycosyltransferase
activity as described in the text

Treatment	GM3 : UDP GalNAc N-acetylgalactos-aminyltransferase	GM2 : UDP Gal Galactosyl-transferase
	cpm/mg protein/20 min	
None	40,000	1,500
Enkephalin (1µM)	6,000	800
PGE$_1$ + IBMX (10µM)	100,000	3,800
Bt$_2$ cAMP (100µM)	33,000	1,800

TABLE V. $\underline{GM_3}$: $\underline{Hexosaminyltransferase\ Activity\ in}$ $\underline{N4TG1\ cells\ exposed\ to\ Fentanyl\ (10\mu M)\ for}$ $\underline{different\ periods\ of\ time}$.

N4TG1 cells were exposed to fentanyl ($10\mu M$) for
varying periods of time and the homogenates
assayed for GM_3 : hexosaminyltransferase activity
with UDP ^3H GalNAc as described in the text

Time (hrs)	Control	+ Fentanyl
	cpm transferred/mg protein/50 min	
4	1,400	500
24	2,630	850
48	5,300	5,400
54	7,200	3,450
72	3,400	2,900

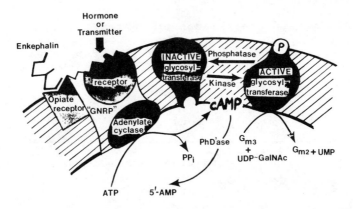

Figure 3.   Proposed model for the action of opiates and opioid peptides on glyco-syltransferase activity. It is postulated that receptors for both opiates and neuro-transmitters are linked to adenylate cyclase by a guanylnucleotide regulatory protein (GNRP). The presence of opiates or opioid peptides inactivates the cyclase, which normally activates a protein kinase–glycosyltransferase system, thereby initiating glycosylation.

cyclic AMP-dependent protein kinase system. Circumstantial evidence exists in that synaptosomal membrane fractions are rich in both gangliosides and glycosyltransferase activities and form an excellent substrate for kinase-catalysed phosphorylation of membrane proteins (18). Studies are currently underway to determine if the phosphorylation of purified glycosyltransferases such as xylosyltransferase (19) causes an increase in activity or if the addition of purified protein kinase and cyclic AMP to membrane preparations can enhance N-acetylgalactosaminyl- or galactosyltransferase activities. We therefore propose the following model to explain some of the actions of opiate peptides on neurons. (Fig. 3). Many confirmatory studies remain to be carried out but we feel that opiate-induced reductions in the level of glycosylated sphingolipids and membrane proteins may explain the chronic effects of opiates, namely the altered sensitivity to other drugs and neurotransmitters such as serotonin. It is also possible that localized small changes in surface carbohydrate in the vicinity of the receptor could be responsible for the initial (acute) effects of opiates which involve rapid changes in cyclic AMP levels. In this context it is interesting to note a recent report (20) that gangliosides stimulate adenylate cyclase in synaptosomal fractions. This apparent reverse correlation appears to have its explanation in the fact that the ganglioside may simply be restoring full β-adrenergic receptor function. In addition, another recent report that ion and metabolite transport may depend on protein glycosylation (21) indicate that the changes we have observed may also explain some of the other observed physiological effects of opiates.

Acknowledgements

We would like to acknowledge the excellent technical assistance of Mr. John Oh and Stewart M. Kernes. Supported in part by USPHS Grants HD-06426, HD-04583 and HD-09402. G. Dawson is the recipient of USPHS Research Career Development Award NS-00029.

Literature Cited

1. Cuatrecasas, P.; Biochem. 1973, 12, 3577.
2. Fishman, P.H.; Brady, R.O.; Science, 1976, 194,906.
3. Fishman, P.H.; Atikkan, E.E.; J. Biol. Chem.,
    1979, 254, 4342.

4.   Peacock, J.; Minna, J.; Nelson, P.G.;
     Nirenberg, M.W.; Exp. Cell. Res., 1972, 73, 367.
5.   Sharma, S.K.; Nirenberg, M.W.; Klee, W.A.;   Proc.
     Natl. Acad. Sci. U.S.A., 1975, 74, 3365.
6.   Hamprecht, B.; Angew. Chem., 1976, 15, 196.
7.   Miller, R.J.; Chang, K-J.; Cooper, B.;
     Cuatrecasas, P.; J. Biol. Chem., 1978, 253, 531.
8.   Dawson, G.; McLawhon, R.; Miller, R.J.;   Proc.
     Natl. Acad. Sci. U.S.A., 1979, 76, 605.
9.   Barker, J.L.; Smith, T.G.;   "The Role of Peptides
     in Neuronal Function", Marcel Dekker, New York,
     1980.
10. North, R.A.; Zieglgansberger, W.; Brain Res.,1978,
     144, 208.
11. Sharma, S.K.; Nirenberg, M.W.; Klee, W.A.; Proc.
     Natl. Acad. Sci. U.S.A., 1975, 74, 590.
12. Dawson, G.; Matalon, R.; Dorfman, A.;   J. Biol.
     Chem., 1972, 247, 5944.
13. Dawson, G.;   J. Biol. Chem., 1979, 254, 154.
14. Steiner, A.L.; Parker, C.W.; Kipnis, D.M.;   J.
     Biol. Chem., 1972, 247, 1106.
15. Dawson, G.; McLawhon, R.; Miller, R.J.;   J. Biol.
     Chem.,   in press, 1979.
16. Brady, R.O.; Fishman, P.H.; Biochem. Biophys.
     Acta, 1974, 355, 121.
17. Johnson, G.S.; Friedman, R.M.; Pastan, I.; Proc.
     Natl. Acad. Sci. U.S.A., 1971, 68, 425.
18. Lohmann, S.M.; Veda, T.; Greengard, P.; Proc.
     Natl. Acad. Sci. U.S.A., 1978, 75, 4037.
19. Schwartz, N.B.; J. Biol. Chem., 1979, 252.
20. Partington, C.R.; Daly, J.W.;   Molec. Pharmacol.,
     1978, 15, 484.
21. Olden, K.; Pratt, R.M.; Jaworski, C.; Yamada, K.M;
     Proc. Natl. Acad. Sci. U.S.A., 1979, 76, 791.

RECEIVED December 10, 1979.

Gangliosides as Receptors for Cholera Toxin, Tetanus Toxin, and Sendai Virus

LARS SVENNERHOLM and P. FREDMAN
Department of Neurochemistry, St. Jörgen Hospital, University of Göteborg, S-422 03 Hisings, Backa, Sweden

H. ELWING, J. HOLMGREN, and Ö. STRANNEGÅRD
Institute of Medical Microbiology, University of Göteborg, S-413 46 Göteborg, Sweden

The carbohydrate residues of cell surfaces have been implicated in many cell functions such as cell-cell interactions,
immunological specificity, and receptors for bacterial toxins,
viruses and hormones. Cell surface carbohydrates reside in both
glycoproteins and glycolipids, and it is conceivable that some
cell surface properties are determined by carbohydrate groups
whether these groups are components of glycolipids and glycoproteins or not. The ABO-blood group substances are an example
of glycoconjugates which are both glycolipids and glycoproteins.
The gangliosides have recently attracted considerable interest as
receptors, not only for bacterial toxins, but also for Sendai
virus, interferon and some glycoprotein hormones (1). The major
basic carbohydrate structure of gangliosides, the gangliotetraose moiety, has been considered unique for this group of
lipids, but future work might show that the same structure might
reside in glycoproteins, and then explain findings which are at
present difficult to understand.

The possibility of gangliosides serving as receptors for
toxins was speculated upon as far back as the 1950s, but was only
recently confirmed by work on cholera toxin. This toxin is readily
isolated, its structure is fairly well known, and it is relatively
stable during storage or when radioactively labelled. The receptor
ganglioside, GM1, is the parent ganglioside of the major brain
gangliosides. Except for ganglioside GM2 it is the most stable of
all known gangliosides, and it occurs abundantly in almost all
mammalian plasma membranes. The studies of the interaction between
ganglioside GM1 and cholera toxin might therefore be a prototype
for ganglioside studies. In our own studies the use of highly
purified gangliosides and newly developed quantitative binding
assays produced results which allowed detailed prediction of the
recognition-specific structures of ganglioside GM1 for cholera
toxin.

0-8412-0556-6/80/47-128-373$5.00/0
© 1980 American Chemical Society

With this system as a point of departure we adopted the same
procedures for our studies of the interactions between gangliosides
and tetanus toxin and Sendai virus.

Isolation of gangliosides

Homogenized human brain tissue was extracted twice with
twenty volumes of chloroform-methanol-water 4:8:3 (final solvent
ratio). The gangliosides were separated from other lipids by phase
partition – water was added to the total lipid extract to give a
final chloroform-methanol-water volume ratio of 4:8:5.6. The upper
phase was evaporated to dryness, dissolved in water, dialysed
against several changes of water for three days, and evaporated to
dryness in a rotating evaporator. The gangliosides dissolved in
chloroform-methanol-water 60:30:4.5 were separated according to
their number of sialic acids on a new anion exchange resin,
Spherosil-DEAE-Dextran, consisting of porous glass beads covered
with cross-linked DEAE-Dextran, with a discontinuous gradient of
potassium acetate in methanol. From the five major ganglioside
fractions obtained – with one to five sialic acids – individual
gangliosides were isolated by silica gel chromatography on columns
or thin-layer plates. The individual gangliosides were purified to
have a carbohydrate composition, which was at least 99% pure.
Ganglioside GTla and GQlb were at least 95% pure, and GPlc contaminated with 10% GPlb. The methods used for the determination of the
purity of individual gangliosides have recently been described
(2).

Ligand methods

Ganglioside-ELISA method. The glycolipid (0.1 or 0.2 ml) is
attached to the polystyrene test tubes (or to the wells of microtiter trays) by incubation at room temperature overnight. Unoccupied binding sites on the plastic surface are blocked by incubation with 1% serum albumin. The tubes are then incubated for
1-5 h at room temperature with dilutions of the test samples
(usually 2 h for non-antibody ligands and 5 h for antibodies) in
buffer pH 7.0-7.2 supplemented with 1% serum albumin or in the
case of antibodies with 0.05% Tween 20. Unbound material is removed by rinsing the tubes with buffer containing 0.05% Tween 20.
Bound non-antibody ligands are thereafter demonstrated by sequential incubations of the tubes with: (1) specific antibody, (2) anti-
immunoglobulin coupled to alkaline phosphatase, and (3) nitro-
phenyl-phosphate substrate (3). The technique is illustrated in
Figure 1.

The water condensation-on-surface (VCS) method. In this
method, originally elaborated for studying antigen-antibody
reactions (4) the ganglioside is attached to the inner bottom

surface of polystyrene Petri dishes (diameter 3.5 cm; Heger
Plastics AB), and unoccupied binding sites blocked with serum
albumin in the same manner as for the ganglioside-ELISA test.
After rinsing with 0.15 M NaCl, melted agar at a concentration of
1% containing 0.01% serum albumin is poured into the dishes forming, after congelation, a 2.5 mm thick agar layer. Wells of 5 mm
diameter are then punched in the gel and filled (50 µl) with
different concentrations of the test ligand (e.g. cholera toxin).
After diffusion for 20 h at room temperature the agar gel is
rinsed off and the plastic surface washed with distilled water and
dried. The dish is then placed upside down over a container filled
with water at 60°C. After exposure to water vapour for one minute
the dish is removed and covered. A condensation pattern consisting
of large confluent water drops is then formed in a circular zone
where a specific ganglioside-ligand reaction has taken place. This
hydrophilic pattern contrasts to that for the remainder of the
surface where only small condensation drops are formed indicating
a distinctly less hydrophilic surface (Figure 2).

Hemadsorption binding assay. Identification of ganglioside-
bound virus by hemadsorption is performed in Petri dishes incubated with gangliosides, egg albumin and virus as for the water condensation method. After removing unbound virus, to the dish is added
5 ml of a 1% suspension of guinea pig erythrocytes in 0.15 M
NaCl. After 15 minutes of sedimentation the unadsorbed erythrocytes are washed off and the hemadsorption pattern inspected.
Permanent recording of the hemadsorption is obtained by contact
copying of the surface on photographic paper with the use of
monochromatic light at 405 nm.

Interaction between cholera toxin and ganglioside GM1

In 1971 van Heyningen and co-workers (5) described that a
crude ganglioside mixture inactivated cholera toxin. We demonstrated that the inactivation was caused by a specific reaction between
the toxin and a single ganglioside GM1 (6). GM1 inhibited the biological effects of cholera toxin down to equimolar with toxin, and
in contrast to all other substances tested, GM1 ganglioside gave a
specific precipitation band with cholera toxin in Ouchterlony
double diffusion-in-gel tests. Independently, Cuatrecasas (7) and
King and van Heyningen (8) described GM1 ganglioside as the substance reacting strongest with cholera toxin but with less
specificity than in our study, probably because the ganglioside
preparations they worked with were not pure.

Subsequent studies in several laboratories have provided
further evidence that ganglioside GM1 is the natural biological
receptor for cholera toxin:
1) Studies of various cell types, including small intestinal
mucosal cells of different species, demonstrated a direct relationship between the cell content of GM1 and the number of toxin

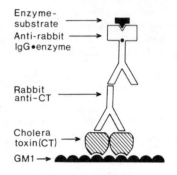

Enzyme-
substrate

Anti-rabbit
IgG•enzyme

Rabbit
anti–CT

Cholera
toxin(CT)

GM1 →

*Figure 1.   Schematic of the ganglioside–
ELISA method*

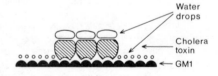

Water
drops

Cholera
toxin

← GM1

*Figure 2.   Schematic of the vapor con-
densation-on-surface method*

molecules that the cells can bind (9, 10).

2) Exogenous GM1 ganglioside can be incorporated into the cell membrane and then act as a functional receptor. This was first shown by Cuatrecasas (11) who observed an increased binding capacity and lipolytic responsiveness of fat cells which had been soaked in GM1. Using tritium-labelled GM1-ganglioside, Holmgren et al. (9) demonstrated incorporation of GM1 into small intestinal epithelial membrane and showed that the increase in GM1 was associated with a corresponding increase in the capacity of the intestine to bind cholera toxin as well as in an increased susceptibility of the gut to the diarrheogenic action of the toxin (9). Incorporation of ganglioside GM1 into transformed cells deficient on this ganglioside restored cell responsiveness to cholera toxin (12).

3) Pretreatment of cell membranes with cholera toxin specifically blocked the membrane GM1-ganglioside from reacting with galactose oxidase (13).

4) Incubation of enzyme-susceptible tissues with *Vibrio cholerae* sialidase increased the number of binding sites proportional to the increase in GM1 ganglioside, as well as increased cellular sensitivity to the toxin (9).

5) Chemical modifications of cholera toxins by means of various reagents were consistently found to affect binding to cells and to GM1 ganglioside to the same extent (14).

Acetylsphingosine-GM1, in which the fatty acid of GM1 was replaced by an acetyl group, had roughly the same ability as intact GM1 to react with cholera toxin including formation of a precipitation line in agar gel. Oligosaccharide-GM1 (gangliotetraose) devoid of both the fatty acid and the sphingosine and therefore unable to form micelles could not precipitate cholera toxin but effectively inhibited the precipitation reactions between toxin and GM1, provided that it was in a 5-fold molar excess to the toxin (15). Essentially similar results were obtained by Staerk et al. (16). The cholera toxin molecule is composed of probably 5 light receptor-binding subunits (B) and a structurally unrelated "heavy" effector subunit (A) (17, 18). Since the five light B-subunits are identical each toxin molecule can be expected to bind up to 5 molecules of GM1, an assumption supported by Sattler et al. (19) and Fishman et al. (20) who found that a cholera toxin molecule binds at most 4-6 gangliotetraose molecules.

The first suggestion that gangliosides may play a role as receptor for tetanus toxin was made before the biochemical structure of the gangliosides were fully known (21, 22). To demonstrate fixation of tetanus toxin by gangliosides van Heyningen and Miller (22) used boundary electrophoresis and ultracentrifugation. They found that the gangliosides GD1b and GT1b had the greatest toxin binding capacity. The same specificity of ganglioside structure was also found for binding at low toxin and ganglioside concentrations. In this case fixation was studied by following the binding of toxin to ganglioside made insoluble by complexing it with cerebroside. The insoluble receptor and the toxin were centrifuged out. Later work using this technique and described in Ledeen and Mellanby (23) showed that the toxin-binding capacity of ganglioside GDlb was about 8 times that of ganglioside GTlb and 20-80 times better than GM1 and GDla.

Table I.

A comparison of the ability of different gangliosides in cerebroside-ganglioside complexes to bind tetanus toxin (23)

Ganglioside	Smallest amount of ganglioside necessary to bind 10 LD_{50} ng
GM1	20
GDla	5.5
GDlb	0.25
GTlb	2.0

Ledley et al. (24), who measured the capacity of various gangliosides to inhibit tetanus toxin binding to thyroid membranes, found the order of reactivity of the various tested gangliosides to be GT1≈GDlb>GM1≈GDla>GM2. In contrast, Helting et al. (25) reported GM1 to be equally effective as GDlb in binding tetanus toxin. The aim of our studies was to define the degree of structure specificity in the binding of tetanus toxin to gangliosides and thus identify the presumed oligosaccharide recognition structure of the receptor. The binding of tetanus toxin to plastic-attached gangliosides was determined using either ganglioside-ELISA or VCS techniques for quantitation of the toxin bound.

Figure 3 shows the results obtained with the ganglioside-ELISA method. In contrast to cholera toxin, which in the same system was bound only to GM1, tetanus toxin bound significantly to several gangliosides. The strongest reactions were obtained with GTlb, GQlb and GDlb closely followed by GTla. Distinctly lower but still strong activity was seen with GM1 (difference to e.g. GDlb: p <0.01), and also GDla and GD3 bound tetanus toxin even less (Figure 3). A decrease in ganglioside concentration used for coating, from e.g. 2 to 0.2 µM, resulted in a markedly reduced binding of tetanus toxin (Figure 3); this differed from cholera toxin the binding of which to GM1 as studied with this method was practically unchanged within the concentration range 0.02 to 2 µM. Also a plot for adsorbance vs toxin concentration

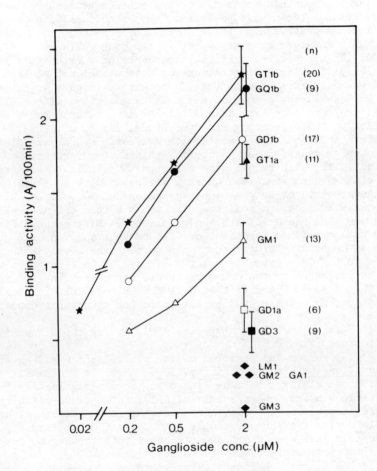

Figure 3. Binding of tetanus toxin to various gangliosides as studied by the gan-glioside–ELISA method. Mean and SEM (bars) values of n experiments, each performed in duplicate or triplicate.

differed markedly between tetanus and cholera toxin in that considerably more material was required for detectable toxin and the
slope was less steep for tetanus toxin.

Essentially the same reaction patterns were demonstrated with
the VCS method where the bound toxin was directly visualized, thus
avoiding the possible influence of any subsequent immunoreaction
steps. Tetanus toxin gave positive, "wet" reactions with the same
gangliosides that were positive in the ELISA method, i.e. GT1b,
GQ1b, GD1b and less strongly with, in the order GT1a, GM1, GD1a
and GD3 (Figure 4). The critical dependence on a relatively high
ganglioside coating concentration for tetanus but not cholera
toxin was evident also with the VCS technique (Figure 4 insert).

Our results by either method that gangliosides of the Glb
series are more effective than GM1 in binding tetanus toxin are in
contrast to those recently reported by Helting et al. (25). We
therefore undertook experiments with enzymatic hydrolysis of the
more complex gangliosides to clarify this matter further. Gangliosides GT1b, GD1b or GM1 were attached to plastic tubes, and
half of the tubes were subsequently incubated with *V. cholerae*
sialidase to hydrolyse the higher gangliosides to GM1. The binding
of tetanus and cholera toxins to the tubes before and after hydrolysis was then measured with the ganglioside-ELISA method. Figure 5
shows that sialidase treatment of GT1b as well as with GD1b decreased the binding of tetanus toxin to the same level as that obtained in the GM1-coated tubes; the binding of cholera toxin was
instead increased to the GM1 tube level, supporting the assumption
that the presumed hydrolysis of GT1b and GD1b to GM1 did really
take place. From these studies the following structural conclusions
may be drawn:

1) Tetanus toxin has a special affinity for the Glb series of
gangliosides (Figure 6).

2) The number and position of the sialic acid residues are
critical for the binding affinity. Thus, there is a minimal requirement for one sialosyl residue linked to the inner galactose for
detectable binding (compare GM1 and GA1), but optimal affinity
requires a disialosyl group, linked to the inner galactose, as in
the Glb series (compare e.g. GD1b with GM1 and GD1a). Additional
sialic acid residues do not seem to contribute further to the
recognition structure (GQ1b≈GT1b≈GD1b).

3) The oligosaccharide backbone is also of critical importance.
GD3 which lacks the terminal galactose and N-acetylgalactosamine
residues has little binding activity despite the proper disialosyl
linkage to a galactose residue, and GM2 in contrast to GM1, has no
binding activity supporting the important role of the terminal
galactose.

The original studies of toxin binding using cerebroside-
ganglioside complexes (26) had the serious disadvantage that the
fixation ability of the ganglioside was critically dependent on the
ganglioside:cerebroside ratio of the insoluble complexes. In the

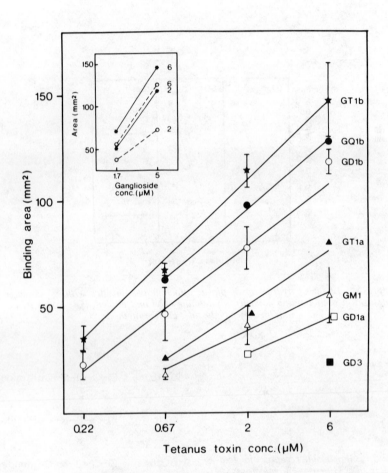

Figure 4. Binding of tetanus toxin to various gangliosides as determined with the VCS method; micromolar ganglioside concentrations were used for coating. Insert shows the effect of decreasing the coating concentrations of tetanus toxin (6 and 2 μM). Mean values of two or three experiments (here SEM values also shown) performed in duplicate.

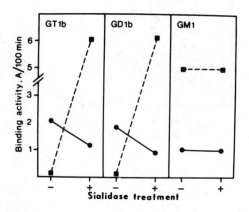

Figure 5. Effect of hydrolysis of plastic-attached gangliosides with V. cholerae sialidase on their binding capacity for tetanus (●) and cholera (■) toxins as determined by ganglioside–ELISA method

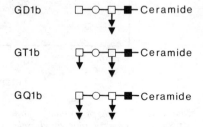

Figure 6. Ganglioside receptors for tetanus toxin. ■, Glucose; □, galactose; ○, galactosamine; ▼, NeuAc.

complexes containing 25% ganglioside, the amount of ganglioside
necessary to bind a certain amount of tetanus toxin is only one-
fiftieth needed by complexes containing 2% or 50% ganglioside. It
seems reasonable to assume that the ganglioside-cerebroside ratio
will vary with the number of sialic acids of the ganglioside and
the composition of the lipophilic portion of cerebroside and ganglioside. If so, it might explain why Mellanby and van Heyningen
(cited in the review by Ledeen and Mellanby (23)) found a higher
binding capacity for ganglioside GDla than GM1. In the present
new methods the influence of the lipophilic portion on the binding seems to be negligible.

The crucial question is then whether the gangliosides of the
Glb series are the biological receptors for tetanus toxin or not.
This question remains to be answered by experiments of the type
described for cholera toxin. Toxin-binding of various brain subcellular fractions has shown that the highest toxin-fixing
capacity resides in a fraction containing small fragments of
synaptic membranes (27). Further purification of the synaptic
membranes resulting in a severalfold enrichment of the gangliosides did not affect the toxin-binding (28). The binding studies
suggest, however, that only minute amounts of the gangliosides of
the Glb series in the synaptic membranes are involved in binding
of toxin. The binding capacity of the synaptic membranes was
drastically reduced when they were hydrolysed by neuraminidase
(28). It is not known whether the binding of the tetanus toxin to
gangliosides leads to the transfer of the toxin or an active fragment inside the cell. Tetanus toxin can block transmission by a
presynaptic action both at peripheral neuromuscular junctions and
at central synapses. We also know that the toxin is able to enter
the axoplasm at peripheral synapses, since retrograde axonal
transport seems to be the mechanism by which the toxin reaches the
central nervous system.

Sendai virus

The initial event in the entry of viruses into cells is the
attachment of the virus to specific receptors on the cell membrane.
The chemical structures of most receptors for animal viruses are
poorly defined. Cell surface glycoprotein, glycolipids, and
phospholipids have been implicated. Very recently Helenius et al.
(29) could identify human HLA and murine H-2 histocompatibility
antigens as receptors for Semliki Forest virus; these antigens are
well-defined membrane glycoproteins.

Sendai virus, like other myxo- and paramyxovirus, has surface glycoprotein spikes which adsorb to specific receptors on
erythrocytes of most mammalian and fowl species and cause
hemagglutination. The receptors on erythrocyte membranes contain
neuraminic acid, as indicated by the fact that they are destroyed
by neuraminidase. Haywood (30) demonstrated that liposomes containing gangliosides could inhibit the agglutination of erythro-

cytes by Sendai virus. Liposomes without ganglioside had no inhibitory effect, suggesting that the Sendai virus receptor might
contain or resemble gangliosides. She further noticed that the
effect seemed to require ganglioside with more than one sialic
acid since commercial di- and trisialoganglioside preparations
could inhibit Sendai virus hemagglutination while a monosialoganglioside preparation could not.

Serial dilutions of Sendai virus, purified after growth in
the allantoic cells of fertilized eggs, was applied spotwise onto
the surface of polystyrene Petri dishes coated with the various
gangliosides in different concentrations (virus spot assay). After
incubation for three hours the dishes were washed and exposed to
water vapour (vapour condensation-on-surface (VCS) method). This
resulted in characteristic "wet" spots with a few of the gangliosides (Figure 7 a). The strongest reactions were obtained with
ganglioside GTla, GQlb and GPlc which gave positive results in
concentrations as low as 0.1 μM (Table II). Two other gangliosides,
GDla and GTlb, also had some binding capacity but only when they
were used in about a 100-fold higher concentration. Other gangliosides tested were completely negative (Table II).

Sendai virus attaches to erythrocytes, and to check that the
wet spots really represented specifically bound virus it was in-

Table II.

High-affinity binding of Sendai virus to Gl gangliosides
with a terminal disialosyl group

Ganglioside	Minimal effective concentration (μM)		
	Virus spot VCS assay	Ganglioside spot assay VCS	Hemadsorption
GAl	>5	>25	>25
GMl	>5	>25	>25
GM2	>5	>25	25
GM3	>5	>25	n.t.[1]
GDla	5	2.8	0.9
GDlb	>5	>25	25
GD3	>5	>25	25
GTla	0.1	n.t.	n.t.
GTlb	5	2.8	2.8
GQlb	0.1	0.05	0.01
GPlc[2]	0.1	0.05	0.02

[1]n.t. = not tested
[2]from dogfish. Approx. 10% of NeuAc was also O-acetylated

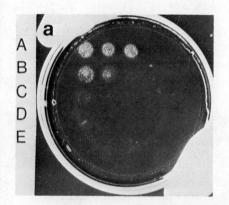 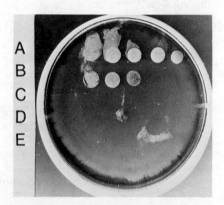

Figure 7. *Specific binding of Sendai virus to GQ1b ganglioside and its inhibition by antiserum as demonstrated with (a) the VCS method and (b) hemadsorption (see text). A: virus in fourfold serial dilutions in buffer; B: in lower-titer antiserum, 1%; C: in the same serum, 10%; D: in higher titer antiserum, 1%; this serum, 10%.*

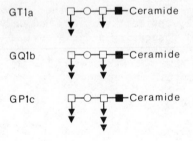

Figure 8. *Ganglioside receptors for Sendai virus.* ■, *Glucose;* □, *galactose;* ○, *galactosamine;* ▼, *NeuAc.*

vestigated whether the spot-forming material would bind erythrocytes secondarily (hemadsorption). This was shown to be the case.
In each instance in which the water condensation method gave
positive results a parallel   titrated plate displayed specific
hemadsorption for the same positions (Figure 7 b). The hemadsorption method was slightly more sensitive than the water condensation technique allowing detection of a 4- to 16-fold higher virus
dilution. Immune serum was shown to specifically inhibit the binding of Sendai virus to e.g. GQlb as examined by either of the two
methods.

A quantitatively more precise method to compare the binding
affinity of Sendai virus for the various gangliosides was to
attach the gangliosides spotwise in serial dilutions to the
plastic and then incubate the whole plate with virus (ganglioside
spot assay). The minimal effective coating concentration of each
ganglioside for binding Sendai virus could then be determined with
either the water condensation or the hemadsorption methods
(Table II). The results confirmed those obtained with the virus
spot assay showing that the virus affinity for GQlb and GPlc was
about 50- to 100-fold higher than for GDla and GTlb and >500-fold
higher than for any of the other tested substances (Table II).

Thus it is clear that Sendai virus has a very strong binding
tendency to GQlb which seems to exceed that of tetanus toxin to
its "receptor" gangliosides and actually approach the binding
strength of cholera toxin to GMl. Conversely to the situation
with tetanus toxin the sialic acid residues extending from the
terminal galactose are the critical ones for binding (Figure 8).
One such residue is an absolute requirement (compare GTlb and
GDla with GDlb) and a disialosyl group in this position apparently confers maximal binding capacity (GQlb≈GTla≈GPlc). However,
also the N-acetylgalactosamine residue (or the chain length as
such) in the backbone seems to contribute markedly to the
"receptor" structure, as indicated by the fact that GD3 had only
minimal binding capacity in spite of possessing a disialosyl
group linked to a terminal galactose.

Sendai virus has been shown to have the strongest affinity
for gangliosides with the common terminal end sequence:

NeuAcα2→8NeuAcα2→3Galβl→3GalNAc→

Some affinity was also shown by gangliosides GDla and GTlb, with
the same carbohydrate sequence but lacking the terminal NeuAc.
The sequence NeuAcα2→3Galβl→3GalNAc also exists in some glycoproteins, i.e. glycophorin, the predominant glycoprotein of human
erythrocytes.

In a recent paper (31) it was demonstrated that the removal
of sialic acid from human erythrocytes with *Vibrio cholerae*
sialidase abolished hemagglutination by Sendai virus. Hemagglutination titers were restored selectively by the incorporation of
NeuAc with β-galactoside α2→3sialyltransferase which has a strict

substrate specificity for the Galβ1→3GalNAc sequence.
It is conceivable that the natural receptor binding structure
for Sendai virus has only one terminal NeuAc residue, but it is
also possible that some of the oligosaccharide moieties received
a disialosyl linkage during the incubation with the specific
sialyltransferase.

Glycophorin is generally assumed to be the erythrocyte
receptor of myxoviruses, primarily because purified glycophorin
effectively inhibits agglutination of erythrocytes by most viruses
(32), and it is the major glycoprotein of red cell membranes. This
membrane may contain also minute amounts of ganglioside GQlb or
one of the other two high affinity gangliosides for Sendai virus.
In this case they may serve as the Sendai virus receptor. The
study by Paulson et al. (31) has not ruled out this possibility,
since their analytical procedure for products from reaction with
specific sialyltransferases does not exclude the existence of
gangliosides.

A clear understanding of the interaction between a virus
and the cell surface receptors will require an exact knowledge
about the oligosaccharide structure. A glycoprotein has in general
a relatively large number of oligosaccharide moities, which might
differ considerably in composition. This disadvantage does not
exist for glycolipids. By the development of the new methods for
the separation of gangliosides with homogenous carbohydrate
moieties and sensitive ligand methods, a sensitive tool has been
created for the elucidation of the receptor structure, irrespective of whether the receptor is a glycoprotein or a glycolipid.

Acknowledgements. The costs of the studies were defrayed by
grants from the Swedish Medical Research Council (3X-627 and
16X-3382).

Literature Cited

1. Fishman, P.H.; Brady, R.O.: Biosynthesis and function of gangliosides. Science, 1976, 194, 906-915.

2. Svennerholm, L. Structure and biology of cell membrane gangliosides, in 43rd Novel Symposium, Ouchterlony, Ö. and Holmgren, J., Eds. "Cholera and Related Diarrheas. Molecular
Aspects on a Global Health Problem"; Karger: Basel, 1979,
in press.

3. Holmgren, J.; Svennerholm, A.-M.: Enzyme-linked immunosorbent
assays for cholera serology. Infect. Immun., 1973, 5, 662-667.

4. Elwing, H.; Nilsson, L.-Å.; Ouchterlony, Ö.: Visualization
principles in thin-layer immunoassays (TIA) on plastic surfaces. Int. Archs. Allergy Appl. Immun., 1976, 51, 757-762.

5. Van Heyningen, W.E.; Carpenter, C.C.J.; Pierce, N.F.;
   Greenough, W.B.: Deactivation of cholera toxin by ganglioside.
   J. Infect. Dis., 1971, 124, 415-418.

6. Holmgren, J.; Lönnroth, I.; Svennerholm, L.: Tissue receptor
   for cholera exotoxin: Postulated structure from studies with
   GM1-ganglioside and related glycolipids. Infect. Immun.,
   1973, 8, 208-214.

7. Cuatrecasas, P.: Interaction of Vibrio cholerae enterotoxin
   with cell membranes. Biochemistry, 1973, 12, 3457-3558.

8. King, C.A.; van Heyningen, W.E.: Deactivation of cholera
   toxin by a sialidase-resistant monosialosyl-ganglioside.
   J. Infect. Dis., 1973, 127, 639-647.

9. Holmgren, J.; Lönnroth, I.; Månsson, J.-E.; Svennerholm, L.:
   Interaction of cholera toxin and membrane GM1 ganglioside of
   small intestine. Proc. Natl. Acad. Sci., (U.S.A.), 1975, 72,
   2520-2524.

10. Hansson, H.-A.; Holmgren, J.; Svennerholm, L.: Ultrastructural
    localization of cell membrane GM1 ganglioside by cholera
    toxin. Proc. Natl. Acad. Sci., (U.S.A.), 1977, 74, 3782-3786.

11. Cuatrecasas, P.: Gangliosides and membrane receptors for
    cholera toxin. Biochemistry, 1973, 12, 3558-3566.

12. Moss, J.; Fishman, P.H.; Manganiello, V.C.; Vaughan, M.;
    Brady, R.O.: Functional incorporation of ganglioside into
    intact cells: Induction of choleragen responsiveness. Proc.
    Natl. Acad. Sci., (U.S.A.), 1976, 73, 1034-1037.

13. Mullin, B.R.; Aloj, S.M.; Fishman, P.H.; Lee, G.; Kohn, L.D.;
    Brady, R.O.: Cholera toxin interactions with thyrotropin
    receptors on thyroid plasma membranes. Proc. Natl. Acad. Sci.,
    (U.S.A.), 1976, 73, 1679-1683.

14. Holmgren, J.; Lönnroth, I.: Cholera toxin and the adenylate
    cyclase-activating signal. J. Infect. Dis., 1976, 133, 64-74.

15. Holmgren, J.; Månsson, J.-E.; Svennerholm, L.: Tissue receptor
    for cholera enterotoxin: Structural requirements of GM1 gang-
    lioside in toxin binding and inactivation. Medical Biology,
    1974, 52, 229-233.

16. Staerk, J.; Ronneberger, H.J.; Wiegandt, H.; Ziegler, W.:
    Interaction of ganglioside G$_{Gtet1}$ and its derivatives with
    choleragen. Eur. J. Biochem., 1974, 48, 103-110.

17. Lönnroth, I.; Holmgren, J.: Subunit structure of cholera toxin. J. Gen. Microbiol., 1973, 76, 417-427.

18. Holmgren, J.; Lönnroth, I.: Structure and function of entero-toxins and their receptors, in 43rd Nobel Symposium, Ouchter-lony, Ö. and Holmgren, J., Eds. "Cholera and Related Diarrheas. Molecular Aspects on a Global Health Problem"; Karger: Basel, 1979, in press.

19. Sattler, J.; Schwarzmann, G.; Staerk, J.; Ziegler, W.; Wie-gandt, H.: Studies of the ligand binding to cholera toxin. Hoppe-Seyler's Z. physiol. Chem., 1977, 385, 159-163.

20. Fishman, P.H.; Moss, J.; Osborne, J.C.: Interaction of cholera-gen with the oligosaccharide of ganglioside GM1: Evidence for multiple oligosaccharide binding sites. Biochemistry, 1978, 17, 711-716.

21. Van Heyningen, W.E.: Tentative identification of the tetanus toxin receptor in nervous tissue. J. Gen Microbiol., 1959, 20, 310-320.

22. Van Heyningen, W.E.; Miller, P.A.: The fixation of tetanus toxin by ganglioside. J. Gen. Microbiol., 1961, 24, 107-119.

23. Ledeen, R.W.; Mellanby, J.: Gangliosides as receptors for bacterial toxins, in "Perspectives in Toxicology", Bernheimer, A., Ed.; John Wiley and Sons (New York), 1977, pp. 15-42.

24. Ledley, F.D.; Lee, G.; Kuhn, L.D.; Habig, W.H.; Hardegree, M.C. Tetanus toxin interactions with thyroid plasma membranes. Im-plications for structure and function of tetanus toxin re-ceptors and potential pathophysiological significance. J. Biol. Chem., 1977, 252, 4049-4055.

25. Helting, T.B.; Zwisler, O.; Wiegandt, H.: Structure of tetanus toxin. II. Toxin binding to ganglioside. J. Biol. Chem., 1977, 194-198.

26. Van Heyningen, W.E.; Mellanby, J.: The effect of cerebrosides and other lipids on the fixation of tetanus toxin by gang-lioside. J. Gen. Microbiol., 1968, 52, 447-454.

27. Mellanby, J.; Whittaker, V.P.: The fixation of tetanus toxin by synaptic membranes. J. Neurochem., 1968, 15, 205-208.

28. Mellanby, J.; Morgan, I.G., cited by Ledeen R.W. and Mellanby, J.: Gangliosides as receptors for bacterial toxins, in "Per-spectives in Toxicology", Bernheimer, A., Ed.; John Wiley and Sons (New York), 1977, pp. 15-42.

29. Helenius, A.; Morein, B.; Fries, E.; Simons, K.; Robinson, P.; Shirrmacher, V.; Terhorst, C.; Strominger, J.L.: Human (HLA-A and HLA-B) and murine (H-2K and H2-D) histocompatibility antigens are cell surface receptors for Semliki Forest virus. Proc. Natl. Acad. Sci., (U.S.A.), 1978, 75, 3846-3850.

30. Haywood, A.M.: Characteristics of Sendai virus receptors in a model membrane. J. Mol. Biol., 1974, 83, 427-436.

31. Paulson, J.C.; Sadler, J.E.; Hill, R.L.: Restoration of specific myxovirus receptors to asialoerythrocytes by incorporation of sialic acid with pure sialyltransferases. J. Biol. Chem., 1979, 254, 2120-2124.

32. Bächi, T.; Deas, J.E.; Howe, C.: Virus-erythrocyte membrane interactions, in "Cell Surface Reviews, Virus Infection and the Cell Surface", Poste, G., and Nicholson, G.L., Eds.; North-Holland (Amsterdam), 1977, Vol. 2, pp. 83-127.

RECEIVED December 10, 1979.

Interferon–Carbohydrate Interaction

HELMUT ANKEL, FRANÇOISE BESANCON[1], and CHITA KRISHNAMURTI

Department of Biochemistry, Medical College of Wisconsin, Milwaukee, WI 53226

Previous investigations suggest that mouse fibroblast inter-feron interacts with carbohydrate-containing cell membrane consti-tuents. Its antiviral action is blocked by certain plant lectins, such as those from Phaeseolus vulgaris (PHA, ref. 1) or the non-toxic agglutinin from Abrus precatorius (2), and after preincuba-tion with gangliosides (3,4). Gangliosides covalently attached to Sepharose avidly bind mouse fibroblast interferon, which is reversed in the presence of N-acetylneuraminyl lactose, the tri-saccharide common to many gangliosides (4). Furthermore, prein-cubation of SV/ALN cells with gangliosides under conditions that lead to incorporation into the cell membrane of these cells, in-creases their sensitivity to the antiviral action of mouse fibro-blast interferon as described by Vengris, Reynolds, Hollenberg and Pitha (5).

In this communication we extend our earlier observations, which primarily dealt with the antiviral action of mouse fibro-blast interferon, to its antigrowth activity and to antiviral and antigrowth activities of mouse T-cell interferon. We will show that inhibition by common gangliosides is restricted to both activities of fibroblast interferon alone. T-cell interferon, al-though its biological activities are analogous to those of fibro-blast interferon, neither binds to nor is inhibited by these gly-colipids. Furthermore we demonstrate that mouse leukemia L-1210 cells that were selected for resistance to fibroblast interferon (6), respond equally well to T-cell interferon as the parent cells which are responsive to both interferons.

Materials and Methods

Cells and Virus. Encephalomyocarditis virus (EMC) and mouse L929 fibroblasts were obtained from Dr. Sidney Grossberg. L929 cells were routinely propagated in MEM containing 10% fetal bovine serum (Gibco). Fibroblast interferon-sensitive and resistant

* Current address: Laboratoire de Biochimie Physique, I.N.R.A. Université Paris-Sud, 91 405 Orsay, France

0-8412-0556-6/80/47-128-391$5.00/0

mouse leukemia L-1210 cells (L-1210S and L-1210R) were provided
by Dr. Ion Gresser and were cultured in RPMI 2310 medium (Gibco)
supplemented with 5% fetal bovine serum.

Reagents. Partially purified mouse fibroblast interferon
was supplied by Dr. Kurt Paucker and had a specific activity of
2.4×10^7 NIH Reference Units per mg protein (IU/mg). Mouse T-
cell interferon was produced in PHA stimulated mouse spleen cells
in culture and purified by Dr. Ernesto Falcoff (7,8). Its specific activity was 1.6×10^5 IU/mg. Mono- and oligosaccharides,
mixed bovine brain gangliosides and polylysine (MW 30,000) were
from Sigma, CNBr-activated Sepharose from Pharmacia, and individual gangliosides from Supelco. Gangliosides G_{M3} and G_{M2} were
kindly provided by Dr. Subhash Basu. Individual gangliosides are
designated according to Svennerholm (9). Their purities were
analysed by TLC (see below). G_{M1}, G_{M2} and G_{M3} showed single
spots after exposure to resorcinol spray or iodine vapor. G_{D1a}
was slightly contaminated with a resorcinol-positive spot with
the mobility of G_{M2}, G_{T1b} showed one additional spot which moved
identically to G_{D1b}. Both contaminants were present in amounts of
10% or less as judged from the relative intensities of the spots.
The mixed ganglioside fraction from mouse brain was prepared
according to Folch et al. (10) as described by Brunngraber et al.
(11). Ganglioside affinity columns were prepared as previously
described (4). The procedure involved coupling of polylysine to
CNBr-activated Sepharose (12), followed by the attachment of
mixed bovine brain gangliosides with carbodiimide (13).

Antiviral Assay. Antiviral activity was determined in mouse
L_{929} cells. Approximately 10^5 cells in 1 ml MEM plus 10% fetal
bovine serum were seeded into 16 mm wells of MultiWell tissue culture plates (Falcon) and kept at 37° in a CO_2 incubator. The following day medium was removed and cells were incubated with appropriate dilutions of interferon in serum-free MEM containing 50
µg/ml bovine serum albumin (0.5 ml/well). Where indicated, interferon solutions were preincubated with glycolipids or carbohydrates at 37° for 30 minutes prior to addition to the cell monolayers. Control cells were incubated under identical conditions,
but in the absence of interferon. After 24 hours at 37°, medium
was removed and the cells were infected with EMC at a multiplicity
of infection of 0.1, adding 0.5 ml of an appropriate virus suspension in MEM-2% fetal bovine serum to each well. After 1 hour
at 37° medium with non-adsorbed virus was removed and the cells
were incubated in 0.5 ml fresh medium plus 2% fetal bovine serum
for 16 to 17 hours at 37°. They were then placed at -80° for at
least 30 minutes, then thawed and virus yield in each well was
determined by hemagglutination of human red blood cells of type 0
in serial two-fold dilutions of the virus suspensions (14). EMC
titers are expressed as the reciprocals of the highest dilutions
that still showed hemagglutination. The accuracy of this assay is

about ± one dilution. When the effects of fibroblast and T-cell interferons and those of different glycosides on these interferons were compared, experiments were performed simultaneously using the same batch of L-cells under identical conditions. This was necessary since interferon sensitivity and viral yield were somewhat variable from culture to culture. Concentrations of interferon used in individual experiments refer to appropriate dilutions of the original solutions, whose interferon titers were determined by comparing their antiviral activities to those of NIH standard G 002-49-511, one IU/ml referring to an interferon concentration that results in 50% inhibition of viral yield as compared to the standard.

Growth Studies. L-1210R and L-1210S cells were grown without agitation in plastic vials in RPMI medium containing 5% fetal bovine serum at 37° in a CO_2 incubator, using between 0.3 and 1 ml culture medium in different experiments. Cell counts were done in a Coulter counter. When L-1210R and L-1210S cells were compared, cultures were investigated simultaneously and under identical conditions.

Column Chromatography. Sepharose beads containing covalently linked gangliosides (0.2 ml packed volume) were placed into a pasteur pipette containing a small amount of glass wool. Columns were washed with MEM containing 50 μg/ml bovine serum albumin (3 ml). Interferon solutions in MEM-albumin (1 ml) were placed on the columns, which were eluted with MEM-albumin at a flow rate of no more than one drop per minute. Fractions of 1 ml were collected and interferon titers determined in each fraction after serial two-fold dilution. Columns onto which mouse fibroblast interferon had been loaded, were eluted with MEM-albumin first, then with 0.07 M N-acetylneuraminyl lactose at pH 2.

Other Analytical Procedures. Thin layer chromatography was carried out with silica gel plates G60 (Merck), using either chloroform-methanol-water (65:45:9) or n-propanol-0.2% $CaCl_2$ in H_2O (80:20). The plates were developed with resorcinol spray (15). Sialic acid was determined with thiobarbituric acid after hydrolysis in 0.1 sulfuric acid (16).

Results

Effects of Glycolipids on Antiviral Activity of Fibroblast Interferon.

Since interferons appear to be species-specific, we investigated whether the ganglioside fraction from mouse brain was more potent in inhibiting antiviral activity of mouse fibroblast interferon than that obtained from heterologous brain extracts. As seen in Figure 1 bovine brain gangliosides were almost as potent

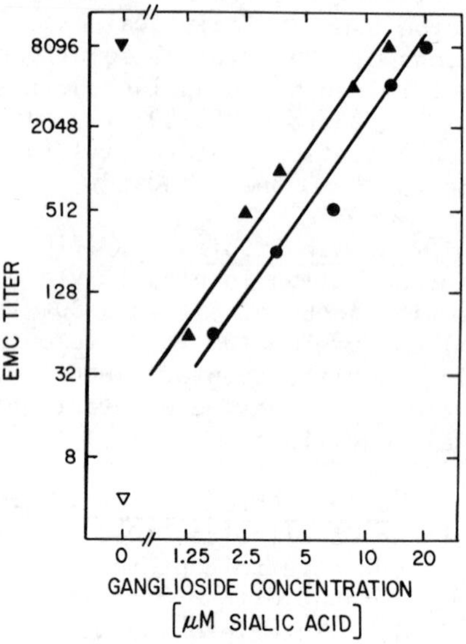

Figure 1. Effects of mouse (▲) and bovine (●) brain gangliosides on antiviral activity of mouse fibroblast interferon

Interferon solutions in MEM (30 IU/mL) were preincubated with the indicated ganglioside concentrations at 37°C for 30 min before addition to the L-cell monolayer. Antiviral assays are described in Materials and Methods. ▼: EMC titer in the absence of interferon. This titer was unchanged in the presence of both ganglioside preparations up to a concentration corresponding to 100 μM sialic acid.

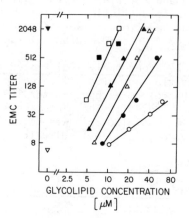

Figure 2. Effects of individual glycolipids on antiviral activity of mouse fibroblast interferon

Experimental conditions as in Figure 1. ○, Globoside; ●, G_{M3}; △, G_{M1}; ▲, G_{D1a}; □, G_{T1b}; ■, G_{M2}. ▼: EMC titer in the absence of interferon. This titer was unchanged in the presence of each individual glycolipid up to a concentration of 100 μM.

in inhibiting antiviral activity as those derived from murine brain. The lines extrapolate to ganglioside concentrations corresponding to 12 and 18 µM sialic acid, respectively, for complete inhibition of antiviral activity. We doubt that this apparent 1.5 fold difference is significant, since the viral assay cannot be carried out with enough accuracy to support differences of less than a factor of 2. Both preparations had very similar patterns of major gangliosides when analyzed by TLC, in accordance with observations by others, who find similar ratios of gangliosides G_{M1}, G_{D1a}, G_{D1b} and G_{T1b} in brain extracts from bovine and murine origin (17).

In earlier work we used preincubation of Sepharose-bound mouse fibroblast interferon with solutions of individual gangliosides to demonstrate their effect on antiviral activity (3,4). Comparison of potency of inhibition by individual glycolipids under these semi-quantitative conditions indicated that G_{M2} and G_{T1b} were equally good inhibitors, and that G_{D1a} and G_{M1} were somewhat less inhibitory. G_{M3} was only slightly inhibitory, whereas ganglio-trihexaosyl ceramide, globo-trihexaosyl ceramide and globoside had no effect. Since ganglioside G_{M3} and gangliotriaosylceramide were much less inhibitory than G_{M2} or did not inhibit at all, it appeared that terminal N-acetylgalactosaminyl and (or) N-acetylneuraminyl residues are important constituents for inhibition of antiviral activity.

In order to obtain more quantitative data on the relative inhibitory potencies of individual gangliosides, we subsequently preincubated underivatized mouse fibroblast interferon with ganglioside solutions prior to the addition to the target cells. The data shown in Figure 2 corroborate our earlier observations, indicating that under these conditions individual glycolipids will cause complete inhibition of antiviral activity of fibroblast interferon in the following order: G_{T1b} and G_{M2} > G_{D1a} > G_{M1} > G_{M3} > globoside, requiring individual concentrations of 14, 30, 45, 100 and 1000 µM for complete reversal of the antiviral effect.

Effects of Saccharides on Antiviral Activity of Fibroblast Interferon.

Since the ceramide portions of more and less inhibitory glycolipids are very similar, differential inhibition of antiviral activity of mouse fibroblast interferon must be related to their carbohydrate side chains. We therefore assayed antiviral activity in the presence of various saccharides contained in gangliosides. As seen in Figure 3, both N-acetylneuraminyl lactose and N-acetylneuraminic acid inhibited antiviral activity, requiring approximately equal concentrations to obtain complete inhibition (60 mM). However, in comparison to G_{M3}, 600-fold higher concentrations of these sugars had to be employed to yield complete inhibition of antiviral activity. N-glycolylneuraminic acid and the β-methyl-

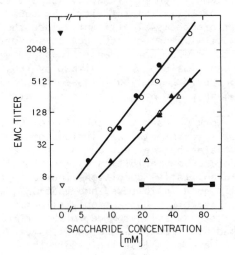

Figure 3.   Effects of mono- and oligosaccharides on antiviral activity of mouse
fibroblast interferon

*Experimental conditions as in Figure 1. ■, Lactose; △, N-glycolylneuraminic acid; ▲,
neuraminic acid β-methyl glycoside; ○, N-acetylneuraminic acid; ●, N-acetylneuraminyl
lactose. ▼: EMC titer in the absence of interferon. This titer was unchanged in the
presence of each saccharide up to a concentration of 100 mM.*

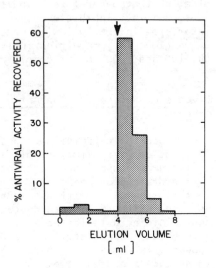

Figure 4.   Binding of mouse fibroblast
interferon to Sepharose–ganglioside col-
ums and elution with N-acetylneuraminyl
lactose

*One mL interferon solution (2 × 10³ IU) in
MEM plus 50 µg/mL bovine serum albumin
was loaded onto a small column containing
0.2 mL of the Sepharose–ganglioside adduct
as described in Materials and Methods. The
column was first eluted with MEM–albumin
alone. At arrow, elution was continued with
a solution of 0.07M N-acetylneuraminyl lactose in MEM–albumin at pH 2. Antiviral
activity in each fraction was determined as
described in Materials and Methods. A small
amount of the antiviral activity (7%) passed
the column unretarded; the remaining portion (89% of that applied) was eluted with
N-acetylneuraminyl lactose.*

glycoside of neuraminic acid also were inhibitory, yet concentrations approximately three times higher than those of the above saccharides resulted in comparable inhibition. In view of the fact that all gangliosides contain substituted lactosyl residues it is interesting that lactose had no effect on antiviral activity up to concentrations of 100 mM. Other sugars that had little or no effect at comparable concentrations were: N-acetylglucosamine, N-acetylgalactosamine, mannose, galactose and L-fucose.

It is surprising that the trisaccharide N-acetylneuraminyl lactose was much less potent in inhibiting antiviral activity than the corresponding glycolipid G_{M3}. This might indicate that either the binding site of mouse fibroblast interferon on G_{M3} includes part of the lipid portion as well, or that arrangement of carbohydrate chains in ganglioside micelles favors a conformation which allows for a much tighter fit of interferon. The free trisaccharide in solution, on the other hand, might assume any number of conformations, of which only one or very few are favorable to interferon binding.

That inhibition of antiviral action is due to binding, and that this involves the carbohydrate side chains on the ganglioside molecule, was clearly indicated by the behavior of fibroblast interferon on affinity columns containing covalently bound gangliosides. As seen in Figure 4, when mouse fibroblast interferon was placed on such a column, less than 10% of the antiviral activity passed through unretarded. The remaining antiviral activity was quantitatively eluted with 70 mM solutions of N-acetylneuraminyl lactose at pH 2. It should be noted that this concentration of the trisaccharide also completely reversed the antiviral effect, as indicated in Figure 3.

Effect of Gangliosides on Antigrowth Activity of Fibroblast Interferon.

Since it has been established that antiviral and antigrowth activities of mouse fibroblast interferon reside in the same molecules (18), one would expect that gangliosides would inhibit both activities in a similar fashion. To investigate the effect of gangliosides on growth inhibition, we used mouse leukemia L-1210 cells, which grow more rapidly than mouse L-cells. This cell line is of additional interest since Gresser, Bandu and Brouty-Boye have isolated a subline (L-1210R) by continuous growth in the presence of mouse fibroblast interferon, which is resistant to its antiviral and antigrowth activities (6). When growth of interferon-sensitive L-1210 cells (L-1210S) was followed for 4 days, the number of cells in control cultures was three times higher than that in cultures which contained mouse fibroblast interferon (Figure 5). Although addition of gangliosides alone inhibited growth to some extent, the effect of fibroblast interferon in the presence of gangliosides was largely reversed and the cell number in these cultures approached that of cultures grown in the

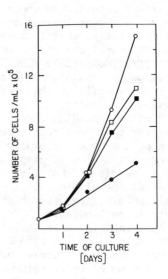

Figure 5. *Effect of bovine brain ganglio-sides on growth inhibition of L-1210S cells by mouse fibroblast interferon*

Cells were grown in RPMI medium plus 5% fetal bovine serum in 1 mL total volume as described in Materials and Methods. At 24-hr intervals the cells were counted in a Coulter counter. Control cells; ●, cells from cultures containing 1000 IU/mL mouse fibroblast interferon; □, cells from cultures containing bovine brain gangliosides at a concentration corresponding to 35 μM sialic acid; ■, cells from cultures containing both interferon (1000 IU/mL) and gangliosides (35 μM sialic acid).

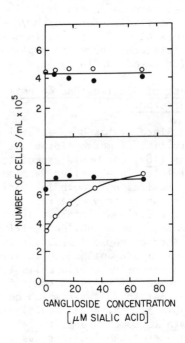

Figure 6. *Effects of mouse fibroblast in-terferon L-1210S and L-1210R cells in the presence of increasing ganglioside concentrations*

Cells were seeded at an original cell density of 8 × 10⁴ L-1210S cells/mL and 7 × 10⁴ L-1210R cells/mL in 0.4 mL total volume. Cells were cultured as in Figure 5 and counted after three days of growth. ○, Cells grown in the presence of 1000 IU/mL mouse fibroblast interferon; ●, those grown in its absence under identical conditions. Top, L-1210R cells; bottom, L-1210S cells.

presence of gangliosides alone.

Since the growth-inhibitory effect of gangliosides was much less pronounced after 3 days of culture, we investigated the effect of different ganglioside concentrations on antigrowth activity by counting the cells after 3 days. A comparison of the effects of gangliosides on antigrowth activity of interferon in both L-1210R and L-1210S cells is shown in Figure 6. As expected from the original observations by Gresser et al. (6) the L-1210R cells were not inhibited by fibroblast interferon and addition of gangliosides had little effect, both in the absence and in the presence of interferon. On the other hand, in the L-1210S cultures interferon produced approximately 50% reduction in cell number. The cell number, however, progressively increased with increasing concentrations of gangliosides. Complete reversal of the interferon effect was observed at a ganglioside concentration corresponding to 70 μM sialic acid. This compares to a concentration of 20 μM for complete reversal of the antiviral effect in the mouse L-cell system (Figure 1 and 7).

Effects of Glycolipids on T-cell Interferon (19).

It has been observed in several laboratories that interferon produced in mitogen-stimulated spleen cells (T-cell interferon) differs from fibroblast interferon in several of its physicochemical properties, although the biological effects are quite similar to those of the fibroblast variety (20). Preliminary studies using crude interferon preparations (appr. 10^3 IU/mg) obtained from cultured mouse spleen cells of BCG sensitized animals after stimulation with old tuberculin (21) indicated that gangliosides were much less inhibitory to this interferon than to mouse fibroblast interferon (22). In an attempt to further elucidate whether affinity for gangliosides is indeed a property not shared by T-cell interferon, we have collaborated with the laboratory of Dr. Ernesto Falcoff and systematically compared the effects of glycolipids on antiviral and antigrowth activities of mouse T-cell and fibroblast interferons under identical experimental conditions using more highly purified preparations of the former (1.6 x 10^5 IU/mg; ref. 7,8). As seen in Figure 7, at ganglioside concentrations where antiviral activity of fibroblast interferon was completely inhibited, that of T-cell interferon remained unchanged.

Individual glycolipids that inhibited fibroblast interferon (Figure 2) had no effect when tested with T-cell interferon under identical conditions at concentrations up to 100 μM (19). These included G_{M3}, G_{M2}, G_{M1}, G_{D1a}, G_{T1b} and G_{L4}.

That T-cell interferon does not bind to gangliosides is demonstrated by its behavior on ganglioside affinity columns: Under conditions where over 90% of mouse fibroblast interferon was retained (as shown in Figure 4) T-cell interferon quantitatively eluted in the breakthrough of the column (Figure 8). T-cell interferon, after passage through the affinity column, was still insensitive to ganglioside inhibition, excluding the possi-

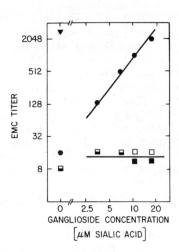

Figure 7. Effects of bovine brain gangliosides on antiviral activities of mouse fibroblast and T-cell interferons

Experiment was carried out as described in Figure 1, using 20 IU/mL of both interferons. ●, Mouse fibroblast interferon; □, T-cell interferon; ■, T-cell interferon after passage through a Sepharose–ganglioside column (see Figure 8). ▼: EMC titer in the absence of interferon (19).

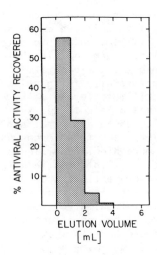

Figure 8. Lack of binding of mouse T-cell interferon to Sepharose–ganglioside columns

Experiment was carried out at the same time and under the same conditions as described in Figure 4, applying 1 mL of a T-cell interferon solution containing 10^3 IU. At least 90% of the applied antiviral activity passed the column unretarded (19).

bility that a non-interferon contaminant with high affinity for
gangliosides had been removed by this procedure (See Figure 7).
To exclude enzymatic destruction of gangliosides we incubated a
solution of gangliosides with T-cell interferon at an interferon
concentration 10 times higher than that normally used in the antiviral assay, for 24 hrs at 37°, then heat-inactivated the T-cell
interferon and assayed the treated gangliosides for their inhibitory action on fibroblast interferon. As control we used a
ganglioside solution treated identically, but in the absence of
T-cell interferon. Although there was a small decrease in inhibitory potency of the ganglioside solutions after this treatment
(approximately 30%), there was no significant difference between
the solutions preincubated with T-cell interferon as compared to
those that were preincubated in MEM alone. Thus it appears that
lack of binding to and inhibition by gangliosides is due to the
T-cell interferon molecule itself and not to contaminating factors that either compete for binding to gangliosides or degrade
them to non-inhibitory breakdown products.

Effect of T-cell Interferon on Growth of L-1210S and L-1210R Cells (19).

If the affinity of mouse fibroblast interferon for gangliosides relates to its functional interaction with mouse cells,
then clearly T-cell interferon must interact with different components of these cells, as it does not bind to gangliosides.
Therefore, if the L-1210R cells that were selected for their resistance to fibroblast interferon (6), had altered or inaccessible sites on the membrane that no longer allowed productive
interaction with fibroblast interferon, then T-cell interferon
might still be active with these cells. That this is indeed the
case is shown in Figure 9. L-1210S cells were found to be equally
sensitive to the antigrowth activities of both interferons. However, L-1210R cells, although insensitive to fibroblast interferon, were as sensitive to T-cell interferon as L-1210S cells. In
addition and as expected, antigrowth activity of T-cell interferon, like antiviral activity, was found to be resistant to inhibition by gangliosides at concentrations that completely reversed
the antigrowth effect of fibroblast interferon (Figure 8). That
resistance of L-1210R cells to fibroblast interferon was not due
to gangliosides (or other inhibitors specific for the fibroblast
variety) shed from these cells into the medium, was confirmed by
assaying the antigrowth activity of fibroblast interferon on
L-1210S cells suspended in 4-day old culture medium of L-1210R
cells. In comparison to control L-1210S cells suspended in 4-day
old culture medium from L-1210S cells, the antigrowth effect of
fibroblast interferon was unchanged, indicating that different
responses to both interferons by L-1210R cells is due to the cells
themselves and not to fibroblast interferon-specific inhibitors
shed into the medium by L-1210R cells, in accordance with results

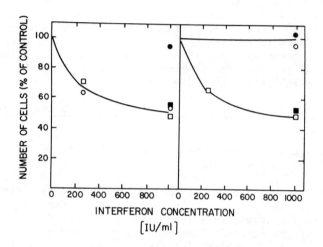

Figure 9. Antigrowth activities of mouse fibroblast and T-cell interferons on L-1210S (left) and L1210R cells (right) and the effects of gangliosides on antigrowth activity

Cells were seeded at an original density of 8 × 10⁴ L-1210S and L-1210R cells/mL in 0.3 mL total volume. Cells were cultured as in Figure 5 and counted after three days of growth. The cell number in control cultures (100%) was 3.1 × 10⁵ L-1210S cells/mL and 2.6 × 10⁵ L-1210R cells/mL. ○, Mouse fibroblast interferon; ●, mouse fibroblast interferon plus gangliosides (52 μM sialic acid); □, T-cell interferon; ■, T-cell interferon plus gangliosides (52 μM sialic acid) (19).

previously reported (6).

Discussion

Data presented in this communication provide the following evidence:
a) both antiviral and antigrowth activities of mouse fibroblast interferon are inhibited by gangliosides;
b) inhibition is due to binding of interferon to gangliosides;
c) binding involves the carbohydrate side chains on the ganglioside molecules, and is at least in part directed towards sialic acid residues;
d) neither antiviral nor antigrowth activities of T-cell interferon are inhibited by gangliosides, and T-cell interferon does not bind to gangliosides;
e) mouse leukemia L-1210R cells selected for resistance to fibroblast interferon retain unchanged sensitivity to T-cell interferon.

It is not known whether fibroblast or T-cell interferons or parts of them have to enter target cells in order to result in antiviral or antigrowth responses. The fact that mouse fibroblast interferon interacts with carbohydrate constituents of ganglioside molecules and that some transformed mouse cells gain increased sensitivity to its antiviral effect after uptake of exogenous gangliosides into the cell membrane (5) tempts us to speculate that interaction of this type of interferon with cell membrane gangliosides is of functional significance. Clearly, if this were the case, then T-cell interferon, although producing the same biological effects as fibroblast interferon, must have a different mechanism by which it interacts with its target cells, since it does not bind to gangliosides and is active with cells selected for resistance to fibroblast interferon.

It is possible that there are two classes of interferon-binding sites on the cell membrane, each specific for productive interaction with only one type of interferon. Thus prolonged growth of L-1210 cells in the presence of fibroblast interferon could select for those cells that have no or non-functional binding sites for fibroblast interferon, but still carry unaltered sites for binding of T-cell interferon. Alternatively, uptake mechanisms for both interferons or their active fragments might be different, one involving gangliosides, the other one a different type of glycolipid or none at all. Thirdly, although the biological responses to both types of interferon appear to be identical, there might be two (or more) different mechanisms by which these might be triggered, involving activation of different enzymes or enzymatic steps, each specific for one type of interferon. At the present time there is no direct evidence to decide which of these possibilities is the correct one.

There are two aspects of medical significance related to our

observations: Firstly it is known that cancer patients ofte have
elevated levels of circulating gangliosides which might reflect
increased concentrations of these glycolipids in the tumor-sur-
rounding tissue (23,24). Therefore, treatment of such patients
with human fibroblast or leucocyte interferon might not be very
effective, as these two also bind to gangliosides (5,25).
Thus using human T-cell interferon as an alternative treatment in
cases where fibroblast or leukocyte interferons fail to show the
desired effects could have obvious advantages, provided that indeed the former is comparable to mouse T-cell interferon in its
resistance to inhibition by gangliosides. Secondly, the observations of Gresser et al. (6) concerning the selection of fibroblast interferon-resistant leukemia cells might be of relevance
in interferon therapy of leukemic patients, which likewise might
select for resistant cells that would escape from the desired
growth inhibition. Our data suggest that alternation between
fibroblast and T-cell interferons might be a useful approach to
prevent such selection.

Acknowledgements

This work was supported by grants from the National Science
Foundation (PCM 76-84125) and from the National Institutes of
Health (AI-15007).

Literature Cited:

1. Besançon, F., and H. Ankel, Nature (London) (1974) 250,
    784-786.
2. Besançon, F., and H. Ankel, unpublished observation.
3. Besançon, F., and H. Ankel, Nature (London) (1974) 252,
    478-480.
4. Besançon, F., H. Ankel, and S. Basu, Nature (London) (1976)
    259, 576-578.
5. Vengris, V.E., F.H. Reynolds, Jr., M.D. Hollenberg, and P.M.
    Pitha, Virology (1976) 72, 486-493.
6. Gresser, I., M.T. Bandu, and D. Brouty-Boye, J. Natl. Cancer
    Inst. (1974) 52, 553-559.
7. Wietzerbin, J., S. Stefanos, M. Lucero, E. Falcoff, D.C.
    Thang, and M.N. Thang, Biochem. Biophys. Res. Commun. (1978)
    85, 480.
8. Wietzerbin, J., S. Stefanos, M. Lucero, E. Falcoff, J.
    O'Malley, and E. Sulkowski, Gen. Virol., in press.
9. Svennerholm, L.J., Neurochem. (1963) 10, 613-623.
10. Folch, J., M.B. Lees, and G.H. Sloane Stanley, J. Biol. Chem.
    (1957) 226, 497.

11. Brunngraber, E.G., G. Tettamanti, and B. Berra, in "Glyco-
    lipid Methodology" (L.A. Witting, ed.) American Oil Chemists'
    Society, pp. 159-186 (1976).
12. Sica, V., I. Parikh, E. Nola, G.A. Puca, and P. Cuatrecasas,
    J. Biol. Chem. (1973) 248, 6543-6558.
13. Cuatrecasas, P., Biochemistry (1973) 12, 4253-4264.
14. Craighead, J.E. and A. Shelokov, Proc. Soc. Exp. Biol. Med.
    (1961) 108, 823-826.
15. Svennerholm, L., Biochim. Biophys. Acta (1957) 24, 604-611.
16. Warren, L., J. Biol. Chem. (1959) 234, 1971-1975.
17. Ledeen, R.W., and R.K. Yu, in "Glycolipid Methodology" (L.A.
    Witting, ed.) American Oil Chemists' Society, pp. 187-214
    (1976).
18. De Maeyer-Guignard, J., M.G. Tovey, I. Gresser, and E.
    De Maeyer, Nature (London) (1978) 271, 622-625.
19. Besançon, F., H. Ankel, C. Krishnamurti, and E. Falcoff,
    Manuscript in Preparation (1979).
20. Epstein, L.P. in "Interferons and their Actions", W.E.
    Stewart II, ed., CRC Press, Inc. pp. 91-132 (1977).
21. Sonnenfeld, G., A.D. Mandel, and T.C. Merrigan, Cellular
    Immunology (1977) 34, 193-206.
22. Besançon, F., H. Ankel, G. Sonnenfeld, and C.T. Merrigan,
    unpublished observation.
23. Kloppel, T.M., T.W. Keenan, M.J. Freeman, and D.J. Morré,
    Proc. Natl. Acad. Sci. USA, (1977) 74, 3011-3013.
24. Portoukalian, J., G. Zwingelstein, N. Abdul-Malak, and J.F.
    Doré, Biochem. Biophys. Res. Comm. (1978) 85, 916-920.
25. Besançon, F., and H. Ankel, Texas Rep. on Biol. and Med.
    (1977) 35, 282-292.

RECEIVED December 10, 1979.

Perturbation of Behavior and Other CNS Functions by Antibodies to Ganglioside

MAURICE M. RAPPORT, STEPHEN E. KARPIAK, and SAHEBARAO P. MAHADIK

Departments of Psychiatry and Biochemistry, Columbia University, College of Physicians and Surgeons, New York, NY 10032

From the viewpoint of the chemist, the brain presents an almost limitless frontier. The brain, as a center for communication control, has been shown by anatomists and physiologists to be composed of a network of neurons that make contact with one another mostly by release of chemicals at synaptic junctions (neurotransmission). There are astronomical numbers of these synaptic junctions, and there is also a complex array of chemical transmitters and chemical modulators involved in neurotransmission. Many of these transmitters and modulators have not yet been identified. The physiological actions of these substances are diverse (they both excite and depress activity) so we must also postulate that many different molecular structures are involved in receptor functions even for the very same transmitter or modulator.

In this extensive array of synaptic connections lie the mechanisms for plastic adaptations of the brain to the external environment -- modifications that subserve the processes of sensory reception, memory and learning, emotional responses, and abstract thought. A major task of neurochemists is to sort out molecules participating in the myriad synaptic connections and to identify them.

Antibodies As A Bridge Between Structure And Function

Methods of separating and characterizing molecules have developed rapidly in the past twenty years and new ones are appear-

0-8412-0556-6/80/47-128-407$5.00/0

ing with regularity. Among these, immunological methods are
unique in representing relatively mature methods that still retain
a large measure of unexploited potential. The specificity of antibodies for both large and small chemical structures has been
well established during the last 30 to 40 years, and has proven
of inestimable value, especially in the field of endocrinology.
However, the application of immunological techniques to study synaptic differences is still in its infancy.

Our laboratory has been addressing itself for more than a
decade to developing an immunological bridge that will make a
connection between specific chemical structures in the synapse
and various CNS functions. The basis for this effort lies in
earlier demonstrations that antibodies or antisera can serve as
interventive agents that will perturb CNS functions -- for example by inducing alterations in the EEG or inhibiting performance on various tasks (1,2,3). We have studied the interventive
action of antibodies against an array of different antigens using
EEG as well as a number of behavioral paradigms (4,5,6) and have
demonstrated to our own satisfaction that an encouraging degree
of specificity is associated with the actions of these different
types of antibodies. For example, we found that functional alterations were induced with antisera to gangliosides, to S-100 protein (a brain specific protein found mainly in glial cells), and
to synaptic membranes. No such effects were seen with antisera
to galactocerebroside, to 14-3-2 protein (a neuron-specific protein), or to erythrocyte membranes. Furthermore, the antibodies
to G_{M1} ganglioside following intracortical injection induced EEG
spiking (7) and inhibited learning (8,9) whereas antiserum to
S-100 protein inhibited learning but did not alter the EEG (10).
If we accept this evidence that passive administration of antibodies is capable of at least some degree of discrimination among
different CNS functions, the specificity of antibodies for molecular structure provides a bridge between structure and function.

G_{M1} Ganglioside As A Synaptic Target For Antibodies

We will make one further assumption in order to reduce the area of investigation to reasonable proportions. We will assume that synaptic contacts are the site of action of these passively transferred antibodies that are able to disrupt CNS functions. Therefore molecules that are directly involved in such contacts and are accessible to the extracellular space (synaptic cleft) become priority targets of our efforts. G_{M1} ganglioside fits the category well. It is a small stable molecule whose chemistry is well-established. It can be prepared in workable quantities by reproducible procedures, and criteria of purity are available and readily met. Gangliosides are present in substantial amounts in synaptic membranes, and the G_{M1} ganglioside molecule is accessible to the synaptic cleft as we have shown by labeling intact synaptosomes using enzymic oxidation (galactose oxidase) followed by reduction with tritium-labelled borohydride (11).

However, available knowledge suggests that gangliosides are present in all synaptic connections, and if the hypothesis is correct that disturbances of CNS functions by antibodies result from perturbation of synaptic contacts, we might then expect that all CNS functions would be susceptible to disturbance by these antiganglioside antibodies. If this were true, it would limit the usefulness of these antibodies as an interventive agent. Antiganglioside serum would still offer two major advantages stemming from 1) the relative ease and reproducibility of the methods for preparing it and characterizing its antibody content (titer) and specificity and 2) its provision of a rigorous control reagent in the form of the antiserum from which antibodies to G_{M1} are absorbed (removed) with pure G_{M1} ganglioside.

Antibodies To Ganglioside Interfere With CNS Functions Selectively

We have now subjected rats to a number of behavioral tests in which we could show inhibition of learning by small quantities of antiganglioside serum and no inhibition by the absorbed serum.

Among these may be listed inhibition of passive avoidance learning (9), inhibition of morphine analgesia (12) and blockade of sedation induced by reserpine (unpublished data). Schupf and Williams (13) have added blockade of cholinergic stimulation of drinking. However, and quite unexpectedly, we found the anti-ganglioside reagent was not effective in a number of other tests, such as pattern discrimination, fixed-ratio conditioning, self-stimulation, pain threshold and activity levels (all unpublished results), and in the experiments of Schupf and Williams, eating and drinking (personal communication). These results, summarized in Table I, indicate that despite the widespread distribution of G_{M1} ganglioside in the brain, one cannot yet predict whether a particular behavior will or will not be affected by administration of antibodies to G_{M1} ganglioside. One infers from these results that G_{M1} ganglioside receptors may provide a chemical basis for discriminating among different behaviors. If this proves to be true for G_{M1} ganglioside, it may also be true for other ganglioside species as well as for other molecules in synaptic membranes that can serve as "receptors" for antibody ligands.

Mechanisms By Which Antibodies May Perturb CNS Functions

How might one account for this discriminatory capability? One explanation might be found in the differences in topography of "antigenic receptors" in different synaptic contacts, differences both in the number of receptors and in their distribution. Another explanation might be found in differences among synaptic contacts with respect to the type of membrane process that is altered by the binding of antibody molecules. The number of such processes is substantial and continues to grow.

If we consider that the binding of an antibody ligand to an antigenic site in the membrane may alter membrane conformation and/or membrane fluidity, then as a consequence of such alterations a number of properties of the membrane may change including its permeability to ions, its enzyme activities, and the distribution

TABLE I

BIOLOGICAL EFFECTIVENESS OF ANTIGANGLIOSIDE SERUM

EFFECTIVE		INEFFECTIVE	
Test Procedure	Injection Site	Test Procedure	Injection Site
1.EEG seizures	cortex: sensori-motor, frontal, visual; hippo-campus; amygdala	1.EEG seizures	hypothalamus
2.Inhibition of learning (passive avoidance)	i.vc.	2.Pattern discrimi-nation	visual cortex lateral geniculate
3.Inhibition of morphine analgesia	periacqueductal grey	3.Activity levels	i.vc.
4.Blockade of reserpine sedation	i.vc.	4.Fixed-ratio conditioning	i.vc.
5.Developmental interference (DRL behavior; dendrogenesis of pyramidal cells in cortex)	i.cist.	5.Self-stimulation	lateral hypothalamus
		6.Pain threshold	a) PAG b) i.vc.
6.Blockade of cholinergic stimulation of drinking	lateral hypothalamus	7.Eating and drinking	lateral hypothalamus

of membrane components, including receptor sites. These in turn
may affect release and uptake of neurotransmitters, cause increased metabolism of receptor sites or trigger endocytosis. For
most of these mechanisms examples are available from studies of
various types of cells. In addition the antibody binding can ac-
tivate the complement system leading to membranolysis. The list
of possible mechanisms, indicated in Figure 1, is by no means
complete. It does, however, suggest a number of experiments that
should be helpful in elucidating the basis for discriminatory capability.

Antibodies To G$_{M1}$ Ganglioside Inhibit Dendritic Development

The effect of antiganglioside serum on development, indicated
in Table I, provides some suggestion that gangliosides may be involved in the signaling mechanisms that regulate the sequential
developmental processes of dendrogenesis and myelinogenesis in the
CNS. It was recently observed (14,15) that intracisternal injection of antiganglioside antibodies into 5 day-old rats caused
chemical, morphological, and behavioral changes in the adult animals. Chemical studies of somatosensory cortex revealed decreases
of about 30% (p<.01) in ganglioside sialic acid, galactocerebroside, and RNA with no change in DNA, total protein, or total
solids.

Quantitative morphological measurements of oblique dendrites
of pyramidal cells in Golgi preparations showed a decrease in
spine density from 1.55 to 1.10 spines per micron without significant decrease in spine length. Thin spines decreased from 74.8%
to 27.8% whereas stubby spines increased from 22.5% to 67.6%. Be-
haviorally the animals injected with antiganglioside serum and
with absorbed antiserum were able as adults to perform equally
well on a passive avoidance learning paradigm and on the early
stages (5 and 7 second intervals) of DRL (differential reinforce-
ment at low rates) learning. However, when the DRL test was made
sufficiently difficult by increasing the interval to 10 and 15

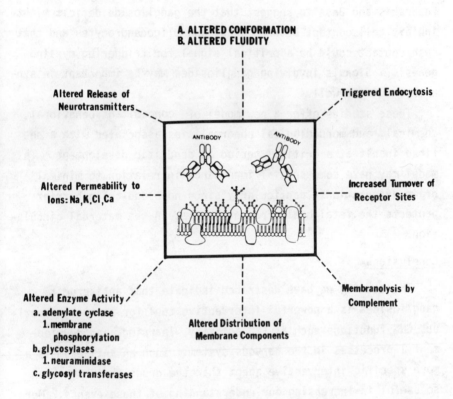

Figure 1.   Possible mechanisms for immunological perturbation of membrane processes

seconds (in this test, the rat must learn to withhold, for a
given number of seconds, the pressing of a lever to obtain a food
reward) the learning deficit in the animals treated with antiserum
compared with those receiving absorbed antiserum was measurable
and significant at the $p<.05$ and $p<.01$ levels, respectively. The
loss of galactocerebroside indicating hypomyelination led Drs.
Kasarskis and Bass to suggest that the ganglioside deficit might
inhibit cell contact between axons and oligodendrocytes and that
such contact could be a critical signal for triggering myelinogenesis. Signals involving gangliosides may be important in syn-
aptogenesis as well.

These studies offer a new model of concomitant behavioral,
chemical, and morphological abnormalities associated with a defined insult at a critical period in dendritic development. This
model may have some special importance in relation to minimal
brain dysfunctions, particularly since no blood-brain barrier
protects the fetal brain from antibodies in the maternal circulation.

Conclusions

The studies we have described indicate that antiserum to
gangliosides is a powerful interventive tool for perturbing various CNS functions such as EEG activity, learning, and developmental processes in the nervous system. Such an antiserum is a
more specific interventive agent than the drugs that have been
so useful in increasing our understanding of these events. Moreover its site of action can probably be better controlled and
more precisely traced. We have focused our attention on gangliosides as "receptor" sites for antibody ligands primarily because
their chemistry is well-established and the evidence for their
localization in synaptic contacts is generally accepted. We
would like to stress the fact that antibodies against other brain
molecules are also effective in disrupting CNS functions and by
mechanisms that do not necessarily involve membrane changes (16,

17). We must therefore recognize the considerable challenge presented in attempting to exploit the discriminatory capabilities of these immunological agents. We are continuing to direct our efforts toward this objective by isolating other synaptic membrane molecules and examining the biological activities of antibodies against them. We believe such experiments will allow us to sort out those molecules in synaptic contacts that have the highest probability of playing some functional role. They may also provide us with criteria for detecting synaptic pathology associated with disorders in man.

Literature Cited

1. Mihailovic, L.; Cupic, D.   Epileptiform activity evoked by intracerebral injection of anti-brain antibodies. Exp. Neurol., 1970, 31, 45-52.
2. Jankovic, B.D.; Rakic, L.; Veskov, R.; Horvat, J.   Effect of intraventricular injection of anti-brain antibody on defensive conditioned reflexes. Nature (Lond.), 1968, 218, 270-271.
3. Hyden, H.; Lange, P.   Correlation of the S-100 brain protein with behavior. Exp. Cell. Res., 1970, 62, 125-132.
4. Rapport, M.M.; Karpiak, S.E.   Discriminative effects of anti-sera to brain constituents on behavior and EEG activity in the rat. Res. Comm. in Psychol., Psychiatr., and Behav., 1976, 1, 115-124.
5. Rapport, M.M.; Karpiak, S.E.   Immunological perturbation of neurological functions, in Nandy, K., Ed. "Senile Dementia & Related Disorders"; Elsevier: New York, 1978, pp. 73-88.
6. Rapport, M.M.; Karpiak, S.E.; Mahadik, S.P.   Biological activities of antibodies injected into the brain. Federation Proceedings (in press).

7. Karpiak, S.E.; Graf, L.; Rapport, M.M. Antiserum to brain gangliosides produces recurrent epileptiform activity. Science, 1976, 121, 735-737.

8. Karpiak, S.E.; Graf, L.; Rapport, M.M. Antibodies to G_{M1} ganglioside inhibit a learned avoidance response. Brain Res., 1978, 151, 637-640.

9. Karpiak, S.E.; Rapport, M.M. Inhibition of consolidation and retrieval stages of passive avoidance learning by antibodies to gangliosides. Behavioral & Neural Biology (in press).

10. Karpiak, S.E.; Serokosz, M.; Rapport, M.M. Effects of antisera to S-100 protein and to synaptic membrane fraction on maze performance and EEG. Brain Res., 1976, 102, 313-321.

11. Rapport, M.M.; Mahadik, S.P. Topographic studies of glycoproteins and gangliosides in the surface of intact synaptosomes, in Roberts, S.; Lajtha, A.; Gispen, W.H., Eds. "Mechanisms, Regulation and Special Functions of Protein Synthesis in the Brain"; Elsevier: Amsterdam, 1977, pp. 221-230.

12. Karpiak, S.E.; Bodnar, R.J.; Hanson, S.; Glusman, M.; Rapport, M.M. Antibodies to G_{M1} ganglioside block morphine analgesia. Neurosci. Abstr., 1978, 4, 460.

13. Schupf, N.; Williams, C.A. Selective blockade of hypothalamic cholinergic pathways by antibody to G_{M1} ganglioside. Neurosci. Abstr., 1978, 4, 521.

14. Kasarskis, E.J.; Karpiak, S.E.; Rapport, M.M.; Bass, N.H. Abnormal cortical maturation after neonatal anti-G_{M1} ganglioside exposure. Trans. Amer. Soc. Neurochem., 1979, 10, 233.

15. Rapport, M.M.; Karpiak, S.E.; Kasarskis, E.J.; Bass, N.H. Behavioral changes after perinatal exposure to anti-G_{M1} ganglioside. Trans. Amer. Soc. Neurochem., 1978, 10, 234.

16. Van Wimersma Greidanus, T.B.; De Wied, D. Modulation of passive-avoidance behavior of rats by intracerebroventricular administration of antivasopressin serum. Behav. Biol.,

1976, 18, 325-333.

17. Shashoua, V.E. Brain protein metabolism and the acquisition of new patterns of behavior. Proc. Natl. Acad. Sci., 1977, 74, 1743-1747.

Supported in part by the USPHS Grant No. NS-13762.

RECEIVED December 10, 1979.

Immunological Properties of Gangliosides

A. J. YATES, C. L. HITCHCOCK, S. S. STEWART, and R. L. WHISLER

Departments of Pathology and Medicine, Ohio State University, Columbus, OH 43210

Gangliosides are complex glycosphingolipids which although present in high concentrations in mammalian brain are also present in many non-neural cells. As yet, no proven physiological role has been shown for them, but considerable evidence exists which indicates that they may function as receptor molecules (1,2). Furthermore, it is speculated that gangliosides might participate in other cell surface phenomena such as cell-cell recognition and growth control(3,4). In relation to the latter function, it is intriguing that concentrations of gangliosides in the serum of patients and animals with some malignancies are elevated(5,6,7). Furthermore, recent evidence indicates that ganglioside added exogenously to in vitro assay systems of immune function can modulate immune reactivity(8,9,10). The present studies were designed to examine and characterize the effects of gangliosides on the in vitro immune reactivity of human peripheral blood mononuclear cells (PBMC).

Effects of Exogenously Added Gangliosides on In Vitro Assays of Cellular Immunity

Regulation of Lymphocyte Activation by Mixtures of Gangliosides From Human Cerebral Cortex.
A mixture of normal human cerebral cortex gangliosides (NHCG) was isolated from human cerebral cortex by the method of Suzuki(11), and further purified by silicic acid column chromatography(12). Purity was assessed by silica gel thin layer chromatography and quantitation was by the resorcinol assay of Svennerholm(13,14). Liposomal membranes with and without incorporated gangliosides were prepared according to the method of Esselman and Miller(15). Human PBMC were obtained from human blood by centrifugation over Ficoll-Hypaque gradients and enumerated by a Coulter counter. The microculture system utilized in these studies has been described previously in detail(16).

The data presented in Table I demonstrate the effects of gangliosides, liposomes, and gangliosides incorporated into lipo-

0-8412-0556-6/80/47-128-419$5.00/0

somal membranes on the lymphocyte reactivity in response to the
nonspecific mitogen concanavalin A (Con-A) as determined by
tritiated thymidine incorporation. The addition of progressively
larger amounts of each of these preparations at the initiation of
cultures resulted in progressively greater suppression of
thymidine incorporation. Although, cholesterol-lecithin liposomes resulted in suppression only at amounts greater than that
used to incorporate 7.88 nanomoles ganglioside, considerably
greater suppression was observed with the addition of 3.94 and
7.88 nanomoles gangliosides alone. Furthermore, a marked suppression occurred with liposomal bound gangliosides resulting in
a synergistic inhibitory effect which was greater than would be
predicted by the additive effects of gangliosides and liposomes
alone.

TABLE I

Effects of Gangliosides, Liposomes, and Liposomal Bound Gangliosides on PBMC ^{3}H-Thymidine Incorporation In Response to 18 microgm Conconavalin A for 96 Hours[a]

Amount added (nanomoles)[b]	Liposomes[c]	Gangliosides	Liposomal Gangliosides
0.98	11.8+1.5	10.2+1.2	9.7+1.3
1.97	10.8+1.3	8.1+1.1	9.7+1.1
3.94	11.2+1.5	5.5+7.7	2.1+0.5
7.88	13.2+1.6	4.5+1.2	2.3+0.7
15.8	6.7+1.1	5.5+0.8	2.6+0.6
31.5	1.0+0.2	2.2+0.5	0.5+0.04

Control (none added 9.8+0.9)

a Results are expressed as CPM x 10^{-4}+S.D. ^{3}H-Thymidine incorporation for triplicate cultures processed on day four of culture
with 18 microgm Con-A. Each culture received 1 microcurie ^{3}H-
Thymidine 18 hours prior to processing.

b At the time of culturing, the indicated amounts of gangliosides, liposomes, or liposomal-bound gangliosides were added to
triplicate cultures with 1.7×10^5 cells per 0.2 ml total volume.

c The amount of liposomes alone is equivalent to that which was
used to incorporate the amount of ganglioside stated.

Effects of Specific Gangliosides on Lymphoblastic Transformation.

Having established that a mixture of gangliosides isolated from human brain can suppress lymphoblastic transformation (as determined by thymidine incorporation) in response to non-specific mitogens, it was of interest to see if there are differences in the inhibitory potencies among different gangliosides. To test this, the four major gangliosides (GM1, GD1a, GD1b, and GT1b) were purified from a crude mixture of normal human brain ganglio-sides. This was accomplished by means of column chromatography using DEAE Sephadex and Iatrobeads (17). The purity of each ganglioside was checked both by silica gel thin-layer chromato-graphy and determinations of sugar and aminosugar ratios by gas-liquid chromatography of their alditol acetate derivatives(18,19). In two TLC solvent systems, each migrated as a single spot and their carbohydrate ratios were within the range of error of the technique to confirm the identities of specific gangliosides.

Four different amounts of the four gangliosides were added to microtiter wells containing human PBMC's and 18 micrograms of concanavalin-A. All four gangliosides markedly inhibited lympho-blastic transformation when 12.6 nanomoles was added to the cul-tures with a total volume of 0.2 ml (Table II). As little as 1.6-nanomoles of GD1b and GT1b resulted in substantial inhibition, while much less inhibition occurred with equivalent amounts of GM1 and GD1a.

TABLE II

Effects of Major Human Brain Gangliosides on ^{3}H-Thymidine Incor-poration (CPMx10^{-3}) Into PBMC In Response to Con-A.

Amount of Ganglioside(nanomoles)
Added to Each Well

	12.6	6.3	3.15	1.57
GM1	7.0	64.0+5.8	66.0+11.0	62.5+14.0
GD1a	5.7+1.6	32.0+2.2	52.1+5.2	63.1+0.5
GD1b	4.9+5.0	10.7+8.7	12.2+2.7	20.4+13.5
GT1b	0.5+0.2	2.1+0.2	3.9+1.2	3.0+1.7

No ganglioside 110+2.0

No Con-A 3.0+1.6

Similar amounts of these gangliosides were added to test for possible effects on mixed leukocyte reactions(MLR). None of these gangliosides inhibited significantly at 1.57 nanomoles and only GM1 inhibited when 3.15 nanomoles was added (Table III). With

6.3 nanomoles all caused inhibition of 50 to 60% of control
values; and with 12.6 nanomoles, GDla was the only one which did
not cause marked inhibition of Con-A reactivity. Thus quantitative differences exist in the inhibitory capacity of specific
gangliosides. Furthermore, the effects are different between
Con-A and MLR reactivity. These differences argue against non-
specific effects and suggest that more than one inhibitory mechanism may be operational.

TABLE III

Effects of Major Human Brain Gangliosides on ^{3}H-Thymidine Incorporation (CPMx10^{-3}) Into PBMC In Mixed Leukocyte Cultures

	Amount (nanomoles) of Each Added			
	12.6	6.3	3.15	1.57
GM1	4.9+1.5	8.4+2.6	9.8+2.4	17.0+2.6
GDla	20.0+6.2	13.4+1.4	15.1+2.4	20.3+5.8
GDlb	0.9+0.2	15.1+3.0	15.6+3.3	24.3+1.8
GTlb	4.0+1.3	9.8+1.8	14.4+1.7	17.8+2.8

No gangliosides 20.5+0.8

No stimulator cells 4.3+1.4

The greater suppression of Con-A reactivity by GDlb and
GTlb suggests that the 2→8 NANA-NANA bond attached to the internal galactose of the oligosaccharide backbone could be responsible for the suppressive effects. To test if these moieties
could be the only structural property causing suppression the
following experiments were performed. GM4, GM3 and GD3 with
only NANA sialic acid residues were prepared from Chicken egg
yolk(20). GD3 with NGNA was isolated from cat erythrocytes(21).
GM2, NANA and NGNA were obtained from Supelco, Inc., Bellefont,
Penn. Prior to their use the purity of each was determined as
described above for the major brain gangliosides.
    The data in Table IV show that GM4 & GM3, resulted in no
inhibition of Con A reactivity even at the highest concentration
of 12.6 nanomoles. In contrast, 12.6 nanomoles of GM2 and GD3
suppressed the Con A reactivity 95% and 40% respectively. Only
GM2 demonstrated significant inhibition at lower concentrations.
Moreover, the inhibition by GM2 was dose related.

TABLE IV

Effects of Minor Human Brain Gangliosides on ^{3}H-Thymidine Incorporation (CPMx10^{-4}) Into PBMC In Response to Con-A.

Amount of Ganglioside (nanomoles)
Added to Each Well

	12.6	6.3	3.15	1.57
GM4	11.4+1.3	9.8+1.7	11.5+0.6	10.8+1.3
GM3	9.0+1.7	7.9+1.7	8.5+0.9	8.2+019
GM2	0.2+0.003	1.7+0.3	3.4+0.3	6.2+0.7
GD3	6.3+0.5	8.4+1.3	9.1+1.1	7.8+1.0

No gangliosides 9.14+0.58

No Con-A 0.21+0.08

TABLE V

Effects of Minor Human Brain Ganglioside on Tritiated Thymidine Incorporation (CPMx10^{-3}) in Mixed Leukocyte Cultures

Amounts (nanomoles) of Each Added

	12.6	6.3	3.15	1.57
GM4	13.9+1.6	15.0+3.9	15.5+2.5	17.5+3.3
GM3	11.8+3.5	11.0+2.5	13.2+3.0	15.3+4.3
GM2	0.8+0.3	8.0+1.4	11.2+3.4	15.1+3.0
GD3	4.9+1.7	4.4+0.6	6.9+0.8	15.8+3.5

No gangliosides 20.5+0.8

No stimulator cells 4.3+1.4

When these same gangliosides were added to mixed leukocyte cultures (MLC), all caused some inhibition at 12.6 nanomoles and none inhibited when 1.6 nanomoles was added (Table V). At intermediate and high concentrations, GM4 and GM3 inhibited less than GD3 and GM2. These results again indicate that their effects on the Con-A and mixed leukocyte reactions are different, and that on a molar basis different gangliosides have differential inhibitory effects.

Furthermore, the inhibitory effect of GD3 supports the concept that the disialo linkage is an important determinant in ganglioside suppression of lymphocyte activity. However, the

consistent suppression by GM2 clearly indicates that certain
structural properties of gangliosides in addition to disialo
linkages can result in suppressive activity. It also should be
noted that large amounts of sialic acids and colominic acid
caused only slight suppression (Table VI). Thus the inhibitory
effects of gangliosides are not solely due to their sialic acid
moieties.

TABLE VI

Effects of Different Sialic Acids on Tritiated Thymidine Incorporation (CPMx10^{-3}) In Mixed Leukocyte Reaction

Amount (nanomoles) of Each Added

	50.4	25.2	12.6	6.3
NANA	9.6+2.3	13.7+2.5	14.3+1.8	16.7+4.0
NGNA	14.4+2.5	17.6+3.4	17.9+1.5	22.0+2.2
Colominic Acid	16.4+1.8	18.5+2.1	18.0+5.0	16.1+2.6

No sialic acid   20.7+0.8

TABLE VII

Con-A Binding[a] to PBMC in the Presence and Absence of Gangliosides

Duration of Incubation	Amount Ganglioside Added	
	None	12.6 nanomoles
One hour	5.0+0.2	4.4+0.1
Four hour	4.8+0.2	4.8+0.2

a 3x10^6 PBMC were incubated with 45,000 CPM of ^{125}I labelled
Con-A in culture media alone or with 20 microgm NHCG. After
the indicated times, the cells were removed, washed twice with
phosphate-buffered saline to remove unbound Con-A and bound
radioactivity determined. The results represent the percent
of the total radioactivity added which was bound to the cells.
The results are the means +S.D. for triplicate samples.

Studies on the Binding of Gangliosides to Peripheral Blood
Mononuclear Cells. Initially it was suspected that the inhibition of lymphocyte activation by gangliosides alone and liposomal-bound gangliosides might merely be due to competition of
these with Con-A binding to PBMC. However, several observations
indicate that factors other than the blocking of Con A binding
to PBMC are responsible.

First, as shown in Table VII, results of experiments under
identical conditions which result in ganglioside suppression of
thymidine incorporation revealed no significant difference in
the percentage of iodinated Con-A which bound to the PBMC in the
presence of gangliosides compared to control cultures. The
percentage of Con-A bound is consistent with previous studies(22).
Second, if the suppressive mechanism involved Con-A binding to
gangliosides, resulting in less available Con-A to bind PBMC,
then it would be expected that a large excess of Con-A should
overcome the ganglioside suppression by making more Con-A available to PBMC receptor sites.

TABLE VIII

Effects of Ganglioside on ^{3}H-Thymidine Incorporation (CPMx10^{-3})
Into PBMC in Response to Variable Doses of Con-A[a]

Dose of Con-A (microgm)	None	12.6 nanomoles
18	16.6+1.3	0.75+0.08
36	94.7+11.5	4.1+1.5
72	1.1+0.01	0.07+0.01
144	0.83+0.06	0.22+0.06

a Similar to Table I except variable amounts of Con-A were used.

The data presented in Table VIII demonstrates that this is
clearly not the situation because gangliosides also suppress
reactivity in the presence of supraoptimal doses of Con-A. It
was difficult to assess adequately the degree of suppression with
the maximal Con-A dose in view of the markedly diminished value of
ganglioside cultures (217+63 CPM) compared to the control response
with an optimal concentration of Con-A (826+56 CPM).

The likelihood that a mechanism other than lectin-ganglio-
side binding contributes to ganglioside suppression is supported
further by the data in Table IX. It demonstrates the results of
adding several different amounts of gangliosides, liposomes, and
liposomal-bound ganglioside at the initiation of mixed leukocyte
cultures (MLC) on subsequent ^{3}H-thymidine incorporation. As with
the Con-A reactivity, the addition of increasing amounts of gangliosides resulted in greater suppression of MLC reactivity.

Although larger amounts of liposomes without gangliosides inhibited MLC reactivity approximately 40%, corresponding amounts of gangliosides alone and liposomal-bound gangliosides suppressed MLC reactivity 60% and 90% respectively. The amount of liposomes used to incorporate 3.94 nanomoles ganglioside had no effect on MLC reactivity, although 3.94 nanomoles of ganglioside alone inhibited thymidine incorporation. The ten-fold reduction in mixed leukocyte reactivity with 3.94 nanomoles liposomal-bound ganglioside indicates that the steric presentation of the ganglioside to the lymphocyte might be an important factor in determining the magnitude of its biological effect. It must be emphasized that the viability of PBMC incubated with and without gangliosides did not differ and, in both cases, exceeded 90% as assessed by trypan blue exclusion. In addition, the inhibition of thymidine incorporation correlated with suppression of lymphoblast transformation as evaluated by morphologic criteria, thus, excluding artifacts by variations in cold thymidine pools.

TABLE IX

Effects of Gangliosides, Liposomes and Liposomal-Bound Gangliosides on ^{3}H-Thymidine Incorporation (CPMx10^{-3}) Into PBMC In Mixed Leukocyte Cultures[a]

Amount Added (nanomoles)	Liposomes	Gangliosides	Liposomal Gangliosides
0.98	28.9+3.9	32.2+1.2	28.4+1.4
1.97	27.1+3.2	19.7+2.7	17.2+0.5
3.94	35.0+2.5	15.3+4.1	0.34+0.07
7.88	27.3+2.2	12.6+0.6	0.20+0.02
15.8	21.1+3.4	11.9+2.5	0.28+0.07
31.5	23.4+2.2	5.8+0.3	0.32+0.02
Control (none added)	34.3+3.3		

a One way mixed leukocyte reactions were performed in flat bottom microtiter plates with 1x10^5 responder cells and 1x10^5 mitomycin-C treated allogeneic stimulator cells per well.

The previous data do not indicate if it is essential that exogenous gangliosides bind to the plasmalemma in order to exert their inhibitory effects or if the inhibition might occur secondary to functional alterations among PBMC not requiring cell binding of gangliosides. To determine if exogenous gangliosides bind to cells and if there is a differential binding of them to different PBMC populations, PBMC were separated into thymus derived lymphocytes (T cells), immunoglobulin bearing

lymphocytes (B cells) and adherent monocytes. ^{14}C-labeled gangliosides were isolated from rat brains after intracranial
injection of 2 week old animals with ^{14}C glucosamine. The
radioactive gangliosides were incubated 24 hours at 37°C in 5%
CO_2 with each subpopulation followed by harvesting of the cells
and determining the degree of radioactivity bound to the cells.
Table X shows that T and B cells bind and/or incorporate 1.9+0.2
and 1.7+0.1 percent of total gangliosides respectively whereas
monocytes result in the significantly higher binding of 4.3+0.1%
(p<.01). These results are interesting in light of the fact that
the cells were separated on the basis of different surface membrane properties.

Studies currently in progress are critically analyzing the
characteristics of ganglioside binding to PBMC and relating them
to functional activity.

TABLE X

Uptake of Exogenous Gangliosides by PBMC Subpopulations[a]

Subpopulation	% Binding
T-lymphocytes	1.9+0.2
B-lymphocytes	1.7+0.1
Monocytes	4.3+0.1

a PBMC were separated into T cells, B cells, and adherent
   monocytes followed by 24 hour incubation with 2,000 CPM ^{14}C-
   labelled gangliosides. Unbound gangliosides were removed by
   washing the cells twice with medium. Percent binding represents the counts bound by the cell subpopulations divided by
   total CPM available x 100.

Studies to Determine During Which Phase of Lymphocyte
Activation That Gangliosides Act. The ganglioside inhibition
of PBMC proliferation might be occurring throughout all stages
of lymphocyte activation, or their inhibitory activity could
depend on the state of lymphocyte activation. To investigate
this question, PBMC were cultured with optimal doses of either
Con-A or Poke Weed Mitogen (PWM). Twenty-five microgms of
ganglioside was added to each of triplicate wells at the point
of culture initiation (time 0). The same amount was added to
different cultures either four hours or eighteen hours later with
control cultures receiving no gangliosides.

TABLE XI

Effects of Adding Gangliosides at Various Times After Culture
Initiation on Incorporation of ^{3}H-Thymidine (CPMx10^{-4}) Into
PBMC in Response to Con-A and Poke Weed Mitogen[a]

Time of Addition	Con-A	PWM
0 hours	1.3+0.05	0.18+0.01
4 hours	4.8+0.9	0.54+0.01
18 hours	11.3+2.2	1.76+0.14
Control	14.3+0.3	3.1+0.15

a PBMC were cultured with optimal doses of Con-A or PWM. 15.8
   nanomoles of gangliosides was added to triplicate samples at
   the time of culturing(Time 0), 4 hours, and 18 hours later.
   Control cultures received no gangliosides.

     As shown in Table XI, the inhibitory effects of gangliosides
was maximal when added at the time of culturing, with less inhibition at the later times of addition. For the Con-A response,
no inhibition was observed when gangliosides were added after
eighteen hours. When added after eighteen hours of culturing,
gangliosides continued to inhibit PWM reactivity 60%. This might
reflect either differences in PBMC subpopulations sensitivity to
ganglioside inhibition or differing kinetics of activation by
Con-A and PWM.
     The data presented in Table XII show the effects of exo-
genous gangliosides, liposomes, and liposomal-bound gangliosides
when added at initiation of MLC and 48 and 96 hours later. The
effects of gangliosides on MLC reactivity with respect to the
time of addition were similar to the effects on Con-A reactivity.
When added at the time of culture initiation, both gangliosides
and liposomal-bound gangliosides resulted in maximal inhibition
with less effect noted when added 48 hours and 96 hours later.
The amount of liposomes added had no effect at any of the times
they were added. It is possible that the inhibition of PBMC reactivity by exogenous gangliosides might merely represent non-
specific interactions between PBMC membranes with gangliosides
being required in the medium throughout the period of culturing
for inhibition to be manifest. However, another possibility is
that exogenous gangliosides might either be incorporated into the
membrane or perturb the membrane such that inhibition occurs
without the necessity of gangliosides in the culture medium. The
data presented in Table XIII and XIV support the latter possibility.

TABLE XII

Comparison of the Effects of Gangliosides, Liposomal-Bound
Gangliosides, and Liposomes Added at Various Time Intervals on
^{3}H-Thymidine Incorporation (CPMx10^{-3}) Into PBMC in Mixed
Leukocyte Cultures

Time of Addition	Liposomes	Gangliosides	Liposomal-Bound Gangliosides
0	27.1+1.8	3.05+0.66	5.1+0.9
48	17.9+2.8	10.9+1.7	11.8+4.1
96	22.8+3.5	15.1+0.8	12.3+2.2
Control (nothing added) 22.9+3.6			

TABLE XIII

Effects of Preincubating PBMC With Gangliosides for 18 Hours
on ^{3}H-Thymidine Incorporation (CPMx10^{-3}) Into PBMC[a]

	Gangliosides	Media
Control	1.46+0.22	2.3+0.7
Con-A	86.3+3.6	79.7+2.4

a   PBMC were preincubated for 18 hours in RPMI 1640 plus 6% fetal
bovine serum (FBS) with 31.5 nanomoles/ml gangliosides.
Control cultures consisted of PBMC preincubated in RPMI 1640
plus 6% FBS (Media). The PBMC were then removed, washed twice
and recultured with or without 18 microgms Con-A in microtiter
plates. 96 hours later, ^{3}H-thymidine incorporation was
measured.

Preincubation of PBMC with exogenous gangliosides for
eighteen hours, (Table XIII) followed by washing twice in media
and then culturing with Con-A resulted in no inhibition of the
Con-A reactivity. In contrast, identical experiments after 72
hours of preincubation (Table XIV) resulted in approximately
60% inhibition of Con-A reactivity compared to control cultures regardless of the concentration of Con-A. Preincubation with gangliosides also diminished the spontaneous ^{3}H-
thymidine incorporation of the control cultures from 3,883 CPM
to 710 CPM. Similar results were obtained in preincubation
experiments with the MLR (data not shown).

TABLE XIV

Effects of Ganglioside Preincubation on ^{3}H-Thymidine Incorporation (CPMx10[-4]) Into PBMC in Response to Con-A

| Amount of Con-A | Preincubation[a] | |
(microgm	Ganglioside	Media
0	0.07+0.01	0.38+0.09
9	7.07+0.7	15.4+1.7
18	5.8+1.06	13.7+2.3
36	0.89+0.32	1.57+2.8

a   Culture conditions were identical to those of Table XIII
    except that PBMC were preincubated 72 hours with gangliosides.

Discussion

    It is now well established that ganglioside added to culture
media can modulate lymphocyte reactivity in vitro. Previous
studies demonstrated this phenomenon in murine systems(8,9,10),
and the present studies show that human peripheral blood mononuclear cell responses to non-specific mitogens and allogeneic
stimuli are similarly sensitive to exogenous gangliosides. In
none of these systems does the effect appear to be due to a non-
specific toxic effect as cell viability and several biological
phenomena remain intact(9). Furthermore, the degree of inhibition is dose dependent.
    The mechanisms through which gangliosides exert their
effects are unknown. Several plant lectins can bind to glycolipids, including ganglioside(23,24). However, our data indicate that the inhibitory effect of gangliosides on PBMC
mitogenesis is not simply due to the prevention of lectin
binding to cells. Firstly, the amount of Con-A which binds to
cells is not altered by the addition of ganglioside to the
culture medium. Secondly, gangliosides inhibit the MLR which
does not involve lectins. Nevertheless, it appears that they
exert their effects on human PBMC in the early phase of activation. Whether or not this involves binding of gangliosides
to the cells is unknown, but our results show that gangliosides
are taken up by T-lymphocytes, B-lymphocytes, and macrophages.
The studies in which cells were preincubated with gangliosides
and then washed prior to adding Con-A indicate that there is a
minimum period of time that the cells must be in contact with
the ganglioside before the inhibitory effect is irreversibly
established. Perhaps during this time period, the ganglioside
is being incorporated into the plasmalemma.

The nature of the uptake of exogenously added gangliosides by PBMC is not known, but results of investigations by others indicate that they are bound to the plasmalemma of several different types of cells. Keenan et al.(26) found that exogenous gangliosides bind to the surface of canine erythrocytes and mouse 3T3 cells, but their association with the membrane was different than that of endogenously synthesized gangliosides. Callies et al.(27) extended these studies and concluded that gangliosides taken up from the culture medium by chicken erythrocytes, erythrocyte ghosts and fibroblasts were mostly bound to plasmalemma protein. The nature of this binding is not known but O'Keefe and Cuatrecacas(28) found that once GM1 is incorporated into the membrane of mouse 3T3 fibroblasts that it persists over several cell divisions without being degraded. Furthermore, the manner in which GM1 binds to mouse fibroblasts allows it to function as a cholera toxin receptor(29). Thus it is reasonable to suspect that gangliosides are modulating lymphocyte reactivity by interacting with cell surface components.

Not all gangliosides have similar inhibitory effects. The ones with the simplest oligosaccharide units (GM4 and GM3) have no inhibitory effect, while some with more complex oligosaccharide chains have greater inhibitory effects. With the exception of GM2, those which inhibited the most have 2→8 disialo linkages. However, the inhibitory effect is not solely due to sialic acid as NANA, NGNA, and colominic acid caused very little inhibition; and that which was seen only occurred with relatively large amounts. Therefore, their biological effects appear to be related more to their overall molecular structure rather than to one specific moiety.

Although many questions remain to be investigated, currently it appears that specific gangliosides may be important in lymphocyte subpopulation interactions. Indeed, some recent studies have demonstrated that soluble lymphocyte-derived regulatory materials might be glycolipids (15,25), and these immunoregulatory glycolipids appear not only to inhibit but also augment immune reactivity. These observations support a possible role of gangliosides as soluble mediators participating in normal immunoregulation. Furthermore, the presence of elevated serum levels of sialic acid containing glycolipids in both experimental animals and humans bearing malignancies (5,6,7) suggests a causal role in the immunodeficiency states frequently observed in malignancy. Thus further studies of interactions between gangliosides and the immune system should prove valuable as a tool to investigate immune mechanisms both in healthy and disease states.

ACKNOWLEDGEMENTS

This work was supported by American Cancer Society Grant IM-199 and N.I.H. Grant NS-10165.

LITERATURE CITED

1. Fishman, P.H.; Brady, R.O.; Science, 1976, 194, 906-915.

2. van Heyningen, W.E.; Nature, 1974, 249, 415-417.

3. Roseman, S.; Chem. Phys. Lipids, 1970, 5, 270-297.

4. Langenbach, R.; Kennedy, S.; Exp. Cell Res., 1978, 112, 361-372.

5. Portoukalian, J.; Zeingelstein, G.; Abdul-Malik, N.; Dore, J-F.; Biochem. Biophys. Res. Comm., 1978, 85, 916-920.

6. Skipski, V.P.; Katopodis, N.; Prendergast, J.S.; Stock, C.C.; Biochem. Biophys. Res. Comm., 1975, 67, 1122-1127.

7. Kloppel, T.M.; Keenan, T.W.; Freeman, M.J.; Moore, D.J.; Proc. Natl. Acad. Sci. U.S.A., 1977, 74, 3011-3013.

8. Ryan, J.L.; Shinitzky, M.; Eur. J. Immunol., 1979, 9, 171-175.

9. Lengle, E.E.; Krishnaraj, R.; Kemp, R.G.; Canc. Res., 1979, 39, 817-822.

10. Miller, H.C.; Esselman, W.J.; J. Immunol., 1975, 115, 839-843.

11. Suzuki, K.; J. Neurochem, 1965, 12, 629-638.

12. Yates, A.J.; Wherrett, J.R.; J. Neurochem., 1974, 23, 993-1003.

13. Svennerholm, L.; Biochim Biophys. Acta, 1957, 24, 604-611.

14. Miettinen, T.; Takki-Luukkainen, T.T.; Acta. Chem. Scand., 1959, 13, 856-857.

15. Esselman, W.J.; Miller, H.C.; J. Immunol., 1977, 119, 1994-2000.

16. Stobo, J.D.; Paul, S.; van Scoy, R.E.; Hermans, P.E.; J. Clin. Invest., 1976, 57, 319.

17. Momoi, T.; Ando, S.; Nagai, Y.; Biochim. Biophys. Acta, 1976, 441, 488-497.

18. Holm, M.; Mansson, J.-E.; Vanier, M.-T; Svennerholm, L.; Biochim. Biophys. Acta, 1972, 280, 356-364.

19. Griggs, L.J.; Post, A.; White, E.R.; Finkelstein, J.A.;
    Moeckel, W.E.; Holden, K.G.; Zarembo, J.E.; Weisbach, J.A.;
    Anal. Biochem.; 1971, 43, 369-381.

20. Li, S.-C.; Chien, J.-L.; Wan, C.C.; Li, Y.-T.; Biochem J.,
    1978, 173, 697-699.

21. Handa, N.; Handa, S.; Japan J. Exp. Med., 1965, 35, 331-341.

22. Stobo, J.D.; Rosenthal, A.S.; Paul, W.E.; J. Immunol., 1972,
    108, 1-17.

23. Deleers, M.; Poss, A.; Ruysschaert, J.-M; Biochem Biophsy.
    Res. Comm., 1976, 72, 709-713.

24. Boldt, D.H.; Speckart, S.F.; Richards, R.L.; Alving, C.R.;
    Biochem. Biophys. Res. Comm., 1977, 74, 208-214.

25. Wolf, R.L.; Merler, E.; J. Immunol., 1979, 123, 1169-1180.

26. Keenan, T.W.; Franke, W.W.; Wiegandt, H.; H.-S., Z. Physiol.
    Chem., 1974, 355, 1543-1558.

27. Callies, R.; Schwarzmann, G.; Radsak, K.; Siegert, R.;
    Wiegandt, H.; Eur. J. Biochem., 1977, 80, 425-432.

28. O'Keefe, E.; Cuatrecacas, P.; Life Sciences, 1977, 21,
    1649-1654.

29. Fishman, P.H.; Moss, J.; Vanghan, M.; J. Biol. Chem., 1976,
    251, 4490-4494.

RECEIVED December 10, 1979.

Ganglioside Composition of Rabbit Thymus
Molecular Specificities Compared with Those of Various Organs

YOSHITAKA NAGAI and MASAO IWAMORI

Institute of Medical Science, University of Tokyo, Tokyo, Japan

It is well established that thymus plays an important role for T cell differentiation, but the molecular basis for the differentiation-related surface change is still incompletely understood. The possible involvement of glyco-lipids and glycolipid biosynthesis correlated with the altered phenotypic properties has been described in the induction of Thy-1 antigen by thymic humoral factor (3) and in the stimulation of thymocytes (4) and peripheral lymphocytes (5, 6) by T cell mitogen. During the course of the study of membrane specificity with special attention to ganglioside molecular species, we found unique characteristics of thymus gangliosides. Although at present many experiments must accumulate to clarify the correlation of such membrane specificity and cellular differentiation, the findings seem to be important to analyze the thymus function and the role of gangliosides.

Materials And Methods

Materials. Male rabbits (New Zealand White) were obtained from the Department of Experimental Animals of this Institute. DEAE-Sepharose, CL-6B, was a product of Pharmacia Fine Chemicals, Uppsala, Sweden and was converted to acetate form as described previously (7). Iatrobeads (6RS 8060) for column chromatography were supplied from Iatron Laboratory, Tokyo and were washed with chloroform/methanol/5N ammonia (3 : 2 : 1, by vol.) until the solution becomes colorless.

Preparation Of Gangliosides. The pooled tissues were homogenized in 7 volumes of cold acetone to remove neutral lipids and water. Gangliosides were extracted from the acetone-dried powder with 3 volumes of chloroform/

The nomenclature of Svennerholm for gangliosides (1) was used through-out. The term "lacto-N-norhexaose" for a straight chain hexasaccharide, Gal(β,1-4)GlcNAc(β,1-3)Gal(β,1-4)GlcNAc(β,1-3)Gal(β,1-4)Glc, was used to discriminate a straight chain from a branched chain (2).

0-8412-0556-6/80/47-128-435$5.00/0

methanol (1 : 1, by vol.), (1 : 2, by vol.), (2 : 1, by vol.) and (1 : 1, by vol.) at 45°C. The lipid solution was applied to a DEAE-Sephadex (A-25, acetate form) column, and the neutral lipids were eluted with 5 column volumes of chloroform/methanol (1 : 1, by vol.) and 1 column volume of methanol. Acidic lipids including gangliosides were then eluted from the column with 10 column volumes of 0.2M sodium acetate in methanol. After the mild alkaline saponification with 0.5N NaOH and neutralization with 1N acetic acid, the solution was dialyzed against distilled water   and evaporated to dryness. The dried residue was dissolved in chloroform/methanol (4 : 1, by vol.) and applied on Silica Gel 60 column. The column was eluted successively with chloroform/methanol (4 : 1, by vol.), (7 : 3, by vol.), (3 : 1, by vol.), (1 : 1, by vol.), (1 : 2, by vol.) and (1 : 4, by vol.). Gangliosides were eluted from the column with chloroform/methanol (1 : 1, by vol.), (1 : 2, by vol.) and (1 : 4, by vol.).

Ganglioside-Mapping And Isolation Of Individual Gangliosides. The gangliosides were fractionated on the basis of the differences in charge by a column packed with DEAE-Sepharose (acetate form) as described previously (7) and the ganglioside-maps were made to observe the whole components and to decide the fraction to be collected. The gangliosides fractionated by DEAE-Sepharose were further fractionated by Silica Gel 60 or Iatrobeads column chromatography (8).

Structural Analyses Of Gangliosides. Carbohydrate composition, methylation analysis, enzyme treatment and partial acid hydrolysis were performed as described previously (9).

Quantitation Of Gangliosides. The gangliosides separated on thin-layer plates were located with resorcinol reagent and the plates were scanned with a double wavelength chromatoscanner (Model CS-900, Shimazu, Kyoto) (10). The reference standards, GM3, GM1, GD1a, GD1b and GT1b, of known amounts were developed together on the same plates and determined in the same manner.

Results And Discussion

Rabbit Thymus Gangliosides. The ganglioside pattern developed with chloroform/methanol/0.5% $CaCl_2 \cdot 2H_2O$ (55 : 45 : 10, by vol.) showed that the gangliosides with high polarity were present in significant amounts (Fig. 1), which contrasted from the other extraneural organs that contained GM3 as a predominant constituent. As shown in Fig. 1, four gangliosides were recognized as the major constituents in rabbit thymus and the ganglioside composition of male and female rabbits were essentially similar. One was a disialoganglioside and the other three were monosialogangliosides.

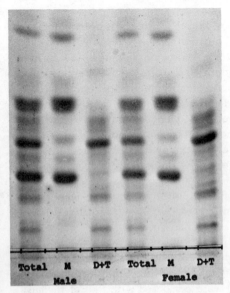

Figure 1. Thin-layer chromatogram of rabbit thymus gangliosides. Plate developed with chloroform/methanol/0.5% $CaCl_2 \cdot 2H_2O$ (55:45:10, by volume); spots located with resorcinol–HCl reagent. Total: total gangliosides; M: monosialogangliosides; D + T: di- and trisialogangliosides.

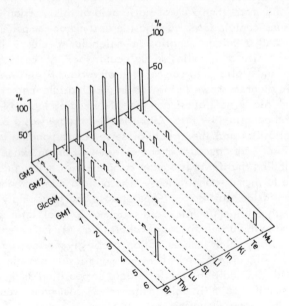

Figure 2. Sialic acid distribution in monosialoganglioside components of various rabbit tissues. Br: brain; Thy: thymus; Lu: lung; St: stomach; Li: liver; In: intestine; Ki: kidney; Te: testis; Mu: muscle; Column height shows percentage distribution of sialic acid in monosialoganglioside fraction. Numbers 1–6: unidentified monosialogangliosides.

The structure of the most polar monosialoganglioside was determined as NeuNGc(α,2-3)Gal (β,1-4) GlcNAc(β,1-3)Gal (β,1-4)GlcNAc(β,1-3)Gal (β,1-4)Glc(β,1-1) ceramide based on the following findings: (a) carbohydrate analysis showed 3 moles of galactose, 2 moles of N-acetylglucosamine, one mole each of glucose and N-glycolylneuraminic acid, (b) sequential enzymatic hydrolysis by neuraminidase, β-galactosidase and β-N-acetylhexosaminidase showed the alternating sequence of galactose and N-acetylglucosamine, (c) methylation analysis yielded 3 moles of 3-linked galactitol and 1 mole of 4-linked glucitol, and 4-linked N-acetylhexosaminitol. In a similar way, the structures of the three other thymus gangliosides were assigned as follows:

NeuNGc(α,2-3)Gal(β,1-4)GlcNAc(β,1-3)Gal(β,1-4)Glc(β, 1-1)ceramide
NeuNGc(α,2-8)NeuNGc(α,2-3)Gal(β,1-4)Glc(β,1-1)ceramide
NeuNGc(α,2-3)Gal(β,1-4)Glc(β,1-1) ceramide.

Uniqueness Of Thymus Gangliosides. The ganglioside composition of several tissues of rabbit (brain, lung, stomach, liver, intestine, spleen, kidney, testis, muscle and bone marrow) were compared with that of thymus. The concentrations of monosialogangliosides of various tissues are shown in Fig. 2. As is well known, the major monosialoganglioside of brain was GM1 ganglioside, and more than 95% of sialic acid of this fraction were in GM1. The other tissues contain GM3 ganglioside as a major component except for thymus which had a unique ganglioside molecular species. To clarify the membrane specificity of ganglioside molecular species, basic asialo-carbohydrate chain and sialic acid composition of various tissues were expressed as a schematic diagram shown in Fig. 3, in which asialo-carbohydrate chain were classified into three groups: ganglio-N-tetraose, lacto-N-neotetraose, and lactose and ganglio-N-triose. Ganglio-N-tetraose was a basic structure of brain gangliosides and the structural series corresponding to lactose and ganglio-N-triose were abundant in the extraneural organ except for thymus. The general finding that GM3 is a main component of extraneural organs was not applicable to thymus gangliosides. The basic asialo-carbohydrate chain of thymus ganglioside was lacto-N-neotetraose and surprisingly the concentration of N-glycolylneuraminic acid reached 87% of total sialic acid. When compared with various tissues of rabbit (Fig. 4), it is clear that thymus contains particularly high concentration of N-glycolylneuraminic acid. Moreover, all molecular species of thymus ganglioside contain N-glycolylneuraminic acid. In addition, the fatty acid composition of thymus ganglioside is quite simple: palmitic acid is the richest component, and this is also characteristic of thymus gangliosides.

Comparison Of Thymus Gangliosides Of Various Animals. The unique characteristic of thymus ganglioside is its distinct species specificity. We examined thymus of human, calf, rat and mouse in addition to rabbit. As

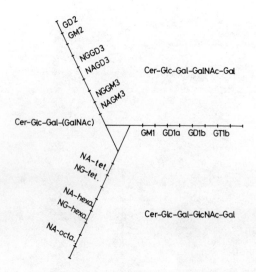

Figure 3.   Procedure for diagramatic representation of ganglioside molecular species

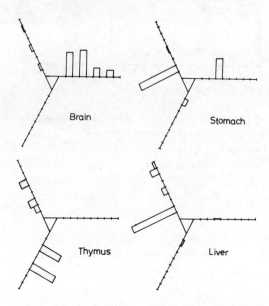

Figure 4.   Diagrams of gangliosides of various tissues. Open column: gangliosides with N-acetylneuraminic acid. Hatched column: gangliosides with N-glycolylneu-raminic acid

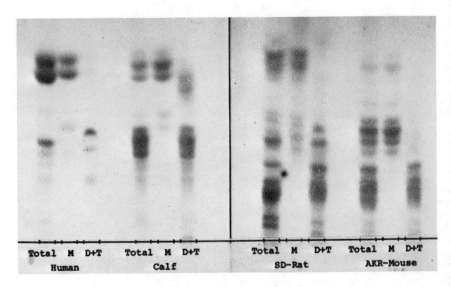

*Figure 5.   Thin-layer chromatogram of thymus gangliosides of various animals.
Abbreviations as in Figure 1.*

shown in Fig. 5, their ganglioside spectra are quite different from that of rabbit. Moreover, the spectrum of each animal is different from each other. Various strains of mice were also examined with regard to thymus ganglio- sides. We found that the inter-strain specificity is not so remarkable as compared with the inter-species specificity. The remarkable species-specific characteristic of thymus is thus in contrast with the case of brain ganglio- sides that exhibits a relatively common composition among various species. Then, what is the implication of this unique character of thymus ganglioside?

Thymus is one of the central immunological organs. Most of the immu- nological events are performed by a dynamic interplay between thymus derived cells (T cells) and bone marrow derived cells (B cells) sometimes with help from macrophage and related cells (accessary cells, A cells). For example, antibody production is initiated by the interaction of antigen- primed B and T cells. T cells themselves are produced first from their stem cells (undifferentiated antecedent cells) in bone marrow as B cells are. Migrating from bone marrow, the immature precursor T cells move to thymus where they are differentiated into mature T cells presumably under the influence of thymic hormone and then leave the thymus for different parts of body where they participate in various immunological events. Recently, different types of T cells (T cell subsets) have been found, which play different important roles in immunological responses (i.e. helper T cells, suppressor T cells, effector T cells, cytotoxic T cells, killer T cells, natural killer T cells etc.). Under certain stimuli these T cell subsets seem to increase in number. It is extremely important to know exactly what type of cells and how many cells of that type are involved both in normal immunol- ogical differentiation and responses, as well as in disease states. For this purpose cell surface markers are a valuable tool, but hitherto known markers are not always present on all the T cell subsets. Gangliosides may be useful to differentiate T cell subsets. Furthermore, their distinct inter-species specificity may allow to obtain efficiently the specific antibody against the marker ganglioside by immunization of animals of different species. It should be noted that antilymphocyte antisera which were obtained by heterologous animals are used as a useful tool in the transplantation medicine.

Marcus et al. reported that anti GM1 ganglioside antibody can be used as a T cell marker (11, 12, 13). In collaboration with Tada's group of the University of Tokyo we very recently found that antibody to asialoganglio- side GM1 specifically labels mouse natural killer cells and that the antibody selectively kills the cells in the presence of complement (14). Natural killer cells are known by their specific ability to attack tumor cells. We also have evidence that certain gangliosides are capable of eliciting auto- immune diseases experimentally which we called the ganglioside syndrome (15). For example, some brain gangliosides produce autoimmune lesions in peripheral nerve in susceptible rabbits. This may correspond to the human disease Landry-Guillain-Barré syndrome, but its responsible antigen is not clear. On the other hand, Sela et al. (16) interestingly reported that anti-body to GM1 ganglioside can initiate cell multiplication of lymphocytes. All these recent observations suggest the important involvement of cell surface gangliosides and their related substances in basic immunological processes.

Abstracts

Gangliosides of rabbit thymus were analyzed by the previously developed ganglioside-mapping procedure. No N-acetylgalactosamine-containing ganglioside was detected in thymus. The following gangliosides were contained in thymus in the following concentrations, expressed in percent as sialic acid distribution:
1. N-glycolylneuraminosyl lacto-N-norhexaosyl ceramide, 30%
2. N-glycolylneuraminosyl lacto-N-neotetraosyl ceramide, 23%
3. GD3 containing N-glycolylneuraminic acid, 28%
4. GM3 containing N-glycolylneuraminic acid, 7%
The concentration of N-glycolylneuraminic acid in thymus reached 87% of total sialic acid which was in contrast with that in the other organs which contained mainly N-acetylneuraminic acid. Furthermore, when the basic asialo-carbohydrate chains of gangliosides of various organs were compared, a uniqueness of thymus was clearly demonstrated. Lacto-N-neotetraose was a basic asialo carbohydrate chain in thymus, whereas ganglio-N-tetraose and lactose were found to be basic asialo-carbohydrate chain in brain and in the other extraneural organs (lung, stomach, liver, spleen, kidney, testis, muscle and bone marrow).

Palmitic acid was present in all thymus gangliosides as a major fatty acid, which is also in contrast to the fatty acid compositions of the various organs.

Acknowledgement

This work was supported in part by a grant from the Ministry of Education, Science and Culture and by a research grant for specific diseases from the Ministry of Health and Welfare, Japan.

Literature Cited

1. Svennerholm, L., "Methods in Carbohydrate Chemistry" Ed. Whistler, R. L. and BeMiller, J. N., Academic Press, New York and London, 1972, Vol.4, pp.464-474.
2. Niemann, N., Watanabe,K., Hakomori, S.,Childs, R.A. and Feizi,T., Biochem. Biophys. Res. Commun., 1978, 81, 1286-1293.

3. Miller, H.C. and Esselman, W.J., Ann. N.Y.Acad. Sci., 1975, 249, 54-60.
4. Rosenfelder, G., Van Eijk, R.V.W., Monner, D.A. and Mühlradt, P. F., Eur. J. Biochem., 1978, 83, 571-580.
5. Inoue, Y., Handa, S. and Osawa,T., J. Biochem., 1974, 76, 791-799.
6. Narasimhan, R., Hay, J.B., Greaves, M.F. and Murray, R. K., Biochim. Biophys. Acta, 1976, 431, 578-591.
7. Iwamori, M. and Nagai, Y., Biochim. Biophys. Acta, 1978, 528, 257-267.
8. Iwamori, M. and Nagai, Y., J. Biochem., 1978, 84, 1601-1608.
9. Iwamori, M. and Nagai, Y., J. Biol. Chem., 1978, 253, 8328-8331.
10. Iwamori, M. and Nagai, Y., J. Neurochem., 1979, 32, 767-777.
11. Stein-Douglas, K.E., Schwarting, G.A., Naiki, M. and Marcus, D. M., J. Exp. Med., 1976, 143, 822-832.
12. Stein-Douglas, K.E., Schwarting, G.A., and Marcus, D. M., J. Immunol., 1978, 120, 676-679.
13. Stein-Douglas, K.E. and Marcus, D.M., Biochemistry, 1977, 16, 5258-5290.
14. Kasai, M., Iwamori, M., Nagai, Y., Okumura, K. and Tada, T., Eur. J. Immunol., 1979, in press.
15. Nagai, Y., Momoi, T., Saito, M., Mitsuzawa, E. and Ohtani, S., Neurosci.Lett., 1976, 2, 107-111.
16. Sela, B.A., Raz, A. and Geiger, B., Eur. J. Immunol., 1978, 8, 268-274.

RECEIVED December 10, 1979.

The Immunology and Immunochemistry of Thy-1 Active Glycolipids

W. J. ESSELMAN, H. C. MILLER, T. J. WANG, K. P. KATO, and W. G. CHANEY

Departments of Surgery and Microbiology, Michigan State University, East Lansing, MI 48824

The Thy-1 antigen is a cell surface antigen found in mice and rats. The antigen is membrane bound and is one of a number of differentiation antigens which appear on bone marrow lympho- cytes and early postnatal development. Thy-1 is the product of codominant alleles on chromosome 9 in mice, and all inbred mouse strains are homozygous at this locus and express either Thy-1.1 or Thy-1.2 antigens (1,2). We previously proposed that the allo- antigenic determinants in brain and thymic lymphocytes were car- ried by glycolipids (gangliosides) as well as glycoproteins (3). Thy-1 antigenicity was assayed with a modified *in vitro* PFC technique originally developed for detecting alloantigenic diff- erences on nucleated cells by Fuji and Milgrom (4). This assay demonstrated the allogenic specificity of both glycoprotein and glycolipid forms of Thy-1.

A glycoprotein with Thy-1.1 activity has been isolated and partially characterized from rat tissue (5). The glycoprotein has a single polypeptide chain and contains about 30% carbohy- drate. The rat antigen, in addition to carrying the Thy-1.1 specificity, also carries determinants recognized by heterologous antisera (which were used in the isolation procedures). Zwerner et al., (6) have isolated glycoproteins of similar molecular weight and carbohydrate composition from two lymphoblastoid cell lines which express either Thy-1.1 or Thy-1.2. The carbohydrate composition consists mainly of mannose, galactose, glucosamine, sialic acid, fucose and galactosamine.

This report concerns the biosynthetic radiolabeling of brain and thymic lymphoma Thy-1 active glycolipids with a sialic acid percursor (N-acetylmannosamine) and a sphingosine percursor (pal- mitic acid), as well as the isolation of Thy-1 glycolipids by two dimensional TLC. The ganglioside nature of Thy-1 glycolipid is suggested by interactions with DEAE cellulose ion-exchange chromatography, and by neuraminidase treatment.

0-8412-0556-6/80/47-128-445$5.00/0

Materials and Methods

Radiolabeling of Brain Glycolipids. Thy-1 glycolipid was
labeled biosynthetically using a previously described method (7).
Five to seven day mice of either AKR/J (H-2^k, Thy-1.1) or ICR
Swiss (Thy-1.2) strain mice were used for each preparation. Each
mouse pup was injected intracranially with 8 µl of sterile saline
solution containing 5 µCi [1-^{14}C]-N-acetylmannosamine (ManNAc)
(54.5 mCi/mmol, New England Nuclear, Boston MA). This solution
was injected into both sides of the head at a point 1-2 mm anter-
ior to the interauricular line and 2-3 mm lateral to the midline
with a 10 µl Hamilton syringe. The tip of the needle was introduced only far enough to completely submerge the bevel of the tip
below the surface of the skull. The pups were then returned to
their mothers. Two days later the brains were removed and
pooled for each group and the glycolipids were isolated as described below. Incorporation of ManNAc exclusively into the
sialic acid of gangliosides was confirmed by neuraminidase hydrolysis.

Radiolabeling of Lymphoma Cells. BW5147 (AKR/J, Thy-1.1,
H-2^k) and S49.1 (BALB/c, Thy-1.2, H-2^d) murine lymphoma cell
lines were obtained from the Salk Institute Cell Distribution
Center (LaJolla, CA). Cells to be labeled for glycolipids in
culture were washed once, then incubated in fresh Dulbecco's modified Eagles medium (D-MEM, Grand Island Biological Co.,) inactivated fetal calf serum (FCS, Grand Island Biological Co.) and
either 2.0 µCi/ml [1-^{14}C]palmitic acid (30 mCi/mmol) (International Chemical and Nuclear Corp., Irvine, CA) or 2 µCi/ml [1-^{14}C]-
N-acetylmannosamine. Cells were cultured at 37°C in flasks at a
concentration of 5 x 10^5 cells/ml. The cells reached a concentration of 2 x 10^6 cells/ml after about 48 hours and were washed
once with phosphate buffered saline (PBS) and the pellet collected by centrifugation was extracted as described below.

Glycolipid Preparations. Biosynthetically radiolabeled
glycolipids were isolated from brain and lymphoma cells with
chloroform-methanol (C:M) mixtures (8). The total lipid extract
was subjected to a Folch partition and the ganglioside rich upper
phase was dried in vacuo and hydrolysed with mild alkali using
0.3 N NaOH in methanol-chloroform (1.1, v/v) for one-half hour at
room temperature. This mixture was neutralized with glacial
acetic acid, evaporated, resuspended in distilled H$_2$O and dialyzed
for 48 hr at 4°C against several changes of distilled water. The
dialyzed samples were checked at this point for radioactive incorporation by liquid scintillation spectrometry. Upper phase
samples from brain were applied to thin layer chromatography (TLC)
plates and column chromatography for further purification. Lymphoma upper phase samples were further purified on TLC plates.

Thin Layer Chromatography. Two dimensional preparative TLC
was used to purify Thy-1 glycolipid. All experiments were performed using Silica Gel 60 TLC plates (E. Merck, Darmstadt, West

Germany). Dialyzed upper phase samples of brain and lymphoma cells with from 30,000 to 50,000 cpm were used for each TLC plate. All radiolabeled samples were chromatographed in parellel with mouse brain ganglioside standards extracted as described above.

A two dimensional TLC system was developed to attempt the purification of Thy-1 glycolipid using only one plate. The radiolabeled glycolipid material (brain or lymphoma) was spotted in a small area in the corner of the plate. The plate was developed twice in one dimension in solvent 1 (C:M:W, 50:40:9, 0.02% $CaCl_2$) and once in solvent 2 (C:M:W, NH_4OH, 60:35:6:14) in the second dimension. Each plate was air dried for one hour then dried *in vacuo* for 45 minutes (between runs) to ensure dryness.

Autoradiography. TLC plates were covered with a 3 x 10 in. sheet of Kodak SB-5 X-ray film (Eastman Kodak, Co., Rochester, NY) and kept at 4°C in the dark until developed. The time of exposure varied between 12 days and one month. The films were developed in Kodak X-ray Developer-Replenisher (#146-5327) and fixed in Kodak Rapid Fix (#146-4106). Selected areas on the TLC plates directly under the spots on the films were eluted from the silica gel and tested in the anti-Thy-1 PFC assay.

Thy-1 Chemical and Enzymatic Treatments. Clostridium perfringens neuraminidase (Sigma, St. Louis, MO) 0.1 units, was added to dried glycolipid in acetate buffer pH 4.5 and incubated for one hour at 37°C. The sample was heated to 100°C for 15 min., dried, extracted with C:M (2:1) and tested as described below. DEAE cellulose chromatography was performed by application of a small volume of glycolipid in C:M (1:1) to a 3 ml column of DEAE cellulose, acetate form, followed by elution with 5 column volumes of C:M:W (60:40:8). Bound glycolipids were eluted with 5 column columes of chloroform-methanol-ammonium acetate. The samples were dried and tested in the PFC assay as described below. Mild HCl treatment was performed as previously described with 0.1 N HCl at 80°C for 30 minutes (8). After hydrolysis the samples were neutralized with 0.1 N NaOH, dried and tested in the PFC assay.

Anti-Thy-1 PFC Assay. Fuji and Milgrom (4) originally developed an *in vitro* PFC assay which detected Thy-1 alloantigen on whole thymocytes. A modified version, used here, has previously been described in detail and found to be effective for measuring the immune response to isolated glycolipid and glycoprotein Thy-1 alloantigens (3).

Thy-1 Glycoprotein. Glycoprotein extracted and highly purified by lectin-affinity chromatography from C57BL/6 mouse thymus (5) was a generous gift from Dr. M. Letarte (Toronto, Canada). Clostridium perfringens neuraminidase 0.1 units was added to the glycoprotein (1 μg) in acetate buffer pH 4.5 and incubated for one hour at 37°C. The sample was heated to 100°C in a water bath to destroy enzyme activity and added to the PFC cultures to test for remaining Thy-1 activity.

Results

Thy-1 Active Glycolipids. We have previously reported that
AKR mouse (Thy-1.1) and C3H mouse (Thy-1.2) gangliosides inhibited anti-Thy-1.1 and anti-Thy-1.2 sera in a hapten inhibition
assay (9). We demonstrated allogenic specificity with gangliosides with some cross reaction. Hapten inhibition studies were
extended to free sugars, oligosaccharides and to oligosaccharides
derived from partially purified AKR and C3H gangliosides (unpublished results). Oligosaccharides were obtained from AKR and
C3H gangliosides by ozonolysis, followed by purification on a
P-10 column. The oligosaccharides were then tested for hapten
inhibition in a microcytotoxicity assay with anti-Thy-1.1 and
anti-Thy-1.2. The oligosaccharides were found to be inhibitors
of the anti-Thy-1 sera and they retained the allogenic specificity for Thy-1.1 or Thy-1.2 (10). There was, however, some
cross reaction. Using this assay, intact gangliosides (as liposomes) were about ten times more inhibitory of the antisera, but
they also demonstrated similar cross reaction.

PFC Assay For Thy-1. Recently we have developed an assay
system for Thy-1 which is based on the immune response to antigen
preparations (3) rather than binding of glycolipid antigen to
antibody. This assay involves addition of the antigen to spleen
cell cultures of the opposite Thy-1 allotype (in Marbrook Chambers). The immune response to the antigen is measured by a
plaque forming cell assay in which the spleen cells, after 4 days
incubation with antigen, are plated in agar on glass slides in a
lawn of either Thy-1.1 or Thy-1.2 thymocytes. Antibody forming
cells, in the presence of complement, produced plaques in the
thymocytes. Thus Thy-1.2 antigen elicited anti-Thy-1.2 PFC response only with Thy-1.1 responder spleen cells: and Thy-1.1
antigen elicited an anti-Thy-1.1 PFC response only with Thy-1.2
responder spleen cells.

This assay is useful because the specificity of the response
to a particular antigen can be determined by using various combinations of responder cells and target cells of the Thy-1.1 or
Thy-1.2 allotype. The specificity of this assay for Thy-1 has
been previously described to us (3,11) and by others (12,13).
Thus we were able to show that partially purified G_{M1} ganglioside
preparations (Thy-1 g1) obtained from Thy-1.1 and Thy-1.2 mouse
brain and thymus exhibited specific Thy-1 antigenicity (Fig. 1a).
A purified glycoprotein (Thy-1.2 gp) preparation, isolated from
Thy-1.2 mouse brain according to the procedure of Williams (5),
was also tested and was found to have comparable Thy-1 activity
and Thy-1 specificity compared to the glycolipids. Absorbtion of
the Thy-1 active glycolipids or glycoprotein with anti-Thy-1
antisera further confirmed the specificity of the antigens (Fig.
1b). Absorbtion of Thy-1.2 active glycolipid or glycoprotein with
anti-Thy-1.2 sera removed the immunological activity. However,
absorbtion with anti-Thy-1.2 sera had no effect. The reverse

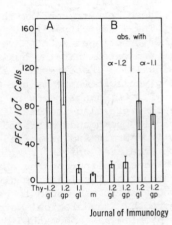

Figure 1. Anti-Thy-1.2 response elicited by glycolipids and glycoprotein

A: Brain glycolipids (gl), (500 ng, and glycoprotein (gp), 100 ng, were added to AKR spleen cells, and resulting PFC were enumerated in a lawn of C3H thymocytes. Culture medium (m) was added to control cultures. Antigens are identified by Thy-1.2 (C3H) or Thy-1.1 (AKR) according to the mice from which they were derived. Control cultures were absorbed before addition to cultures with anti-Thy-1.2 (α-1.2) or anti-Thy-1.1 (α-1.1) antisera. Values are the means and standard errors of five cultures (3).

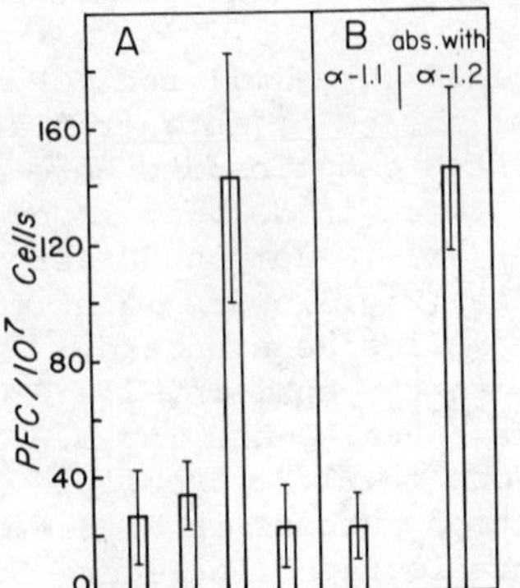

Figure 2. Anti-Thy-1.1 response elicited by glycolipids and glycoprotein

Labels are identical to those in Figure 1. Antigens in this case were added to C3H spleen cells and tested for PFC in a lawn of AKR thymocytes. Values are the means and standard errors of five cultures (15).

Journal of Immunology

assay for Thy-1.1 using the same antigens and absorbtion conditions is shown in Figure 2.   In this assay the Thy-1.1 active
glycolipid (Thy-1.1 gl) gave excellent response and Thy-1.2 active glycolipid gave no response (Fig. 2a).   The Thy-1.2 active
glycoprotein was also negative.   Absorbtion of the Thy-1.1 active
glycolipid with anti-Thy-1.1 sera completely removed the activity;
but absorbtion with anti-Thy-1.2 sera had no effect on the activity of Thy-1.1 (Fig. 2b).

These results confirm that glycolipid and glycoprotein both
carry Thy-1 activity with identical properties in the PFC assay.
Further fractionation of Thy-1 active glycolipids derived from
brain indicated that the activity was seperable from G_{M1} ganglioside (3).   These results show that Thy-1 antigenicity resides
in a glycolipid which is not G_{M1} ganglioside, but which is purified with G_{M1} ganglioside by most procedures.   Thy-1 active glycolipids were also isolated from thymic lymphocytes (thymocytes).
The active compounds migrated with an Rf similar to G_{M1} ganglioside from brain and exhibited Thy-1 activity according to the
mouse strain of origin.

Radiolabeling and the Analysis of Thy-1 Active Glycolipids.
The amount of Thy-1 active glycolipids in brain, thymocytes or
lymphoma cells was found to be extremely small and could only be
detected by immunological methods and not by chemical means.
This problem was approached by incorporating radioactive percursors into the glycolipids of both brain and lymphoma cells of
the Thy-1.2 and 1.1 types.   We have used ^{14}C-palmitate as a percursor of ceramide, and ^{14}C-N-acetylmannosamine as a percursor of
sialic acid (7).   Glycolipids were isolated and the radioactive
gangliosides were resolved by two-dimensional thin layer chromatography in two different solvent systems followed by autoradiography.

AKR (Thy-1.1) and ICR (Thy-1.2) mouse brain gangliosides
were labeled by intracranial injection of ^{14}C-ManNAc and the
isolated gangliosides were applied to thin layer plate and developed twice in solvent 2 on one axis and once in solvent 1 on the
other axis.   Compounds detected by autoradiography of these
plates (Fig.3) were identified by relative TLC mobility compared
to ganglioside standards.   The AKR brain gangliosides were tested with the anti-Thy-1.2 PFC (Fig. 4).   The Thy-1 antigenicity
tests revealed one Thy-1.2 glycolipid (Fraction 4) from AKR brain
and one Thy-1.2 glycolipid (Fraction 4) from ICR brain.   The TLC
fraction numbers in Figure 4 correspond to the numbers on the
autoradiographs in Figure 3.   Thy-1 activity was associated with
the spot directly below G_{D3} and to the right of G_{D1a} in the orientation shown (Fig. 3).   Brain Thy-1.1 glycolipid (Fraction 4)
consistently migrated slightly faster in solvent 1 and 2 than
Thy-1.2 glycolipid (Fraction 4).

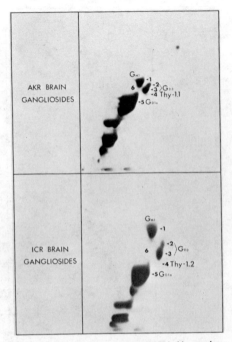

Journal of Immunology

Figure 3. Autoradiograms of Man-NAc- labeled brain gangliosides

Gangliosides labeled by intracranial injection of 1-¹⁴C]ManNAc were extracted from 1 g of brain, applied with solvent 1 (vertically) and solvent 2 (horizontally). Assayed fractions indicated by numbers (and ganglioside abbreviations) correspond to the PFC assay results in Figure 4. Fraction 6 refers to the area surrounding all the assayed spots (15).

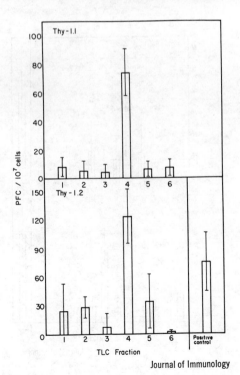

TLC Fraction

Journal of Immunology

Figure 4. Anti-Thy-1 PFC assay of brain gangliosides

Gangliosides assayed were eluted from the plate shown in Figure 3, diluted, and the amount derived from 0.1 g of brain was added to the PFC assay cultures. Values are the mean ± standard error of five cultures. Application of the Student t test to the standard errors for the samples gave p values less than 0.05 when compared with the Thy-1-active fraction. Positive control for the anti-Thy-1.2 PFC assay was a column G_{M1} fraction containing Thy-1 glycolipid (15).

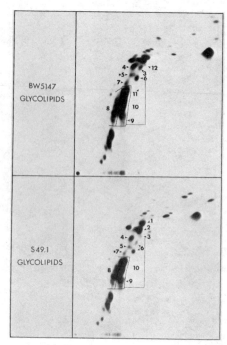

Figure 5.   Autoradiogram of palmitate-labeled lymphoma cell glycolipids

Chromatography was done as described in Figure 3, using the gangliosides derived from 2 × 10⁸ cells. Assayed fractions are labeled with numbers that correspond to the anti-Thy-1 PFC assay in Figure 6. Number 10 refers to the area surrounding the spots that were assayed (15).

Figure 5.   Autoradiogram of palmitate-labeled lymphoma cell glycolipids

Chromatography was done as described in Figure 3, using the gangliosides derived from 2 × 10⁸ cells. Assayed fractions are labeled with numbers that correspond to the anti-Thy-1 PFC assay in Figure 6. Number 10 refers to the area surrounding the spots that were assayed (15).

Journal of Immunology

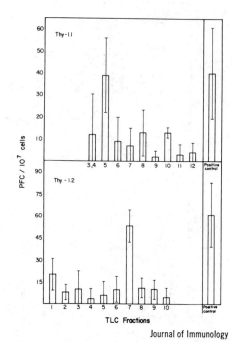

Figure 6.   Anti-Thy-1 PFC assay of lymphoma glycolipids

Glycolipids were eluted from the TLC plate shown in Figure 5. Glycolipids derived from 2 × 10⁷ cells were added to each culture. Values are the mean ± standard error of five cultures. Application of the Student t test to the standard errors for these samples gave p values less than 0.05 when compared with the Thy-1-active fraction. Positive controls are brain Thy-1 glycolipids (15).

Journal of Immunology

BW5147 (Thy-1.1) and S49.1 (Thy-1.2) lymphoma cell lines are
of T cell lineage and express Thy-1 antigen. These cells were
labeled biosynthetically with ^{14}C-palmitate or ^{14}C-ManNAc. Glycolipids from these cells were isolated by extraction, and two-
dimensional TLC and autoradiography. About 50 spots were observed after palmitate incorporation (Fig. 5). BW5147 glycolipids
were tested with the anti-Thy-1.2 PFC assay and S49.1 glycolipids
were tested with the anti-Thy-1.2 PFC assay. One Thy-1.1 glycolipid was detected from BW5147 cells (Fraction 5, Figure 6) and
one Thy-1.2 glycolipid from S49.1 cells (Fraction 7, Figure 6).
The fractions in Figure 6 correspond to the numbered spots of the
autoradiography in Figure 5.

BW5147 and S49.1 lymphoma cell lines were labeled in culture
with ^{14}C-ManNAc and autoradiography of these labeled gangliosides
from two-dimensional TLC plates is shown in Figure 7. Almost all
of the glycolipids labeled with palmitate in Figure 5 in the
lower left quadrant of the plate were labeled with ManNAc. Many
of the other spots in the upper right quadrant of Figure 5 were
neutral glycolipids and hence were not labeled with ManNAc. Lymphoma Thy-1.1 and Thy-1.2 were both labeled with ManNAc as seen
in Figure 7.

Table I: PROPERTIES OF THY-1 ACTIVE GLYCOLIPIDS

		THY-1.1 ANTI-THY-1.1 RESPONSE	THY-1.2 ANTI-THY-1.2 RESPONSE
TREATMENT			
Glycolipid	None	51 ± 8^a	63 ± 6
	C.p. Nase (1 hr)^b	13 ± 6	15 ± 9
	HCl^c	1 ± 1	7 ± 2
	DEAE: bound^d	36 ± 5	45 ± 9
	DEAE: not bound	8 ± 6	1 ± 1

a. The values shown are the mean $\pm$ standard error of five cultures.

b. Clostridium perfringens neuraminidase was incubated with glycolipids for 1 hr at 37°C.

c. 0.1 N HCl was incubated with glycolipids for 30 min at 80°C.

d. DEAE chromatography was performed as described in the text.

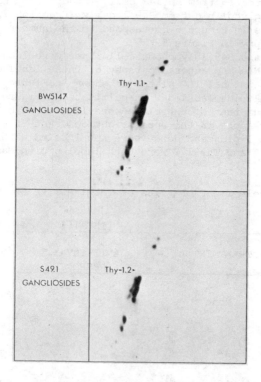

Figure 7.   Autoradiogram of ManNAc-labeled lymphoma cells

Lymphoma cells were labeled with [1-^{14}C]ManNAc in culture. Chromatography was done as in Figure 3 with the extract of 2×10^8 cells. Thy-1 glycolipids are indicated (15).

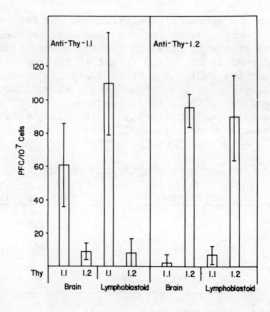

Journal of Immunology

Figure 8. *Anti-Thy-1 PFC assay for allogenic specificity of brain and lymphoma Thy-1 glycolipids*

Thy-1 glycolipids isolated from the two-dimensional TLC plates in Figures 3 and 5 were tested for allogenic specificity with the anti-Thy-1 PFC assays. Values are the mean ± standard error of five cultures. Application of the Student t test to the standard errors for these samples gives p *values less than 0.05 when compared wtih the Thy-1-active fraction (15).*

Thy-1 glycolipids isolated using two-dimensional TLC from
brain (AKR/J and ICR) and lymphoma cells (BW5147 and S49.1) tested for allogenic specificity with the anti-Thy-1 PFC assay. All
four samples were tested in both the anti-Thy-1.1 and anti-Thy-
1.2 PFC assays. AKR/J brain and BW5147 glycolipids (both of
allotype Thy-1.1) elicited a response only in the anti-Thy-1.1
PFC assay (Fig. 8). Thy-1 glycolipids of the Thy-1.2 allotype
showed no response in this assay. ICR brain and S49.1 glycolipids (both of allotype Thy-1.2) elicited a response only in the
anti-Thy-1.2 PFC assay but Thy-1 glycolipids of the Thy-1.1 allotype did not.

Mild acid and neuraminidase treatment and DEAE cellulose
chromatography were used to further characterize the Thy-1 active compounds (Table I, previous page). Neuraminidase treatment
and mild acid conditions, which result in the removal of sialic
acid, destroyed the anti-Thy-1 PFC response to these antigens.
Furthermore, both Thy-1.1 and Thy-1.2 glycolipids bound to DEAE
cellulose, confirming their acidic nature.

The neuraminidase sensitivity to Thy-1 activity of glycoprotein was also assessed (Table II).

Table II: NEURAMINIDASE TREATMENT OF THY-1.2 GLYCOPROTEIN

	TREATMENT	ANTI-THY-1.2 RESPONSE
Glycoprotein[a] (Thy-1.2)	None	58 ± 16[b]
	C.p. Nase (1 hr)[c]	29 ± 16
	C.p. Nase (24 hr)	7 ± 5

a. 100 mg of purified Thy-1.2 glycoprotein was used for these
    experiments.

b. Values shown are the mean $\pm$ standard error of five cultures.

c. Clostridium perfringens neuraminidase was incubated with the
    glycoprotein at 37°C.

Thy-1.2 activity decreased rapidly after 1 hour of treatment and
diminished to background levels after 24 hours. Thus we have
found that the Thy-1 activity of purified glycolipid and glycoprotein was destroyed by neuraminidase treatment.

Further characterization of Thy-1 active glycolipids was
accomplished by fractionating brain gangliosides into mono-,
di- and trisialo-gangliosides by DEAE column chromatography according to the procedure described by Nagai (14). These fractions eluted from the column were tested for Thy-1 activity
(Table III).

Table III: FRACTIONATION OF BRAIN GANGLIOSIDES BY DEAE
COLUMN CHROMATOGRAHPY[a]

GANGLIOSIDE FRACTION ADDED	THY-1 PFC RESPONSE
Unfractionated	32 ± 9.4[b]
Monosialo	9 ± 2.4
Disialo	26 ± 9.0
Trisialo	7 ± 3.4
No Addition	10 ± 5.6

a. Chromatography was performed on a DEAE-Sephacel column eluted
with a gradiant prepared from 0.05M and 0.45M ammonium acetate.

b. Mean $\pm$ standard error of five cultures.

The majority of the activity applied to the column appeared in
the disialoganglioside fraction. Thin layer chromatography
(not shown) verified the composition of the fraction.

Discussion

A sensitive immunological assay permitted the detection of
small quantities of Thy-1 active glycolipids in extracts from
brain, thymocytes and lymphoma cells. The active components were
radiolabeled using carbohydrate percursors (^{14}C-ManNAc) and lipid
percursor (^{14}C-palmitate) and were visualized by autoradiography
of this layer chromatograms. A critical factor in positive identification of alloantigens (Thy-1.1 and Thy-1.2) is the demonstration of activity which is specific for the Thy-1 allotype. These
results indicated that the antigen displayed only the allotype of
the mouse strain from which they were isolated. Thus, the AKR
strain and the BW5147 cell line were positive in the anti-Thy-1.1
PFC assay and the ICR strain and the S49.1 cell line were positive in the anti-Thy-1.2 PFC assay. No cross reaction was observed at the levels tested. Demonstration of the reciprocal
allogenic specificity is important because this supports the suggestion that the glycolipids carry specificities which parellel
the serological specificity of the Thy-1 allotypes.
Thy-1 glycolipids had different TLC R_f values in solvent 1
and solvent 2. The mobility of brain Thy-1 glycolipids was very
similar to G_{D3} ganglioside in that they migrated ahead of G_{M1} in
solvent 2, and behind G_{M1} and slightly ahead of GD1a in solvent

1. We have also found that Thy-1.1 migrates slightly faster
than Thy-1.2 in solvent system 1. Structural differences between
Thy-1.1 and 1.2 presumably result in different mobility in either
solvent system. We found lymphoma cell lines to be a good *in
vitro* source of Thy-1 glycolipid after labeling with either
$[1-^{14}C]$palmitate or $[1-^{14}C]$ManNAc. About 30 compounds were labeled with both plamitate and ManNAc in these cell lines. Additional compounds labeled with palmitate were neutral glycolipids.
We have concluded that Thy-1 glycolipids are gangliosides
because of the following evidence: 1) The glycolipids were isolated in ganglioside fractions after Folch partition and thin
layer chromatography and they were resistent to mild base treatment; 2) The Thy-1 glycolipids were labeled with ManNAc and palmitic acid suggesting the presence of sialic acid and sphingosine
respectively; 3) Ion exchange chromatography with DEAE cellulose
indicated the acidic nature of the glycolipids; and 4) The presence of sialic acid on the active molecule was indicated by
hydrolysis with neuraminidase.
The activity of Thy-1.2 active glycoprotein and glycolipids
were parellel with respect to the magnitude and specificity of
the Thy-1 assay and to the ability of anti-Thy-1.2 sera to absorb
out this activity. Anti-Thy-1.1 sera had no effect on the Thy-1.2
glycoprotein or glycolipid. This suggests that the antigenic
determinant of both of these species is similar. However the
exact nature of the antigenic determinant is not yet known and
will await the complete chemical characterization of both molecules. The activity of both the glycolipid and the glycoprotein
were also destroyed by neuraminidase treatment. This does not
necessarily mean that sialic acid is the antigenic determinant
however because removal of sialic acid from either a glycolipid
or a glycoprotein significantly changes the properties of the
individual compounds. A small ganglioside becomes insoluble in
water after removal of sialic acid, and the charge changes in a
glycoprotein could change the tertiary structure and hydration of
the molecule. Further chemical elucidation of the structure of
the glycoprotein and glycolipid species of Thy-1 are in progress.

Literature Cited

1.   Itakura, K., Muttan, J.J., Boyse, E.A., Old, L.J.   Nature
     New Biol., 1971, 230, (12),126.

2.   Blankenhorn, E.P., Douglas, T.C.   J. Hered., 1972, 63, 259.

3.   Wang, T.J., Freimuth, W.W., Miller, H.C., and Esselman, W.J.
     J. Immunol., 1978, 121, (4), 1361.

4.   Fuji, H., and Milgrom, F.   J. Exp. Med., 1973, 138, 1:16.

5. Barclay, A.N., Letarte-Muirhead, M., Williams, A.F., and Faulkes, R.A., Nature, 1976, 263, 563.

6. Zwerner, R.K., Barstad, P.A., and Acton, R.T., J. Exp. Med., 1977, 146, 986.

7. Kolondy, E.H., Brady, R.O., Quirk, J.M., Kanfer, J.N., J. Lipid Research, 1970, 11, 144.

8. Esselman, W.J., Laine, R.A., Sweeley, C.C., Methods Enzymol., 1972, 28, 140.

9. Miller, H.C., and Esselman, W.J., J. Immunol., 1975, 115, 839.

10. Esselman, W.J., and Kato, K., Fed. Proc., 1976, 35, 1643.

11. Freimuth, W.W., Esselman, W.J., and Miller, H.C., J. Immunol., 1978, 120, 1651.

12. Lake, P., Nature, 1976, 262, 297.

13. Zaleski, M.B. and Klein, J., J. Immunol., 1977, 145, 1602.

14. Iwamori, M. and Nagai, Y., Biochimica et Biophysica Acta., 1978, 528, 257.

15. Kato, K., Wang, T.J., and Esselman, W.J., J. Immunol., in press.

RECEIVED December 10, 1979.

Immune Reactivities of Antibodies against Glycolipids

Natural Antibodies

ROBERTA L. RICHARDS and CARL R. ALVING

Department of Membrane Biochemistry, Walter Reed Army Institute of Research, Washington, DC 20012

Naturally-occurring antibodies against simple glycolipids have been described only in scattered reports in the literature. About 75% of all normal humans have complement-activating anti-Forssman activity (1), and a monoclonal Waldenstrom macroglobulin IgM antibody (McG) having specificity for Forssman glycolipid was derived from the plasma of a patient (2,3). Some normal, or abnormal, human sera have anti-monogalactosyl (4) or anti-digalactosyl diglyceride antibodies (5). Recently we reported the occurrence of "natural" antibodies, apparently autoantibodies, with specificity against di- and trihexosyl ceramide haptens (CDH and CTH), in normal rabbit sera (6). We also found natural anti-ganglioside G_{M1} antibodies in normal human, guinea pig, and rabbit sera (7).

The major purpose of the present study was to describe, and to quantify, the widespread occurrence of natural complement-fixing autoantibodies against numerous simple glycolipids. We show that every individual rabbit and human serum tested had complement-fixing autoantibodies against glycolipids that are widely distributed in circulating blood cells and other tissues.

Numerous studies of specificities of anti-glycolipid antibodies produced by immunizing rabbits have been reported (8,9), and some discrepancies have been noted between laboratories (9). The widespread occurrence of natural antibodies might cause confusion in analyzing specificities produced by experimental immunization. Because of this we have purified both immune and natural anti-glycolipid antibodies from rabbit sera, and we have compared their specifities against the same, and different, glycolipids.

Methods

In all the experiments reported here, antibody activities were determined by antibody-mediated complement-dependent release of trapped marker from liposomes (3,10). The liposomes contained dimyristoylphosphatidylcholine (except where indicated), cholesterol, and dicetyl phosphate in molar ratios of 2:1.5:0.22, and the phospholipid was 10 mM in the final aqueous swelling volume. Glycolipid antigens were incorporated into the liposomes where indicated, generally at the level of 150 µg per µmole of phospholipid. Purified gangliosides and asialo-G_{M2} were kindly provided by Drs. Roscoe O. Brady and Peter H. Fishman. Galactosyl ceramide (CMH) from bovine brain and synthetic lactosyl ceramide (CDH) were from Miles Laboratories, Inc., Elkhart, Indiana. Ceramide trihexoside (CTH) and globoside from human erythrocytes and Forssman glycolipid from sheep erythrocytes were isolated as previously described (3). The sources of synthetic and natural phospholipids, cholesterol, dicetyl phosphate and digalactosyl diglyceride have been given elsewhere (11).

Antibody-dependent complement damage to the liposomes was detected by release of trapped glucose, using a spectrophotometric assay as described previously (10). All sera tested for the presence of antibodies were inactivated at 56° for 30 min., and fresh (unheated) guinea pig serum was used as the complement source. The assays for natural antibodies contained 5 µl of liposomes, 500 µl of glucose assay reagent, either 30 µl of rabbit serum or 50 µl of human serum, 120 µl of guinea pig serum, and sufficient 0.15 M NaCl to bring the final volume to 1 ml. Glucose release was measured after 30 min at room temperature (ca 22°) (3). Glucose release of more than 5% was considered a positive antibody reaction.

Results

Analysis of natural antibodies against galactosyl ceramide (CMH) and lactosyl ceramide (CDH) in sera from several normal humans and from several normal rabbits resulted in the data shown in Figure 1. None of the humans, and very few rabbits, had activity against liposomes when no glycolipid was present. All the data shown for natural anti-glycolipid antibodies were corrected for any activity observed with liposomes lacking glycolipid. Few of the human sera had even a slight level of activity against CMH, but about half of the individual humans did have anti-CDH activity. The rabbits were more reactive against both CMH and CDH. About 40% of the rabbits had activity against CMH, while only 2% lacked anti-CDH activity. The differences in the reactivity of rabbit sera with CMH and CDH suggests that there is little cross reactivity between these two activities in unimmunized animals – a conclusion previously reported by Rapport and colleagues (8).

These same human sera and some of the rabbit sera also were tested against ceramide trihexoside (CTH) and digalactosyl diglyceride (Table I). Several of the human sera reacted with digalactosyl diglyceride, in confirmation of the results of Hirsch and Parks (5). Although only a few human sera reacted with CTH, all of the rabbit sera tested showed reactivity.

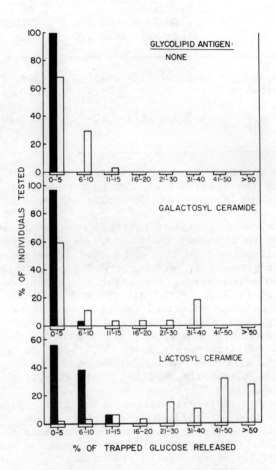

Figure 1. Natural antibodies in normal human and rabbit sera against liposomes containing no glycolipid, galactosylceramide, or lactosylceramide. Glucose release measured from liposomes containing DMPC, CHOL, DCP, and, where indicated, galactosylceramide (150 µg/µmol PC) or lactosylceramide (150 µg/µmol PC). Closed bars, humans; open bars, rabbits.

TABLE I. HUMAN AND RABBIT NATURAL ANTIBODIES AGAINST
DIGALACTOSYL DIGLYCERIDE AND CERAMIDE TRIHEXOSIDE

GLYCOLIPID	HUMAN		RABBIT	
	% OF TRAPPED GLUCOSE RELEASED[a]	% REACTING	% OF TRAPPED GLUCOSE RELEASED[a]	% REACTING
Digalactosyl diglyceride[b]	6.1±1.7 (17)	76	9.8± 8.1(42)	69
Ceramide trihexoside[c]	2.3±3.6 (17)	12	46.7±10.8(12)	100

[a]Expressed as: mean±standard deviation (number of sera tested).
[b]Present in the liposomes at 150 μg per μmole of phospholipid.
[c]Present in the liposomes at 150 nmoles per μmole of phospholipid.

Natural antibodies were purified from rabbit serum by affinity binding to liposomes (6,12). Briefly, this involved adsorbing the antibodies from the serum onto liposomes containing the appropriate glycolipid, washing the liposome-antibody complexes free of unreacted serum, then eluting the antibodies from the liposomes in 1M NaI. Both anti-CDH and anti-CTH were isolated from the same batch of normal rabbit serum and were compared for specificity. As shown in Figure 2, the anti-CDH did not react with CTH-containing liposomes. In contrast, the anti-CTH did react with CDH liposomes (Figure 3), though to a lesser extent than did the anti-CDH. Since no anti-CMH activity was observed in the whole serum, the purified antibodies were not tested against this antigen.

A somewhat different pattern of reactivity was observed with purified antibodies obtained from rabbits immunized with CDH or CTH. As Figure 4 shows, immune anti-CDH antibodies did cross-react with CMH. This observation is in contrast to the lack of cross-reactivity observed with the natural antibody (see above), but is in agreement with the results of Arnon et al. (13) obtained with rabbits that were immunized with lactosylsphingosine conjugated to a polypeptide. The immune anti-CTH studied here showed little or no reactivity with CMH.

The normal human sera shown in Figure 1 also were tested against four glycolipids (globoside, Forssman, asialo-G_{M2}, G_{M2}) having terminal N-acetylgalactosamine residues (Table II). The finding that all the individual human sera had activity against globoside (Table II) was surprising, since globoside is the major glycolipid of human erythrocytes, and only individuals of the rare p and p^k blood types lack globoside (14). Several, but not all, of the individuals tested also had activity against Forssman glycolipid, as reported previously (1,15). This lack of correla-

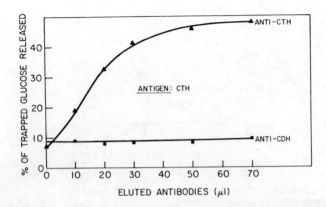

Immunochemistry

Figure 2. Reactivities of purified natural anti-CDH and anti-CTH antibodies against CTH-containing liposomes. Glucose release measured from liposomes containing DMPC, CHOL, DCP, and CTH (50 nmol/μmol PC) (6).

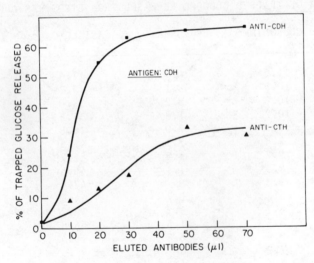

Immunochemistry

Figure 3. Reactivities of purified natural anti-CDH and anti-CTH antibodies against CDH-containing liposomes. Glucose release measured from liposomes containing DMPC, CHOL, DCP, and CDH (154 nmol/μmol PC) (6).

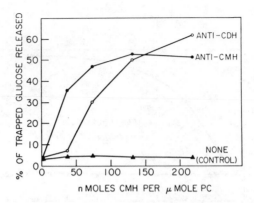

Immunochemistry

Figure 4.   Reactivity of purified immune anti-CDH antibodies against CMH-containing liposomes.   Glucose release measured from liposomes containing DPPC, CHOL, DCP, and the amounts of CMH shown (6).

TABLE II. HUMAN NATURAL ANTIBODIES AGAINST GLYCOLIPIDS HAVING
TERMINAL N-ACETYLGALACTOSAMINE RESIDUES

	% OF TRAPPED GLUCOSE RELEASED FROM LIPOSOMES CONTAINING:			
SERUM	GLOBOSIDE[a]	FORSSMAN[b]	ASIALO-G_{M2}[c]	G_{M2}[c]
1	11.0	0	17.4	0
2	10.9	8.1	35.2	11.7
3	10.8	8.2	33.0	6.4
4	10.7	1.8	54.5	5.6
5	10.1	7.0	18.8	3.2
6	9.8	7.3	46.5	5.2
7	9.7	1.5	81.2	14.9
8	9.5	1.1	45.5	13.4
9	9.3	0	47.4	5.1
10	9.1	10.7	44.6	5.0
11	8.7	12.5	57.2	8.6
12	8.6	11.5	56.8	1.4
13	8.4	6.8	15.4	5.0
14	7.2	0	37.5	0.7
15	6.4	4.2	55.4	1.5
16	5.9	9.6	33.9	3.3

[a] Present in the liposomes at 150 nmoles per μmole of phospholipid.
[b] Present in the liposomes at 10 nmoles per μmole of phospholipid.
[c] Present in the liposomes at 150 μg per μmole of phospholipid.

tion between anti-Forssman and anti-globoside activities suggests a lack of cross-reactivity between them, in agreement with the results of Young et al. (15). These results contrast with those of Naiki and Marcus (16), who found that anti-P antibodies reacted with globoside and Forssman. The reactivity against Forssman apparently also may be an auto-antibody, since Forssman glycolipid reportedly is present in some human colonic mucosa, both normal and malignant (17).

All of the human sera reacted with asialo-G_{M2}, and most gave strong reactions. In contrast, only about half of the individuals reacted with G_{M2}, and these gave only weak reactions. Thus the presence of the sialic acid group on G_{M2} may affect reactivity of the antibodies with the oligosaccharide structure common to asialo-G_{M2} and G_{M2}. Further, since there is no correlation between high reactivity with asialo-G_{M2} and with G_{M2}, the cross-reactivity between these two glycolipids must be limited.

The normal rabbit sera that were tested also had a high degree of reactivity against asialo-G_{M2} (Table III), but of the few rabbits tested against G_{M2}, all reacted with low to moderate activity. Like the humans, all rabbits tested had activity against globoside; but, unlike the humans, most of the rabbits were high reactors.

TABLE III.   RABBIT NATURAL ANTIBODIES AGAINST GLYCOLIPIDS HAVING TERMINAL N-ACETYLGALACTOSAMINE RESIDUES

GLYCOLIPID[a]	% OF TRAPPED GLUCOSE RELEASED[b]
G_{M2}	17.7± 9.5 (4)
Asialo-G_{M2}	54.0±13.4 (24)
Globoside	42.7± 9.7 (21)

[a]Liposomal concentrations are the same as in Table II.
[b]Expressed as: mean±standard deviation (number of sera tested).

Both normal rabbit and human sera have been tested against two gangliosides having terminal galactose residues, namely G_{M1} and G_{D1b}. As seen in Figure 5, most of the human sera tested had at least low levels of anti-G_{M1} antibodies, but none have been observed to have anti-G_{D1b} activity. All normal rabbit sera tested had both activities.

We also investigated the specificities of purified antibodies against G_{M1} and G_{D1b} from the serum of a rabbit immunized with bovine brain gangliosides. As shown in Figure 6, the whole antiserum had substantial activity both against the two ganglio-

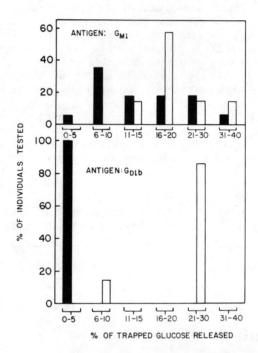

Proceedings of the National Academy of Sciences

Figure 5.   Natural antibodies in normal human and rabbit sera against liposomes containing gangliosides G_{M1} or G_{D1b}. Glucose release measured from liposomes containing DMPC, CHOL, DCP, and G_{M1} (150 μg/μmol PC) or G_{D1b} (150 μg/μmol PC). Human serum was complement source with the G_{M1} liposomes due to high level of anti-G_{M1} activity in guinea pig serum (7).

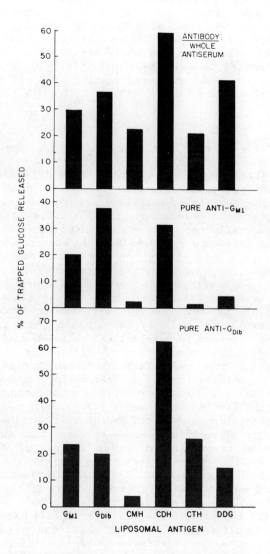

Figure 6.   Reactivities of whole antiserum and purified anti-G_{M1} and anti-G_{D1b} antibodies against liposomes containing glycolipids.   Antibodies purified from rabbit anti-bovine brain ganglioside serum as described previously (12).   Glucose release measured from liposomes containing DMPC, CHOL, DCP, and either G_{M1}, G_{D1b}, CMH, CDH, DDG (each 150 µg/µmol PC) or CTH (150 nmol/µmol PC).

sides, and against four neutral glycolipids, all of which had terminal galactose residues. The anti-G_{M1} antibodies reacted both with G_{M1} and G_{D1b}, and also with CDH, but not with CMH, CTH or digalactosyl diglyceride. The reactivity with CDH was reduced from the level present in the whole serum. Thus this anti-G_{M1} antibody appears to be relatively specific. In contrast, anti-G_{D1b} isolated from the same serum reacted with all of the antigens except CMH (Figure 6).

Discussion

We have demonstrated that every one of the human and rabbit sera tested have natural antibodies against several glycolipids. Most of the glycolipids used as antigens in this study normally occur in human and rabbit tissues (18), and therefore many, if not all, of the activities described are due to autoantibodies.

Previous studies have shown that antibodies against simple glycolipids generally recognize the terminal sugar as the immunodominant group (reviewed in 8 and 9). In the present study, all of the glycolipids used as antigens had either galactose, N-acetylgalactosamine, or N-acetylneuraminic acid as the terminal sugar. Several human sera reacted with CDH (Figure 1), and a few sera reacted with CTH (Table I). Both CDH and CTH are found almost ubiquitously in human red and white cells (19,20). Previously, only rare individuals of the p blood group type (lacking CTH) have been reported to have natural anti-CTH antibodies (21). We have confirmed the observation of Hirsch and Parks (5) that many normal human sera have antibodies against digalactosyl diglyceride. Our observations of natural antibodies against galactosyl glycolipids are consistent with previous reports of antigalactose antibodies in normal serum (22,23,24).

All of the human sera tested had activity against both globoside, the major glycolipid of human erythrocytes, and asialo-G_{M2} (Table II); and several human sera also reacted with Forssman and G_{M2} (Table II). These observations suggest the natural occurrence of antibodies against N-acetylgalactosamine in human sera, and such antibodies might have the potential for reacting with red cells or other tissues.

In humans, ABH (25), Lewis (26), and P (27) blood group substances are glycolipids, at least in part, and glycolipid blood group systems may occur in animals (25). Natural antibodies also have been reported in mice against human blood group A glycolipid (28). Blood group glycolipids in humans are thought to be protected from the immune system, and antibodies against them occur only in individuals lacking the antigen. Anti-blood group P antibodies are directed against glycolipids (such as CTH or globoside) that are ubiquitous in humans (16,21), and generally are considered to be cold agglutinins (reacting only at temperatures below 37° (29) and usually tested at 4°). These antibodies reportedly occur only in those rare humans that lack CTH or glob-

oside (21), although many of the cold agglutinins causing paroxys-
mal cold hemoglobinuria react with the P antigen (30). In our
experiments (at 22°) we found that all normal human or rabbit sera
had anti-globoside antibodies, and all rabbit sera, and approximately 12% of human sera, had anti-CTH antibodies.

Numerous studies have demonstrated that, compared to proteins,
lipids are poor antigens, when judged on a weight basis. Based
on the size of the antigenic site (mainly a mono- or oligosaccharide), most simple glycolipids are good antigens, and antibodies against them are easily raised (9). Occasional studies on
the specificities of experimentally raised antisera have produced
apparently conflicting conclusions. One example is the specificity of antibodies against CDH. Ceramide dihexoside was thought
to be a tumor antigen in humans, in that anti-tumor antibodies
produced in rabbits reacted with it (31). Anti-CDH antibodies
did not cross-react with CMH (galactosylceramide), despite the
presence of a terminal α-galactose (8). The conclusion was drawn
that anti-CDH antibodies were highly specific, noncross-reacting,
antibodies. Our experiments demonstrate that most, and perhaps
all, normal rabbit sera contain naturally-occurring anti-CDH antibodies, and we confirm that such antibodies can be highly specific. In contrast, in an animal that had only a low level of natural anti-CDH, and that was immunized with CDH to raise the titer, anti-CDH antibodies cross-reacted readily with CMH (6). Likewise, anti-CMH and anti-CDH sera appeared to cross-react with CTH,
but antibodies purified from the antisera failed to cross-react
with CTH (6). All normal rabbit sera tested had natural antibodies against CTH (Table I), and these antibodies might give a
false impression of cross-reactivity with CTH. Although the physiological, or pathological, significance of naturally-occurring
anti-glycolipid antibodies is unknown, it is evident that recognition of the existence of such antibodies is important in evaluat-
ing the specificities obtained by experimental immunization.

Abbreviations: DMPC and DPPC, synthetic dimyristoyl and dipal-
mitoyl phosphatidylcholines; CHOL, cholesterol; DCP, dicetyl phosphate; CMH, galactosyl ceramide, $Gal\beta1 \rightarrow ceramide$; CDH, lactosyl
ceramide, $Gal\beta1 \rightarrow 4Glc\beta1 \rightarrow ceramide$; CTH, ceramide trihexoside, $Gal-\alpha1 \rightarrow 3Gal\beta1 \rightarrow 4Glc\beta1 \rightarrow ceramide$; globoside, $GalNAc\alpha1 \rightarrow 4Gal\alpha1 \rightarrow 3Gal1-\beta1 \rightarrow 4Glc\beta1 \rightarrow ceramide$; Forssman, $GalNAc\alpha1 \rightarrow 3GalNAc\beta1 \rightarrow 4Gal\alpha1 \rightarrow 3Gal1-\beta1 \rightarrow 4Glc\beta1 \rightarrow ceramide$; asialo-$G_{M2}$, $GalNAc\beta1 \rightarrow 4Gal\beta1 \rightarrow 4Glc\beta1 \rightarrow cera-$
mide; G_{M2}, $GalNAc\beta1 \rightarrow 4Gal[3 \leftarrow 2\alpha AcNeu]\beta1 \rightarrow 4Glc\beta1 \rightarrow ceramide$; G_{M1}, $Gal-\beta1 \rightarrow 3GalNAc\beta1 \rightarrow 4Gal[3 \leftarrow 2\alpha AcNeu]\beta1 \rightarrow 4Glc\beta1 \rightarrow ceramide$; G_{D1b}, $Gal-\beta1 \rightarrow 3GalNAc\beta1 \rightarrow 4Gal[3 \leftarrow 2\alpha AcNeu8 \leftarrow 2\alpha AcNeu]\beta1 \rightarrow 4Glc\beta1 \rightarrow ceramide$.

472   CELL SURFACE GLYCOLIPIDS

LITERATURE CITED

1. Levine, P. Proc. Natl. Acad. Sci. USA, 1978, 75, 5697-5701.
2. Joseph, K.C.; Alving, C.R.; Wistar, R. J. Immunol., 1974, 112, 1949-1951.
3. Alving, C.R.; Joseph, K.C.; Wistar, R. Biochemistry, 1974, 13, 4818-4824.
4. Dupouey, P. J. Immunol., 1972, 109, 146-153.
5. Hirsch, H.E.; Parks, M.E. Nature, 1976, 264, 785-787.
6. Alving, C.R.; Richards, R.L. Immunochemistry, 1977, 14, 383-389.
7. Moss, J.; Fishman, P.H.; Richards, R.L.; Alving, C.R.; Vaughan, M.; Brady, R.O. Proc. Natl. Acad. Sci. USA, 1976, 73, 3480-3483.
8. Rapport, M.M.; Graf, L. Prog. Allergy, 1969, 13, 273-331.
9. Alving, C.R. (M. Sela, ed.) "The Antigens", Vol. 4, Academic Press, New York, 1977, 1-72.
10. Kinsky, S.C. Methods Enzymol., 1974, 32, 501-513.
11. Alving, C.R.; Fowble, J.W.; Joseph, K.C. Immunochemistry, 1974, 11, 475-481.
12. Alving, C.R.; Richards, R.L. Immunochemistry, 1977, 14, 373-381.
13. Arnon, R.; Sela, M.; Rachaman, E.S.; Shapiro, D. Eur. J. Biochem., 1967, 2, 79-83.
14. Marcus, D.M.; Naiki, M.; Kundu, S.K. Proc. Natl. Acad. Sci. USA, 1976, 73, 3263-3267.
15. Young, W.W.,Jr; Hakomori, S.I.; Levine, P. J. Immunol., 1979, 123, 92-96.
16. Naiki, M.; Marcus, D.M. Biochemistry, 1975, 14, 4837-4841.
17. Hakomori, S.; Wang, S.-M.; Young, W.W.,Jr. Proc. Natl. Acad. Sci. USA, 1977, 74, 3023-3027.
18. Hakomori, S. Adv. Cancer Res., 1973, 18, 265-315.
19. Sweeley, C.C.; Dawson, G. (Jamieson, G.A.; Greenwalt, T.J., ed.) "Red Cell Membrane Structure and Function", J.B. Lippincott Co., Philadelphia, 1969, 172-227.
20. Gottfried, E.L.; (Nelson, G.J., ed.) "Blood Lipids and Lipoproteins: Quantitation, Composition and Metabolism", Wiley-Interscience, New York, 1972, 387-415.
21. Kato, M.; Kubo, S.; Naiki, M. J. Immunogenet., 1978, 5, 31-40.
22. Tsai, C.-M.; Zopf, D.A.; Wistar, R.,Jr.; Ginsburg, V. J. Immunol., 1976, 117, 717-721.
23. Rogentine, G.N.,Jr.; Plocinik, B.A. J. Immunol., 1974, 113, 848-858.
24. Sela, B.-A.; Wang, J.L.; Edelman, G.M. Proc. Natl. Acad. Sci. USA, 1975, 72, 1127-1131.
25. Hakomori, S.; Kobata, A. (M. Sela, ed.)"The Antigens", Vol. 2, Academic Press, New York, 1974, 79-140.
26. Marcus, D.M. N. Engl. J. Med., 1969, 280, 994-1006.

27. Naiki, M.; Marcus, D.M. Biochem. Biophys. Res. Commun., 1974, 60, 1105-1111.
28. Arend, P; Nijssen, J. Nature, 1977, 269 255-257.
29. Roelcke, D. Clin. Immunol. Immunopath., 1974, 2, 266-280.
30. Schwarting, G.A.; Kundu, S.K.; Marcus, D.M. Blood 1979, 53, 186-192.
31. Rapport, M.M.; Graf, L.; Skipski, V.P.; Alonzo, N.J. Cancer, 1959, 12, 438-445.

RECEIVED December 10, 1979.

INDEX

475

D

E